AF167413

Lecture Notes
in Business Information Processing   **526**

Series Editors

Wil van der Aalst , *RWTH Aachen University, Aachen, Germany*
Sudha Ram , *University of Arizona, Tucson, AZ, USA*
Michael Rosemann , *Queensland University of Technology, Brisbane, QLD,
Australia*
Clemens Szyperski, *Microsoft Research, Redmond, WA, USA*
Giancarlo Guizzardi , *University of Twente, Enschede, The Netherlands*

LNBIP reports state-of-the-art results in areas related to business information systems and industrial application software development – timely, at a high level, and in both printed and electronic form.

The type of material published includes

- Proceedings (published in time for the respective event)
- Postproceedings (consisting of thoroughly revised and/or extended final papers)
- Other edited monographs (such as, for example, project reports or invited volumes)
- Tutorials (coherently integrated collections of lectures given at advanced courses, seminars, schools, etc.)
- Award-winning or exceptional theses

LNBIP is abstracted/indexed in DBLP, EI and Scopus. LNBIP volumes are also submitted for the inclusion in ISI Proceedings.

Andrea Marrella · Manuel Resinas · Mieke Jans ·
Michael Rosemann
Editors

Business Process Management Forum

BPM 2024 Forum
Krakow, Poland, September 1–6, 2024
Proceedings

 Springer

Editors
Andrea Marrella ⓘ
Sapienza University of Rome
Rome, Italy

Manuel Resinas ⓘ
University of Seville
Seville, Spain

Mieke Jans ⓘ
Hasselt University
Hasselt, Belgium

Michael Rosemann ⓘ
Queensland University of Technology
Brisbane, QLD, Australia

ISSN 1865-1348 ISSN 1865-1356 (electronic)
Lecture Notes in Business Information Processing
ISBN 978-3-031-70417-8 ISBN 978-3-031-70418-5 (eBook)
https://doi.org/10.1007/978-3-031-70418-5

This Springer imprint is published by the registered company Springer Nature Switzerland AG
The registered company address is: Gewerbestrasse 11, 6330 Cham, Switzerland

If disposing of this product, please recycle the paper.

Preface

This volume comprises all papers presented at the BPM Forum of the 22nd International Conference on Business Process Management (BPM), held during 1–6 September 2024 at the AGH University of Krakow, Poland. The BPM Forum features novel ideas about emerging topics in the field of Business Process Management. As a platform, the BPM Forum provides authors with the possibility to showcase their ongoing research and with this trigger an early scientific debate and critical reflection on its potential and merit.

This year, the conference received a total of 171 submissions, the second highest number of submissions in the history of this conference. 144 papers were selected for a full review process. The review process followed the established high-quality standards of the BPM Conference series, and for the first time saw the conduct of a double-blinded review process. Each paper was reviewed by at least three carefully selected Program Committee members of the respective track. Then, an extensive discussion between the reviewers and a Senior Program Committee member took place. This discussion was captured in a meta-review by the Senior Program Committee who offered their recommendations to the Track Chairs, who were responsible for the final decision.

This thorough review process resulted in 29 papers being accepted at the main conference and 21 papers were included in the BPM Forum. The BPM Forum papers are compiled in these Proceedings and demonstrate the contemporary nature of current BPM research. This is demonstrated by the diverse topics presented at the BPM Forum, including object-centric BPM, predictability, sustainability, AI and the role of data and LLMs, design principles for RPA, and the changing role of humans in fast-transforming business processes.

In addition to the main conference and the Forum, there was a multitude of affiliated events including 9 workshops, five fora including a Central and Eastern Europe forum, tutorials, journal-first and demos and resources tracks, and a doctoral consortium.

Open Science aiming at reproducibility and replicability of research results continued to be a major principle for the BPM Forum. Authors were explicitly requested to link one or more repositories with additional artefacts such as data sets, prototypes, and interview protocols alongside implemented prototypes to their papers. For the first time, papers that made artefacts available are recognised with a related badge in these proceedings.

We also encouraged all participants to consider diversity, equity, and inclusion in their writing, reviews, presentations, and all other interactions related to the BPM conference. We are grateful for the work of the Diversity and Inclusion Chairs, Artur Lesner, Andrea Marrella, and Adela del Rio Ortega.

We would like to thank all authors, both regular and senior members of the Program Committees, and all external reviewers. Their diligent responses and effective collaboration made a rigorous, extensive, and timely review process possible and enabled the high-quality research output reflected by the papers in this volume.

We acknowledge and are grateful to the sponsors of BPM 2024; without their support, such an event would not have been possible: Celonis as the platinum sponsor, SAP Signavio, tecna, and Wyzsza Szkoła Biznesu - National Louis University as gold sponsors, and DCR Solutions, OKNOPLAST Sp. z o.o., and Grupa pbi as bronze sponsors.

Finally, we would like to express our special appreciation to Edyta Brzychczy and Krzysztof Kluza from the AGH University of Krakow as the General Chairs of BPM 2024. Together with the local organizing team, consisting of Katarzyna Gdowska, Renata Gabryelczyk (University of Warsaw), Aneta Napieraj, Marta Podobińska-Staniec, Emilia Mecfel, and conference manager Anna Smyk, they all created a flawless conference and BPM Forum fostering inspirations, intellectual debates, and new friendships.

We hope that the papers presented at the BPM Forum 2024 and captured in this volume will provide fertile stimulation for BPM academics and professionals.

September 2024

Andrea Marrella
Manuel Resinas
Mieke Jans
Michael Rosemann

Organization

Steering Committee

Jan Mendling (Chair)	Humboldt-Universität zu Berlin, Germany
Marlon Dumas	University of Tartu, Estonia
Avigdor Gal	Technion – Israel Institute of Technology, Israel
Chiara Ghidini	Free University of Bozen-Bolzano, Italy
Manfred Reichert	Ulm University, Germany
Hajo A. Reijers	Utrecht University, the Netherlands
Stefanie Rinderle-Ma	Technical University of Munich, Germany
Adela del Río Ortega	University of Seville, Spain
Michael Rosemann	Queensland University of Technology, Australia
Shazia Sadiq	University of Queensland, Australia
Barbara Weber	University of St.Gallen, Switzerland
Matthias Weidlich	Humboldt-Universität zu Berlin, Germany
Mathias Weske	HPI, University of Potsdam, Germany

Executive Committee

General Chairs

Edyta Brzychczy	AGH University of Krakow, Poland
Krzysztof Kluza	AGH University of Krakow, Poland

Main Conference Program Committee Chairs

Andrea Marrella (Track I Chair)	Sapienza University of Rome, Italy
Manuel Resinas (Track II Chair)	University of Seville, Spain
Mieke Jans (Track III Chair)	Hasselt University, Belgium
Michael Rosemann (Consolidation Chair)	Queensland University of Technology, Australia

Workshop Chairs

Katarzyna Gdowska	AGH University of Krakow, Poland
María Teresa Gómez López	University of Seville, Spain
Jana-Rebecca Rehse	University of Mannheim, Germany

Demonstration and Resources Chairs

Weronika T. Adrian   AGH University of Krakow, Poland
Laura Genga   Eindhoven University of Technology,
                               the Netherlands
Sander J.J. Leemans   RWTH Aachen, Germany

Industry Forum Chairs

Iris Beerepoot   Utrecht University, the Netherlands
Renata Gabryelczyk   University of Warsaw, Poland
Ralf Plattfaut   University of Duisburg-Essen, Germany

Blockchain Forum Chairs

Claudio Di Ciccio   Utrecht University, the Netherlands
Walid Fdhila   University of Vienna, Austria

RPA Forum Chairs

Simone Agostinelli   Sapienza University of Rome, Italy
Daniel Amyot   University of Ottawa, Canada
Henrik Leopold   Kühne Logistics University, Germany

Education Forum Chairs

Katarzyna Gdowska   AGH University of Krakow, Poland
Thomas Grisold   University of St. Gallen, Switzerland
Piotr Sliż   University of Gdańsk, Poland

Central and Eastern European Forum Chairs

Michal Krčál   Masaryk University, Czech Republic
Monika Malinova Mandelburger TU Wien, Austria
Gregor Polančič   University of Maribor, Slovenia
Katarina Tomičić-Pupek   University of Zagreb, Croatia

Tutorial Chairs

Maximilian Röglinger   University of Bayreuth, Germany
Inge van de Weerd   Utrecht University, the Netherlands

Doctoral Consortium Chairs

Han van der Aa	University of Vienna, Austria
Benoît Depaire	Hasselt University, Belgium
Marta Indulska	University of Queensland, Australia

Journal-First Track Chairs

Chiara Di Francescomarino	University of Trento, Italy
Felix Mannhardt	Eindhoven University of Technology, the Netherlands

BPM Dissertation Award Chair

Hajo A. Reijers	Utrecht University, the Netherlands

Publicity Chairs

Marco Comuzzi	Ulsan National Institute of Science and Technology, South Korea
Arik Senderovich	York University, Canada
Sarah Winkler	Free University of Bozen-Bolzano, Italy

Proceedings Chair

Simone Agostinelli	Sapienza University of Rome, Italy

Diversity and Inclusion Chairs

Artur Lesner	AGH University of Krakow, Poland
Andrea Marrella	Sapienza University of Rome, Italy
Adela del Río Ortega	University of Seville, Spain

Track I: Foundations

Senior Program Committee

Wil van der Aalst	RWTH Aachen University, Germany
Jörg Desel	Fernuniversität in Hagen, Germany
Claudio Di Ciccio	Utrecht University, the Netherlands
Chiara Di Francescomarino	University of Trento, Italy

Thomas Hildebrandt	University of Copenhagen, Denmark
Arthur ter Hofstede	Queensland University of Technology, Australia
Rick Hull	University of California, Davis, USA
Sander J.J. Leemans	RWTH Aachen, Germany
Fabrizio Maria Maggi	Free University of Bozen-Bolzano, Italy
Marco Montali	Free University of Bozen-Bolzano, Italy
Artem Polyvyanyy	University of Melbourne, Australia
Manfred Reichert	Ulm University, Germany
Hagen Voelzer	University of St.Gallen, Switzerland
Matthias Weidlich	Humboldt-Universität zu Berlin, Germany
Jan Martijn van der Werf	Utrecht University, the Netherlands
Mathias Weske	HPI, University of Potsdam, Germany

Program Committee

Simone Agostinelli	Sapienza University of Rome, Italy
Ahmed Awad	British University in Dubai, United Arab Emirates
Mario Luca Bernardi	University of Sannio, Italy
Marta Cimitile	University of Rome Unitelma Sapienza, Italy
Patrick Delfmann	University of Koblenz-Landau, Germany
Johannes De Smedt	KU Leuven, Belgium
Ivan Donadello	Free University of Bozen-Bolzano, Italy
Rik Eshuis	Eindhoven University of Technology, the Netherlands
Dirk Fahland	Eindhoven University of Technology, the Netherlands
Stephan Fahrenkrog-Petersen	Humboldt-Universität zu Berlin, Germany
Peter Fettke	German Research Center for Artificial Intelligence (DFKI) and Saarland University, Germany
Valeria Fionda	University of Calabria, Italy
Francesco Folino	ICAR-CNR, Italy
Fabiana Fournier	IBM Research, Haifa, Israel
María Teresa Gómez López	University of Seville, Spain
Alessandro Gianola	University of Lisbon, Portugal
Giancarlo Guizzardi	University of Twente, the Netherlands
Ekkart Kindler	Technical University of Denmark, Denmark
Akhil Kumar	Pennsylvania State University, USA
Irina Lomazova	National Research University Higher School of Economics, Russia
Hugo A. López	Technical University of Denmark, Denmark
Qinghua Lu	CSIRO, Australia

Xixi Lu	Utrecht University, the Netherlands
Felix Mannhardt	Eindhoven University of Technology, the Netherlands
Werner Nutt	Free University of Bozen-Bolzano
Chun Ouyang	Queensland University of Technology, Australia
Oscar Pastor	Technical University of Valencia, Spain
Luigi Pontieri	ICAR, National Research Council of Italy (CNR), Italy
Jana-Rebecca Rehse	University of Mannheim, Germany
Daniel Ritter	SAP, Germany
Yara Rizk	IBM Research Cambridge, USA
Andrey Rivkin	Technical University of Denmark, Denmark
Massimiliano Ronzani	Fondazione Bruno Kessler, Italy
Arik Senderovich	York University, Canada
Tijs Slaats	University of Copenhagen, Denmark
Monique Snoeck	KU Leuven, Belgium
Ernest Teniente	Technical University of Catalonia, Spain
Eric Verbeek	Eindhoven University of Technology, the Netherlands
Sarah Winkler	Free University of Bozen-Bolzano, Italy

Track II: Engineering

Senior Program Committee

Boualem Benatallah	University of New South Wales, Australia
Andrea Burattin	Technical University of Denmark, Denmark
Cristina Cabanillas	University of Seville, Spain
Massimiliano de Leoni	University of Padua, Italy
Jochen De Weerdt	KU Leuven, Belgium
Benoît Depaire	Hasselt University, Belgium
Remco Dijkman	Eindhoven University of Technology, the Netherlands
Marlon Dumas	University of Tartu, Estonia
Avigdor Gal	Technion – Israel Institute of Technology, Israel
Henrik Leopold	Kühne Logistics University, Germany
Massimo Mecella	Sapienza University of Rome, Italy
Jorge Munoz-Gama	Pontificia Universidad Católica de Chile, Chile
Luise Pufahl	Technical University of Munich, Germany
Hajo A. Reijers	Utrecht University, the Netherlands
Stefanie Rinderle-Ma	Technical University of Munich, Germany

Shazia Sadiq	University of Queensland, Australia
Pnina Soffer	University of Haifa, Israel
Moe Thandar Wynn	Queensland University of Technology, Australia
Boudewijn Van Dongen	Eindhoven University of Technology, the Netherlands
Ingo Weber	Technical University of Munich, Germany
Barbara Weber	University of St. Gallen, Switzerland

Program Committee

Lars Ackermann	University of Bayreuth, Germany
Robert Andrews	Queensland University of Technology, Australia
Abel Armas Cervantes	University of Melbourne, Australia
Fabio Casati	Servicenow, USA
Claudia Diamantini	Marche Polytechnic University, Italy
Andrea Delgado	Universidad de la República, Uruguay
Joerg Evermann	Memorial University of Newfoundland, Canada
Walid Gaaloul	Télécom SudParis, France
Luciano García-Bañuelos	Tecnológico de Monterrey, Mexico
Laura Genga	Eindhoven University of Technology, the Netherlands
Oscar González-Rojas	University of the Andes, Colombia
Daniela Grigori	Paris Dauphine University, France
Georg Grossmann	University of South Australia, Australia
Gert Janssenswillen	Hasselt University, Belgium
Andrés Jiménez Ramírez	University of Seville, Spain
Anna Kalenkova	University of Adelaide, Australia
Agnes Koschmider	University of Bayreuth, Germany
Manuel Lama Penin	University of Santiago de Compostela, Spain
Francesco Leotta	Sapienza University of Rome, Italy
Orlenys López Pintado	University of Tartu, Estonia
Elisa Marengo	University of Turin, Italy
Alfonso Márquez-Chamorro	University of Sevilla, Spain
Giovanni Meroni	Technical University of Denmark, Denmark
Hye-Young Paik	University of New South Wales, Australia
Cesare Pautasso	University of Lugano, Switzerland
Pierluigi Plebani	Politecnico di Milano, Italy
Pascal Poizat	Université Paris Nanterre and LIP6, France
Barbara Re	University of Camerino, Italy
Stefan Schönig	University of Regensburg, Germany
Marcos Sepúlveda	Pontificia Universidad Católica de Chile, Chile

Natalia Sidorova	Eindhoven University of Technology, the Netherlands
Renuka Sindhgatta	IBM Research, India
Minseok Song	Pohang University of Science and Technology, South Korea
Niek Tax	Meta, UK
Victoria Torres	Universitat Politècnica de València, Spain
Nick van Beest	CSIRO, Australia
Han van der Aa	University of Vienna, Austria
Sebastiaan van Zelst	Celonis, Germany
Seppe Vanden Broucke	KU Leuven, Belgium
Karolin Winter	Eindhoven University of Technology, the Netherlands
Nicola Zannone	Eindhoven University of Technology, the Netherlands

Track III: Management

Senior Program Committee

Daniel Beverungen	Paderborn University, Germany
Adela del Río Ortega	University of Seville, Spain
Paul Grefen	Eindhoven University of Technology, the Netherlands
Thomas Grisold	University of St. Gallen, Switzerland
Mojca Indihar Štemberger	University of Ljubljana, Slovenia
Marta Indulska	University of Queensland, Australia
Christian Janiesch	TU Dortmund University, Germany
Peter Loos	IWi at DFKI, Saarland University, Germany
Niels Martin	Hasselt University, Belgium
Jan Mendling	Humboldt-Universität zu Berlin, Germany
Maximilian Röglinger	University of Bayreuth, Germany
Flavia Santoro	University of the State of Rio de Janeiro, Brazil
Peter Trkman	University of Ljubljana, Slovenia
Amy Van Looy	Ghent University, Belgium

Program Committee

Amine Abbad-Andaloussi	University of St Gallen, Switzerland
Banu Aysolmaz	Eindhoven University of Technology, the Netherlands

Christian Bartelheimer	University of Paderborn, Germany
Markus Becker	University of Southern Denmark, Denmark
Iris Beerepoot	Utrecht University, the Netherlands
Marco Comuzzi	Ulsan National Institute of Science and Technology, South Korea
Barbara Dinter	Chemnitz University of Technology, Germany
Irene Bedilia Estrada Torres	University of Seville, Spain
Renata Gabryelczyk	University of Warsaw, Poland
Kanika Goel	Queensland University of Technology, Australia
Michael Leyer	University of Marburg, Germany
Alexander Mädche	Karlsruhe Institute of Technology, Germany
Monika Malinova Mandelburger	TU Wien, Austria
Martin Matzner	Friedrich-Alexander-Universität Erlangen-Nürnberg, Germany
Patrick Mikalef	Norwegian University of Science and Technology, Norway
Nadine Ostern	Queensland University of Technology, Australia
Geert Poels	Ghent University, Belgium
Gregor Polančič	University of Maribor, Slovenia
Pascal Ravesteyn	Utrecht University of Applied Sciences, the Netherlands
Kate Revoredo	Humboldt-Universität zu Berlin, Germany
Dennis M. Riehle	Koblenz University, Germany
Estefanía Serral	KU Leuven, Belgium
Rehan Syed	Queensland University of Technology, Australia
Oktay Turetken	Eindhoven University of Technology, the Netherlands
Irene Vanderfeesten	KU Leuven, Belgium
Inge van de Weerd	Utrecht University, the Netherlands
Sven Weinzierl	FAU, Germany
Axel Winkelmann	University of Würzburg, Germany
Bastian Wurm	LMU Munich School of Management, Germany
Francesca Zerbato	Eindhoven University of Technology, the Netherlands
Michael Zur Muehlen	Stevens Institute of Technology, USA

Additional Reviewers

Amiri Elyasi, Keyvan
Back, Christoffer Olling
Camargo Chávez, Manuel A.
Christfort, Axel

Corea, Carl
Dong, Liming
Fonger, Frederik
Groefsema, Heerko

Gunklach, Jonas
Harms, Maximilian
Hasel Mehri, Gelareh
Imenkamp, Christian
Janssen, Dominik
Käppel, Martin
Langner, Moritz
Lee, Suhwan
Liessmann, Annina
Liu, Yue
Lupia, Francesco
Mele, Alessandro
Mircoli, Alex
Mitsyuk, Alexey A.
Morgan, Rebecca
Neuberger, Julian
Oriol, Xavier

Potena, Domenico
Rebmann, Adrian
Rossetti, Cristina
Rossi, Lorenzo
Schelhorn, Till Carlo
Steflova, Klara
Susaiyah, Allmin P. Singh
Tang, Willi
Velásquez, Ignacio
Wang, Weixin
Wei, Jia
Wegener, Adrian
Xia, Boming
Yang, Jing
Zilker, Sandra
Zisgen, Yorck
Zschech, Patrick

Contents

Management

Foundations

Expressive Power and Complexity Results for SIGNAL, an Industry-Scale Process Query Language

Timotheus Kampik[1,2] and Cem Okulmus[2(✉)]

[1] SAP, Berlin, Germany
[2] Department of Computing Science, Umeå University, Umeå, Sweden
{tkampik,okulmus}@cs.umu.se

Abstract. With the increased adoption of process mining, there is also a need for practical solutions that work at industry scales. In this context, process querying methods (PQMs) have emerged as an important tool for drawing inferences from event logs. Here, it can be expected that industry approaches differ from academic ones, due to practical engineering and business considerations. To understand what is at the core of industry-scale PQMs, a formal analysis of the underlying languages can provide a solid foundation. To this end, we formally analyse SIGNAL, an industry-scale language for querying business process event logs developed by a large enterprise software vendor. The formal analysis shows that the core capabilities of SIGNAL, which we refer to as the *SIGNAL Conjunctive Core*, are more expressive than relational algebra and thus not captured by standard relational databases. We provide an upper-bound on the expressiveness via a reduction to semi-positive Datalog, which also leads to an upper bound of P-hard for the data complexity of evaluating SIGNAL Conjunctive Core queries. The findings provide first insights into how (real-world) process query languages are fundamentally different from the more generally prevalent structured query languages for querying relational databases and provide a rigorous foundation for extending the existing capabilities of the industry-scale state-of-the-art of process data querying.

Keywords: Process mining · Process querying · Databases

1 Introduction

The increased industry adoption of process mining requires practical solutions that work at industry scales. Process querying methods (PQMs) have emerged as central, laying the foundations of industry-scale tools for drawing inferences from the event logs that collect the process traces as recorded by enterprise systems [11,16,19]. Intuitively, these methods provide structured query languages that focus on querying representations common in business process management, such as *graphs* on modalities that are traditionally considered important

A. Marrella et al. (Eds.): BPM 2024 Forum, LNBIP 526, pp. 3–19, 2024.
https://doi.org/10.1007/978-3-031-70418-5_1

in the domain, such as *logical time*. Despite the importance of PQMs in academia and practice, so far and to the best of our knowledge, none of these industry-scale methods have been subject to a detailed formal analysis. This is understandable, since these tools have been developed using a top-down approach, with the need to find practical solution quickly to meet industry demand for process mining insights. However, the lack of a formal understanding of these industry-scale PQMs, we argue, leads to an interesting and relevant research challenge. First, establishing such an understanding firmly answers the question as to how PQMs can theoretically scale with the size of the underlying data, thus showing whether these tools are indeed an efficient and practical solution beyond individual examples. Second, a complete formal analysis opens the way to more expressive extensions of existing PQMs that are still "well-behaved" in terms of computational complexity. Finally, and most importantly, a formalisation of a PQM allows us to compare its expressive power to other data management approaches and see if PQM methods are indeed adding something new compared to the state of the art in database management, such as relational or NoSQL database systems. We thus see this work as the first step towards the goal of a formal description of, ideally, all PQMs that are publicly documented and can thus be studied. There are already numerous PQMs described in the literature [16], such as PQL [18], Celonis PQL [19], and SIGNAL [11], as well as several graph-based languages for process querying that have gained—while nascent—substantial research attention in recent years [4, 8, 10, 12].

For this paper we focus our attention on SIGNAL, which is actively developed by a large enterprise software vendor in a process mining product[1]. We note that our focus on SIGNAL should be understood as providing the necessary starting point in our endeavour to better understand the formal properties of PQMs, and not as the end goal. Still, we argue that focusing on an industry-scale language *is* desirable to start with, as it helps unearth practical considerations emerging from real-world engineering and business requirements.

Thus, our research questions for this work are as follows:

1. Which well-known query languages can and cannot match the process querying language SIGNAL with respect to expressive power?
2. What are the complexity bounds of the process querying language SIGNAL?

Answering Question 1. tells us whether state-of-the-art process data querying requires fundamentally different capabilities than, e.g., querying a traditional relational database using an SQL dialect; answering Question 2. tells us to what extent it is (theoretically) possible to query large amounts of data fast.

---

[1] Here, we can rely on "openly" available documentation, cf. help.sap.com/docs/ signavio-process-intelligence/signal-guide/syntax (last accessed at 2024-02-20) – as well as on a technical report [11] shared by the vendor.

Our Contributions. In this work we present the following results:

- We provide the first formalisation of the conjunctive core of the SIGNAL query language. This formalisation is based on publicly available documentation and grammar specification; we also had access to a running SIGNAL instance that we could query via a RESTful API.
- We show that the expressive power of the conjunctive core of SIGNAL cannot be captured by relational algebra.
- We provide an upper-bound on the data complexity of evaluating the conjunctive core of SIGNAL by a translation to semi-positive Datalog. It is left as an open question if this also forms the lower-bound, which would lead to a P-completeness result.

The rest of the paper is structured as follows. We provide the formal preliminaries of event logs and cases, presented in the context of the relational model, in Sect. 2. Based on the preliminaries, we provide a formalisation of the core of SIGNAL in Sect. 3. We then show that the SIGNAL core cannot be expressed in relational algebra and translate the language to semi-positive Datalog in order to allow for a straightforward theoretical analysis in Sect. 4. The results of our analysis show that the core of SIGNAL is more expressive than relational algebra and not more expressive than semi-positive Datalog; hence, we can also establish that the upper bound in the data complexity is in polynomial time. The paper concludes in Sect. 5 with a call for future work that will ideally expand the expressive power and complexity results to other PQMs.

2 Preliminaries

We proceed to give a succinct summary of the relational model. For a more detailed introduction of this and related topics on the foundations of databases, we refer to [2]. A domain (or attribute) D is a set of values, such as the natural numbers, alphabetic words of various length, and so on. A tuple $t \subseteq D_1 \times \cdots \times D_n$ is an n-ary combination of values from n many domains. A relation $R \subseteq 2^{D_1 \times \cdots \times D_n}$ is a set of tuples. We distinguish between the *instance* of a relation, which is the set described above, and the *schema* of all relations of this type via the set of domains (or attributes) that make up the values in the tuples in the relation instance. When writing about the schema of a relation, we shall equate it simply with the list of its domains. We assume that any reader is familiar with the relational algebra operations, like projection (π), selection (σ), various joins, and set operations. For readers that are not familiar with these, we again refer to the excellent textbook from Abiteboul et al. [2]. Another concept we need is the notion of conjunctive query (CQ). Formally, a CQ has a body, made up of relational atoms, and a head, a subset of variables occurring in the body, also called the answer variables:

$$q(\mathbf{x}) \leftarrow R_1(\mathbf{y_1}), \ldots, R_n(\mathbf{y_n}),$$

where we have that $\mathbf{x} \subseteq \bigcup_{1 \leq i \leq n} \mathbf{y_i}$. CQs correspond to SELECT-FROM-WHERE queries in SQL where we only allow equations between attributes in the WHERE clause. Another correspondence is to the positive fragment of first-order logic. Despite its structural simplicity, the problem of answering a CQ—defined as the problem of finding all mappings between the variables of the body atoms to instances of the respective relations in a given database—is known to be NP-complete. This combination of structural simplicity while still retaining a high complexity for the basic problem of query answering has made CQs an important object of study in database theory.

Computational Complexity of Query Answering. For the problem of query answering as defined above, there are three notions of computational complexity. The first is to consider both the query and the database instance to be part of the input – this gives us the *combined complexity* of query answering. As discussed, already fairly simple queries, such as CQs, can be hard to answer. However, usually the query is fairly small and thus one might want to investigate how the problem of query evaluation scales with the database size. In this case we treat the query as a constant. This gives rise to the notion of *data complexity*. And finally, if one only wants to study how the problem scales with the size of the query, one can set the database size to constant and this gives the notion of *query complexity.*

Due to the prevalence of temporal data in the form of event logs in process mining, we introduce a special timestamp attribute $\mathcal{T}$. For the sake of simplicity, we assume that a timestamp here refers to a natural number (or 0), representing the UNIX timestamp.

We present here a simplified view on the concepts of event logs and cases, as they play a crucial role in process mining and to understand the query language SIGNAL we wish to present. While there are different notations and terminologies for these concepts in the literature [1], we will present these concepts in the context of the (flat) relational model, which most database theorists and users will likely be familiar with.

Definition 1 (Events and Event Logs). *An* event log L *is a relation with the schema* $(Eid, Cid, \mathcal{T}, A_1, \ldots A_n)$, *where the first three attributes are fixed, namely* Eid *for the* event id, Cid *for the* case id *and* $\mathcal{T}$ *to indicate time, and we allow an n-ary set of attributes for the rest of the event log schema, which we shall call the* event attributes. *We will refer to a single tuple inside* L *as an* event. *We furthermore define the functional dependencies that hold for any event log* L: *both* (Eid, Cid) *and* $(Cid, \mathcal{T})$ *shall be key candidates. Formally, for no two events* $e, e' \in L$ *where* $e \neq e'$, $e = (eid, cid, t, a_1, \ldots a_n)$ *and* $e' = (eid', cid', t', a'_1, \ldots a'_n)$ *such that* $cid = cid'$ *it may hold that* $t = t'$ *or* $eid = eid'$. *Some further notation: we define the function* att *to project an event to its event attributes, e.g.:* $att(e) = (a_1, \ldots, a_n)$. *Due to the timestamp attribute* $\mathcal{T}$ *in events, we can also define a ordering over events of the same case:* $e \succeq e'$ *(resp.* $e \succ e'$) *holds, if we have* $t \geq t'$ *(resp.* $t > t'$).

Table 1. An event log without case attributes.

event_ID	case_ID	timestamp	event_name	status
e0001	0001	1675086864052	Review request	NEW
e0002	0002	1675147138009	Review request	NEW
e0003	0001	1675160180724	Calculate terms	WIP
e0004	0002	1675213914098	Define terms	WIP
e0005	0001	1675220315296	Prepare contract	WIP
e0006	0002	1675282027657	Prepare contract	WIP
e0007	0002	1675414104525	Send quote	SENT

To easily refer to the set of cases in an event log L with schema $(Eid, Cid, \mathcal{T}, A_1, \ldots A_n)$, we introduce the notation $Cases(L) = Cid$. When we need to refer to values inside events, we will by slight abuse of notation use the attribute names as functions: for $e = (eid, cid, t, a_1, \ldots a_n)$, we can then use $\mathcal{T}(e)$ to refer to t, $A_1(e)$ to a_1, and so on.

Let us give a simple example that illustrates how a set of events can be represented.

Example 1. Consider a quote creation process, e.g., for credits in retail banking. First, the request is reviewed (Review request). Then, standard terms are calculated, if applicable (Calculate terms). Otherwise, custom terms are defined (Define terms). In either case, the contract is subsequently prepared (Prepare contract) and finally sent out (Send quote). The events generated during process execution have the following attributes, besides timestamp, case ID, and event ID: *event_name*, giving the event a human-interpretable meaning; and *status*, describing the status change that occurs (i.e., the resulting case status) when the event occurs. We then have the schema $(Eid, Cid, \mathcal{T}, event\_name, status)$. We provide an example for such an event log in Table 1, with events ordered by end timestamp—the assumption that timestamps are unique gives a total order on the set of events. This simplistic example is missing case attributes, as we will introduce another relation to take care of those.

Definition 2 (Cases and Their Events). *A case set C is a relation $(Cid, B_1, \ldots, B_\ell)$, where (Cid) is the sole key candidate and $B_1, \ldots, B_\ell$ are the case attributes. In other words, this means that no two distinct entries of C may share the same case id; case ids are thus unique across the relation C. We refer to tuples $c \in C$ as cases, and by slight abuse of notation we shall identify c via its value for Cid. For a given event log L and case $c \in Cid$, we define its set of events (or event set) $E_c \subseteq L$ as $\{e \mid e \in L, Cid(e) = c\}$.*

Due to the functional dependency of L on $(Cid, \mathcal{T})$, we have that every $e' \in E_c$ has a unique timestamp value for $\mathcal{T}$ and thus $\succeq$ acts as a total order over the elements in E_c for any $c \in Cid$.

Example 2. Let us extend the event collection as specified in Example 1 with case attributes. Our case set is defined as (*Cid, customer_ID, terms*). Due to Definition 2, it follows that the attribute *customer_ID* identifies the case's customer and the attribute *terms* logs the terms that apply for the case (standard or custom terms). In the event log from Example 1, we have two cases; case 0001 has the *customer_ID C*0001 and standard terms, whereas case 0002 has the the *customer_ID C*0002 and custom terms.

Nested Structures in Event Logs. In the literature, and in the implementation of SIGNAL, event logs are in fact nested relations. In order to simplify the presentation, we do not assume such a nested relational model for this paper. The first reason is that the (flat) relational model is conceptually easier to grasp. The second reason is that it allows us to compare the expressiveness of SIGNAL with the expressiveness of relational algebra, without any support for nested relations. Because SIGNAL does not allow arbitrary nested structures this simplification seems justified, as we do not need the full power of nested relational algebras [2].

3   The Conjunctive Core of SIGNAL

We begin by stating the formal syntax of a subset of SIGNAL. We restrict ourselves to a fragment of SIGNAL which is already expressive but hopefully still easy to grasp.

Definition 3 (SIGNAL Conjunctive Core). *To simplify the presentation and focus on the core aspects of the SIGNAL query language, we define a subset of SIGNAL queries, that we shall call* SIGNAL *Conjunctive Core, or SCC for short. We define the syntax of SCC queries in Extended Backus-Naur Form*[2].

```
<scc>              :=  SELECT <varlist>
                         FROM <Eventlog name>
                       [WHERE <conditions>]
<conditions>       :=  <condition> AND <conditions> | <condition>
<condition>        :=  <var> = <var>   | <var> = <const>
                       <var> MATCHES <pattern> |
                       BEHAVIOUR <behaviours> MATCHES <pattern>
<varlist>          :=  <var> |   <var> , <varlist>
<behaviourCond>    :=  <var> = <var> | <behaviourCond> AND <behaviourCond>
<behaviours>       :=  <behaviourCond> AS <var> | <behaviours> , <behaviours>
<pattern>          :=  <pattern> ⤳ <pattern> | <pattern> → <pattern> | <id> |
                       <pattern>* | ANY | START <pattern> | <pattern> END
<id>               :=  string | <var> | <id> OR <id> | NOT ( <id> )
```

We assume the set of variables (<var>) and constants (<const>) to be system defined and omit a formal definition.

---

[2] Recall that as a reference, the complete syntax for SIGNAL, beyond *SCC*, can be found in the official guide provided athelp.sap.com/docs/signavio-process-intelligence/signal-guide/syntax (last accessed at 2024-02-20).

Example 3. The simplest *SCC* queries use only sets of equalities and constant assignments.

```
SELECT case_id, event_name, event_time
  FROM eventlog
 WHERE case_id = 2 AND event_name = "package received"
```

We can also define pattern matching clauses.

```
SELECT case_id, event_name, event_time
  FROM eventlog
 WHERE event_name MATCHES ('package_sent' ⤳ 'package_accepted')
```

Lastly, we give an example for a behaviour pattern formula.

```
SELECT case_id, event_name, event_time
  FROM eventlog
 WHERE BEHAVIOUR A AS event_name = 'Review request' AND status = NEW,
                  B AS event_name = 'Send quote' AND status = SENT
       MATCHES (A ⤳ B)
```

As the name of *SCC* suggests, we are focusing on all SIGNAL queries that correspond to the well known formalism of *conjunctive queries* in database theory, with the only extension being the ability to capture *patterns* over event sets. In the next section, we will also briefly consider more expressive fragments of SIGNAL, such as those that permit to have nested queries (also known as subqueries).

We proceed to define the set of patterns, beginning with simple patterns that can only refer to values of a single event attribute.

Definition 4 (Simple Pattern). *We are given an event log L. A simple pattern over L is a pair $\langle A_i, \mathcal{P}_s \rangle$, where A_i is an event attribute in L and $\mathcal{P}$ is a simple pattern formula. We first define the notion of* event identifiers, *which may be used in simple pattern formulas. We shall define the set of event identifiers inductively.*

- *Every value $a \in A_i$ is an event identifier.*
- *If e and e' are event identifiers, then so are: $e \vee e'$, $not(e')$.*

Now we can define the set of simple event patterns.

- *any is a simple pattern formula.*
- *Every event identifier is a simple pattern formula.*
- *If P' and Q are simple pattern formulas, then so are:*
  $P' \to Q$, $P' \rightsquigarrow Q$, P'^*, $start(P')$, $(P')end$.

Let us highlight that similar patterns are common in process mining approaches utilising regular languages, such as implementations of DECLARE [6] and graph-based PQMs (regular path queries) [4,8].

Example 4. We shall use the schema we presented in Example 1 to give an example of a simple pattern formula. Let us first fix the following constants.

$n_0 = $ 'Review Request' $n_1 = $ 'Calculate terms' $n_2 = $ 'Send quote'

An example of a simple pattern formula is: $p = (n_0 \to ((n_1 \vee n_0) \rightsquigarrow n_2)^* end)$.

Simple patterns cannot address conditions over multiple event attributes. To overcome this limitation, *SCC* also provides patterns that can refer to behaviours over multiple event attributes. We call these behaviour patterns.

Definition 5 (Behaviour Pattern). *We are given an event log L, with schema $(Cid, Eid, \mathcal{T}, A_1, \ldots, A_n)$. A behaviour pattern over L is a pair $\langle B, \mathcal{P}_b \rangle$, where B is a* behaviour matching set *and $\mathcal{P}_b$ is a* behavioural pattern formula. *A* behaviour matching *is a function $\sigma_x : A_1 \times \cdots \times A_n \to \{\top, \bot\}$ from the event attributes to either $\top$ or $\bot$. We can also identify each such function with an identifier x. We first define the set of behaviour identifiers:*

- *Every value x for some behaviour matching σ_x, is a behaviour identifier*
- *If b and b' are behaviour identifiers, then so are $b \vee b'$, $not(b)$.*

With this, we can now present the definition of behavioural pattern formulas.

- *any is a behavioural pattern formula.*
- *Every behavioural identifier is a behavioural pattern formula.*
- *If P' and Q are behavioural pattern formulas, then so are: $P' \to Q, P' \leadsto Q, P'^*, start(P'), (P')end$.*

Example 5. To give some intuition for behaviour pattern formulas, we focus on the behaviour matching function, which is the only difference to simple pattern formulas. We fix the schema to the one in Example 1.

$$b_1 = (\mathbf{status} = \text{NEW} \vee \mathbf{event\_name} = \text{Review Request})$$

In b_1 we see one natural example of how to fix such functions: we define a logical formula and evaluate it over the tuple in the event log. Any tuple that satisfies it is matched to $\top$, otherwise to $\bot$. We note, however, that Definition 5 does not impose a specific formalism on behaviour matching functions, as long as tuples in event logs can be accepted (returning $\top$) or rejected (returning $\bot$).

Note that behaviour patterns are strictly more general than simple patterns, and indeed one could define simple patterns as a special case of behavioural patterns, where the behaviour matching function is only considering the values in a single event attribute. In practice, there are also some limitations: while in Definition 5 we give no limit on the size of the behaviour matching set (i.e. the number of matching functions), the SIGNAL guide – as of the writing of this paper – states that only *eight* such functions are allowed at once. We consider this and other such technical deviations to be of little importance to the effort of formalising SIGNAL.

To define how simple or behavioural patterns are evaluated, we introduce the concept of a segment, which is an interval inside an event set of a case.

Definition 6 (Segment). *Given an event log L, a case $c \in Cases(L)$ and its event set $E_c \subseteq L$, a segment $s \subseteq E_c$ is a subset of E_c that contains for two time points $t_b, t_e \in \mathcal{T}$ all events $e \in E_c$ such that $t_b \leq \mathcal{T}(e) \leq t_e$, where we require that $\exists e', e'' \in s$ such that $\mathcal{T}(e') = t_b$ and $\mathcal{T}(e'') = t_e$. For simplicity, we can identify s simply by $\langle t_b, t_e \rangle$, and we also introduce some needed notation: $b(s) = t_b$, $e(s) = t_e$.*

Since segments are simply sets of events, we are free to use set operations on segments and still get segments as output[3]. As a special case, we also introduce the empty set segment $s_\emptyset$, which will play a technical role in the definition below.

We can now define when a segment satisfies a given simple pattern.

Definition 7 (Simple Pattern Segment Satisfaction). *Given an event log L, a case $c \in Cases(L)$ and its event set $E_c \subseteq L$ and a simple pattern $\langle A_i, \mathcal{P}_s \rangle$ over L, we say that a segment s satisfies $\langle A_i, \mathcal{P}_s \rangle$, if the following holds:*

1. *If $\mathcal{P}_s = a \in A_i$, then $\exists e \in s$ s.t. $A_i(e) = a$;*
2. *if $\mathcal{P}_s = P' \vee Q$, then s satisfies either $\langle A_i, P' \rangle$ or $\langle A_i, Q \rangle$;*
3. *if $\mathcal{P}_s = not(P')$, then s must not satisfy $\langle A_i, P' \rangle$;*
4. *if $\mathcal{P}_s = start(P')$, then s satisfies $\langle A_i, P' \rangle$ and $\not\exists e \in E_c$ with $\mathcal{T}(e) < b(s)$;*
5. *if $\mathcal{P}_s = (P')end$, then s satisfies $\langle A_i, P' \rangle$ and $\not\exists e \in E_c$ with $\mathcal{T}(e) > e(s)$;*
6. *if $\mathcal{P}_s = any$, then s trivially satisfies $\langle A_i, \mathcal{P}_s \rangle$;*
7. *if $\mathcal{P}_s = P' \rightsquigarrow Q$, then $\exists s', s'' \subseteq s$ with $e(s') < b(s'')$ and s' satisfies $\langle A_i, P' \rangle$ and s'' satisfies $\langle A_i, Q \rangle$;*
8. *if $\mathcal{P}_s = P' \rightarrow Q$, then $s' \cup s'' = s$ in addition to all conditions from item 7;*
9. *if $\mathcal{P}_s = P'^*$, then either $\exists s' \subseteq s$ s.t. s' satisfies $\langle A_i, P' \rangle$ and $s \setminus s'$ satisfies $\langle A_i, P'^* \rangle$, or we have $s = s_\emptyset$;*

Note that in condition 8, we require that s' and s'' partition s. In other words, the segment satisfying P' must be directly followed by the one satisfying Q.

Example 6 (Segment Examples). In Fig. 1 we give an example of satisfying segments. For the sake of brevity, we only provide an event set, for some undefined case, and thus omit the definition of a complete event log.

We proceed to give the analogous definition for behavioural patterns. We shall reuse most of the conditions from Definition 7, as the structure of behavioural pattern formulas is mostly the same as for simple pattern formulas, except for the use of behavioural matching functions.

Definition 8 (Behavioural Pattern Segment Satisfaction). *Given an event log L, a case $c \in Cases(L)$ and its event set $E_c \subseteq L$ and a behavioural pattern $\langle B, \mathcal{P}_b \rangle$ over L, we say that a segment s satisfies $\langle B, \mathcal{P}_b \rangle$, if the following conditions hold:*

1. *If $\mathcal{P}_b = x$ where $\sigma_x \in B$, then $\exists e \in s$ s.t. $\sigma_x(att(e)) = \top$;*

in addition to conditions 2 to 9 from Definition 7.

---

[3] There are caveats, though: when using set minus, an expression $s \setminus s'$, can only produce a segment in the sense of Definition 6, if the time interval in s' and s satisfy the Allen's relations [3] *finishedBy* or *startedBy* or $s = s'$. For set union, we require that the two segments involved have a non-empty intersection. Intersection has no special requirements.

Table 2. An example event set.   **Table 3.** Simple pattern formulas.

event_name	timestamp
$e_1 = $ order_received	$t_1 = 15.10.23$
$e_2 = $ package_collected	$t_2 = 16.10.23$
$e_3 = $ package_checked	$t_3 = 17.10.23$
$e_4 = $ package_sent	$t_4 = 23.10.23$

pattern	Satis. segment
$e_2 \rightsquigarrow e_4$	$\langle t_2, t_4 \rangle$
$any \rightsquigarrow e_4$	$\langle t_3, t_4 \rangle$
$start(e_1 \rightarrow e_2)$	$\langle t_1, t_2 \rangle$
$e_1 \rightarrow (e_2 \rightsquigarrow e_4)^*$	$\langle t_1, t_4 \rangle$

event_name	timestamp
$e_1 = $ order_received	$t_1 = 15.10.23$
$e_2 = $ package_collected	$t_2 = 16.10.23$
$e_3 = $ package_checked	$t_3 = 17.10.23$
$e_4 = $ package_sent	$t_4 = 23.10.23$

pattern	Satis. segment
$e_2 \ ^3 \ e_4$	⊠t_2, t_4⊠
$any \ ^3 \ e_4$	⊠t_3, t_4⊠
$start \ (e_1 \ ⊠ \ e_2)$	⊠t_1, t_2⊠
$e_1 \ ⊠ \ (e_2 \ ^3 \ e_4)^⊠$	⊠t_1, t_4⊠

Fig. 1. Table 2 contains an event set consisting of events with a single event attribute (event_name) and a timestamp (shortened and formatted for brevity and human readability). Table 3 contains four simple pattern formulas and their satisfying segments over the event set of Table 2.

With this machinery, we can now clearly define how an SCC query is to be evaluated. For the parts of SCC that correspond to regular SQL, we use the familiar relational semantics. This only leaves the issue of how to deal with patterns. Informally speaking, patterns serve to filter out certain cases from our event log, where patterns are evaluated against the event set of each case individually. If no satisfying segment according to Definition 7 (or resp. Definition 8) exists, then all events of this case are to be removed.

Our plan is to translate SCC to an extended form of RA, by introducing a new selection operator that takes as input an event log, and is given a (simple or behavioural) pattern as a parameter. Analogous to normal selection, it will filter out parts of the input table, namely those events in our event log that are from cases that do not have a satisfying segment for the pattern.

Definition 9 (Pattern Selection Operator). *Given a simple or behavioural pattern $\mathcal{P}$, and an event log L, we define a pattern selection operator $\sigma_\mathcal{P}$ as follows:*

$$\sigma_\mathcal{P}(L) = \{e \mid e \in L \wedge E_{Cid(e)} \subseteq L \text{ has satisfying segment for } \mathcal{P}\}$$

We shall call the extension of RA with this new operator simply $RA^\mathcal{P}$. Formally, from the definitions for patterns and segment satisfaction, combined with Definition 9, we get the following corollary.

Proposition 1. *Given an SCC query q, there exists an $RA^\mathcal{P}$ expression φ, s.t. for every event log L it holds that:*

$$\varphi(L) \equiv q(L).$$

Note that in Proposition 1 above, we present both the query and the relational algebra expression as a function from an event log to a set of events.

The translation of SCC to $RA^{\mathcal{P}}$ follows almost exactly the translation of CQs to RA – recall that we selected SCC as a subset of SIGNAL that structurally matches CQs, with the sole addition of MATCHES clauses with pattern formulas. As we consider the translation of CQs to RA to be folklore, we omit technical details, and simply give an example below.

Example 7 (SCC to $RA^{\mathcal{P}}$). Consider the following SCC query from Example 3:

```
SELECT case_id, event_name, event_time
  FROM eventlog
 WHERE event_name MATCHES ('package_sent' ⤳ 'package_accepted')
```

The query corresponds to the following $RA^{\mathcal{P}}$ expression:

$$\pi_{xtcase\_id,event\_name,event\_time}\left(\sigma_{\mathcal{P}}(\text{eventlog})\right)$$

where $\mathcal{P} = \langle \text{event\_name}, \text{package\_sent} \rightsquigarrow \text{package\_accepted} \rangle$.

SIGNAL Queries with Nesting and Aggregation. In this section, we focused deliberately on a subset of SIGNAL which already captures its abilities to express complex patterns over event sets. We limit our study to such simple queries and deliberately leave out more expressive fragments, such as SIGNAL queries with nesting or the ability to aggregate over groups of attributes. The primary reason for this is that we believe that the ability to express patterns over event logs is already enough to set SIGNAL apart from standard relational algebra, as we shall show in the rest of this paper. Exploring how nesting further adds expressive power and affects data complexity is left for future work.

4 Towards Lower and Upper Bounds on the Data Complexity of SCC Evaluation

After having presented the semantics of SCC in Sect. 3, we next consider how it fits into the complexity landscape of existing query languages. We will first show that SCC is indeed more expressive than relational algebra, and cannot be captured by it without the use of extensions, as were introduced in Sect. 3. Next, we provide some preliminary work towards an exact characterisation of the evaluation problem by stating an upper bound via a reduction from SCC to semi-positive Datalog.

4.1 Inexpressibility of SCC in Relational Algebra

In Sect. 3 we provide a translation of SCC to an extension of RA. It remains to show that this extension is actually necessary, i.e., that it is beyond the expressive capabilities of regular RA to capture the semantics of SCC. We can achieve this by largely relying on well-known "text-book"-level results.

Theorem 1 (Inexpressibility of *SCC* in RA). *There exists an SCC query q, such that it is not possible to find an RA expression φ for which we have that:*

$$q(L) \equiv \varphi(L)$$

for any given event log L.

Proof of Theorem 1. Our argument will make use of Codd's Theorem [7], which states that the expressive power of relational algebra is the same as the domain-independent domain calculus. This in turn is equivalent in expressive power to first-order logic (FO). Hence, all we need to show is that the semantics of *SCC* cannot be captured in FO. To this end, we make use of the fact that FO cannot recognise *evenness* in its models, i.e., models which only have an even number of elements. In the context of the relational model, this would correspond to the number of tuples in a relation. For an easy to follow proof of this inexpressibility, we refer interested readers to [14], where the desired result is captured by Proposition 3.3. What is left to show is that there exists some *SCC* query that can express evenness. Specifically, this means that it will distinguish between event logs of even size and those that have an uneven size. We will make use of simple pattern matching for this, but we first need to establish that in *SCC* we can concatenate the events of all cases into one, single event set. This is possible by simply projecting out the case_id, and replacing it with a single ID across the event log. This gives us a new event log, with the same number of events but where all events share the same case_id. There are some technical complexities here, such as making sure the timestamps are still unique. We can accomplish this by also replacing the timestamps with stand-holder values, using for example the row number to indicate number of seconds. Afterwards, we filter this new event log by a simple pattern formula : $start(any \rightarrow any)^* end$. We now have two possible outputs: either the event set (which corresponds to all events in this event log) has even size and thus has a satisfying segment: in this case we return every event in our event log. Alternatively, if the event log is of uneven size and thus permits no satisfying segment, we return nothing. Note the use of the *start* and *end* constructors, which are critical to capture this property: they ensure that the satisfying segment must span the entire event set.   □

As a minor comment to the proof above, it should be noted that most implementations of SQL can effectively capture evenness via basic arithmetic. However, from a complexity standpoint, SQL with arithmetic is trivially undecidable, just as FO extended with (Peano) arithmetic is undecidable, so equating SQL with RA is sensible when talking about computational complexity.

On the Role of the Kleene Star in Pattern Formulas. To give some more insights into which components of a pattern formula are responsible for its expressive power, let us consider a subset of simple pattern formulas that do not involve the * constructor. In this case, patterns could not express repetitions, and every pattern formula would simply correspond to some fixed ordering over the event set. While we do have disjunction and negation, they are limited to event identifiers; more complex pattern constructors, such as →, do not appear inside disjunctions. Thus, if we consider a pattern formula without the Kleene star, we could rewrite it into an FOL formula which captures the fixed ordering expressed in it, and so this fragment would stay within the expressive power of RA. Thus, it is not a coincidence that the counter-example we use in the proof above makes use if it, it is the crucial component that allows us to escape capture by FOL. Let us also highlight that the Kleene star is apparently central to well-established approaches to reasoning about event log data. In particular DECLARE, a declarative language for process mining frequently used and studied by academics, features the Kleene star in all of its common expression templates when expressed in POSIX regular expressions [6] and many of the most central DECLARE expressions feature the Kleene star when translated to SIGNAL [5].

Let us note that in contrast, the any constructor is *not* at the heart of the problem. Intuitively, the example we use in the proof – $start(any \rightarrow any)^*end$ – is equivalent to the regular expression ^(..)*$, and hence can be verified with a run-off-the-mill regular expression solver. Analogously, $start('a' \rightarrow 'a')^*end$ (^(aa)*$ as a standard regular expression) still provides the needed counter-example, albeit not for *any* event log, but for a specific event containing only events with name "a"; for SIGNAL, the example can be trivially generalised to behaviours that match every event in an event log.

4.2   Translation of *SCC* to Semi-positive Datalog

Following the result that RA itself cannot capture every *SCC* query, we next want to find an existing formalism that does capture the semantics of *SCC* and thus get an upper bound on the complexity of evaluating *SCC*. We proceed to show the following result.

Theorem 2 (Reduction of *SCC* to Semi-positive Datalog). *Given a SCC query q, and an event log L, we can define a semi-positive Datalog program P with a linear ordering over timestamps, such that:*

$$q(L) \equiv P(L),$$

where we understand $P(L)$ to be the output (generated tuples) of the Datalog program P when evaluated with facts from the event log L.

Proof of Theorem 2. To simplify the presentation of the Datalog program, we make use of the functional dependencies of event logs, and identify each event with just the pair of its event id and case id, *Eid*, *Cid*, in addition to the timestamp as it plays a crucial role in capturing the semantics of patterns. We shall make use of the following extensional predicates. The first is $event(c, e, t)$, where the three arguments are, respectively, the case id, the event id, and the timestamp. We encode with $start(c, t)$ (resp. $end(c, t)$) the timestamp of the chronologically first (resp. last) event in the event log of the case with id c. We also have the ability to refer to individual attribute values via predicates of the form $A_i(c, e, a)$, which indicate that an event with case id c and event id e has value

a for attribute A_i. To ensure safety of any rules with negation, we require predicates *time* and *events* which identify, resp., all timestamps and event ids.

In order to capture when any given simple pattern has a satisfying segment, we use an intensional predicate $P_\varphi(t_s, t_e, c)$, which captures the satisfying segments (t_s, t_e) of a given pattern formula φ inside the events set E_c. We use the convention of using upper case for variables and lower case for constants. Here are the rules for intensional predicates P_φ:

$$P_a(T, T, C) \leftarrow segment(T, T, C), event(C, E, T), A_i(C, E, a)$$
$$P_{\neg a}(T, T, C) \leftarrow segment(T, T, C), event(C, E, T), \neg A_i(C, E, a)$$
$$P_{A \vee B}(T_s, T_e, C) \leftarrow P_A(T_s, T_e, C) \vee P_B(T_s, T_e, C)$$
$$P_{\neg(A \vee B)}(T_s, T_e, C) \leftarrow P_{\neg A}(T_s, T_e, C), P_{\neg B}(T_s, T_e, C)$$
$$P_{\neg\neg A}(T_s, T_e, C) \leftarrow P_A(T_s, T_e, C) \text{ }^4$$
$$P_{start(A)}(T_s, T_e, C) \leftarrow P_A(T_s, T_e, C), start(C, T_s)$$
$$P_{(A)end}(T_s, T_e, C) \leftarrow P_A(T_s, T_e, C), end(C, T_e)$$
$$P_{A \rightsquigarrow B}(T_s, T_e', C) \leftarrow segment(T_s, T_e', C), P_A(T_s, T_e, C), P_B(T_s', T_e', C), T_e < T_s'$$
$$P_{A \rightarrow B}(T_s, T_e', C) \leftarrow segment(T_s, T_e', C), P_A(T_s, T_e, C), P_B(T_s', T_e', C), T_e < T_s',$$
$$\neg event(C, E, T''), T_s < T'', T'' < T_s', time(T''), events(E)$$
$$P_{A^*}(T_s, T_e', C) \leftarrow segment(T_s, T_e', C) \vee \big(P_A(T_s, T_e, C), P_{A^*}(T_s', T_e', C), T_e < T_s',$$
$$\neg event(C, E, T''), T_e < T'', T'' < T_s', time(T''), events(E)\big)$$

In order for this translation to capture the intended meaning, we also require the following two rules to be included in the final Datalog program.

$$segment(T, T, C) \leftarrow event(C, E, T).$$
$$segment(T_s, T_e', C) \leftarrow segment(T_s, T_e, C), segment(T_s', T_e', C), T_e < T_s',$$
$$\neg event(C, E, T''), T_e < T'', T'' < T_s', events(E), time(T'')$$

Note that in the translation above, we introduce a disjunction connective in the body of Datalog rules. This is simply "syntactic sugar".

$$Atom_1(A, B, C) \leftarrow \big(Atom_2(A, B) \vee Atom_3(A, B)\big), Atom_4(B, C).$$

is a space-efficient way to encode the following:

$$Atom_1(A, B, C) \leftarrow Atom_2(A, B), Atom_4(B, C).$$
$$Atom_1(A, B, C) \leftarrow Atom_3(A, B), Atom_4(B, C).$$

Equipped with these tools, we shall now present the translation of SCC into Datalog. Let us first consider the case of SCC queries without any pattern matching. We can then model the query as a conjunctive query and the translation to Datalog is trivial: as we only have a single table—the event log—and can only join

---

[4] Note that negation in simple pattern formulas can only occur for event identifiers so that in the translation, we need not consider the negation of more complex patterns.

its event attributes or set them to constants, we can express the query inside a single Datalog rule:

$$Output(X_1, \ldots X_n) \leftarrow event(C, E, T), A_1(E, V_1), \ldots A_n(E, V_\ell), V_1 = c_1, \ldots, V_m = c_m.$$

where $\bigcup_i X_i \subseteq \{C, E, T, V_0, \ldots, V_\ell\}$ are the output variables, as determined in the SELECT clause. The joining of two event attributes is done by simply using the same variable symbol and when specifying the attribute via A_i.

As such, we need only consider how to deal with pattern matching. For this, we make use of the intensional predicate P: for any given pattern formula φ, we add P_φ to the body of the rule above. For every subexpression ϕ that occurs inside φ, we also add the rule P_ϕ to our Datalog program.

Given Definition 7 and the construction above, it should not be difficult to see that our translation of simple patterns captures the intended meaning exactly. Each rule for a given pattern is intended to specify all the necessary conditions for the satisfying segment of that pattern to be fulfilled. We leave out the case for behaviour patterns, as it is analogous, safe that instead of event identifiers, we have behaviour identifiers.

What is left is to show is that our produced Datalog program after applying this translation, falls into the class of semi-positive Datalog. We note that negation can only occur in the translation of patterns and in this translation we only negate the EDB predicates *event* and A_i, which come from the event log.   □

The reduction to semi-positive Datalog to capture *SCC* gives us an upper bound on the data complexity of evaluating *SCC* queries, namely P. This would directly lead to a completeness result, if one could show that the lower bound for *SCC* evaluation is also in P, in other words, that the bound is tight.

Open Question. Is the data complexity of SCC evaluation P-hard?

5   Conclusion and Future Work

We formally presented and analysed a subset of the SIGNAL query language, which forms the heart of a commercial process mining software product. We provided the intended semantics for the pattern matching formulas that are used in SIGNAL to then show that SIGNAL cannot be captured in relational algebra. Also, we conducted a preliminary analysis of the data complexity of the problem of evaluating *SCC* queries, establishing an upper bound of P-hard. The findings highlight that there is more to process querying methods than "just relational algebra" and indicate that the temporal reasoning capabilities, which increase expressive power and raise the complexity bound, may indeed be good motivators of domain-specific implementations. Future work could answer the question posed in this paper regarding the tight upper bound of the SIGNAL core and conduct analogous analyses of other process querying methods such as PQL [15]. For future work, we also leave the formal analysis of new process mining approaches such as object-centric process mining [13], graph-based process querying [4,8,10,12], or the modern wide-column store and OLAP-style systems (such as Apache Arrow) that underlie SIGNAL. Another promising future research direction is the integration of our results with expressive power analyses of "academic" declarative process querying approaches, such as behavioural profiles [17] and DECLARE [9].

References

1. van der Aalst, W.M.P.: Process Mining - Data Science in Action. Springer, Cham (2016)
2. Abiteboul, S., Hull, R., Vianu, V.: Foundations of Databases. A.-W. (1995)
3. Allen, J.F.: Maintaining knowledge about temporal intervals. Commun. ACM **26**(11), 832–843 (1983)
4. Beheshti, A., Benatallah, B., Motahari-Nezhad, H.R.: Processatlas: a scalable and extensible platform for business process analytics. Softw. Pract. Exp. **48**(4), 842–866 (2018). https://doi.org/10.1002/SPE.2558
5. Bergmann, A., Rebmann, A., Kampik, T.: BPMN2Constraints: breaking down BPMN diagrams into declarative process query constraints. In: Proceedings of BPM. CEUR Workshop Proceedings, vol. 3469, pp. 137–141. CEUR-WS.org (2023)
6. Di Ciccio, C., Bernardi, M.L., Cimitile, M., Maggi, F.M.: Generating event logs through the simulation of declare models. In: Barjis, J., Pergl, R., Babkin, E. (eds.) EOMAS 2015. LNBIP, vol. 231, pp. 20–36. Springer, Cham (2015). https://doi.org/10.1007/978-3-319-24626-0_2
7. Codd, E.F.: Relational completeness of data base sublanguages. Research Report/RJ/IBM RJ987 (1972)
8. Esser, S., Fahland, D.: Multi-dimensional event data in graph databases. J. Data Semant. **10**(1–2), 109–141 (2021). https://doi.org/10.1007/S13740-021-00122-1
9. De Giacomo, G., De Masellis, R., Grasso, M., Maggi, F.M., Montali, M.: Monitoring business metaconstraints based on LTL and LDL for finite traces. In: Sadiq, S., Soffer, P., Völzer, H. (eds.) BPM 2014. LNCS, vol. 8659, pp. 1–17. Springer, Cham (2014). https://doi.org/10.1007/978-3-319-10172-9_1
10. Jalali, A.: Graph-based process mining. In: Leemans, S., Leopold, H. (eds.) ICPM 2020. LNBIP, vol. 406, pp. 273–285. Springer, Cham (2021). https://doi.org/10.1007/978-3-030-72693-5_21
11. Kampik, T., Lücke, A., Horstmann, J., Wheeler, M., Eickhoff, D.: SIGNAL – The SAP Signavio analytics query language (2023). https://arxiv.org/abs/2304.06811
12. Khayatbashi, S., Hartig, O., Jalali, A.: Transforming event knowledge graph to object-centric event logs: a comparative study for multi-dimensional process analysis. In: Almeida, J.P.A., Borbinha, J., Guizzardi, G., Link, S., Zdravkovic, J. (eds.) ER 2023. LNCS, vol. 14320, pp. 220–238. Springer, Cham (2023). https://doi.org/10.1007/978-3-031-47262-6_12
13. Li, G., de Murillas, E.G.L., de Carvalho, R.M., van der Aalst, W.M.P.: Extracting object-centric event logs to support process mining on databases. In: Mendling, J., Mouratidis, H. (eds.) CAiSE 2018. LNBIP, vol. 317, pp. 182–199. Springer, Cham (2018). https://doi.org/10.1007/978-3-319-92901-9_16
14. Libkin, L.: Elements of Finite Model Theory. Texts in Theoretical Computer Science. An EATCS Series, Springer, Cham (2004). https://doi.org/10.1007/978-3-662-07003-1. http://www.cs.toronto.edu/%7Elibkin/fmt

15. Polyvyanyy, A.: Process query language. In: Polyvyanyy, A. (ed.) Process Querying Methods, pp. 313–341. Springer, Cham (2022). https://doi.org/10.1007/978-3-030-92875-9_11

16. Polyvyanyy, A. (ed.): Process Querying Methods. Springer, Cham (2022)

17. Polyvyanyy, A., Armas-Cervantes, A., Dumas, M., García-Bañuelos, L.: On the expressive power of behavioral profiles. Form. Asp. Comput. **28**(4), 597–613 (2016)

18. Polyvyanyy, A., ter Hofstede, A.H., La Rosa, M., Ouyang, C., Pika, A.: Process query language: design, implementation, and evaluation. Inf. Syst. **122**, 102337 (2024)

19. Vogelgesang, T., Ambrosy, J., Becher, D., Seilbeck, R., Geyer-Klingeberg, J., Klenk, M.: Celonis PQL: a query language for process mining. In: Polyvyanyy, A. (ed.) Process Querying Methods, pp. 377–408. Springer, Cham (2022). https://doi.org/10.1007/978-3-030-92875-9_13

Multivariate Anomaly Detection in Object-Centric Event Data

Luka Abb[ID] and Jana-Rebecca Rehse[(✉)][ID]

University of Mannheim, Mannheim, Germany
{luka.abb,rehse}@uni-mannheim.de

Abstract. Statistical anomaly detection is a powerful tool for identifying irregularities and improvement potential in process execution data. Existing research in this area has predominantly focused on applications in case-centric event data, considering a limited set of attributes. In this paper, we systematically study event-level anomaly detection in an object-centric context. We define and categorize the anomalies that can occur and derive detection strategies and features for object-related anomalies. We also introduce two novel object-centric datasets with labeled anomalies of varying complexity and empirically evaluate an existing detection approach on them. We find that methods developed for case-centric event logs can be effective in identifying complex and object-related anomalies, but suffer from robustness issues.

Keywords: Anomaly Detection · Object-Centric Process Mining · Multi-Perspective Process Mining

1 Introduction

Identifying irregularities in process executions is one of the principal motivations for process mining, because each irregularity points to a potential for process improvement [22]. For a manual identification of irregularities, expert analysts can inspect a discovered process model [3] to identify characteristics that do not conform to their expectations about the intended process. If a model of this process is available, the identification can be automated by means of conformance checking [10]. Often, however, such a model is not available, especially not a complete one that describes the intended behavior across all observable attributes. In these situations, anomaly detection techniques [11,17] provide a viable alternative for identifying irregularities in an automated way. These techniques are able to find statistically unusual process behavior, hinging on the (intuitive) assumption that behavior that significantly deviates from the typical patterns observed within an event log is more likely to be erroneous.

Process execution data is typically captured in event logs, i.e., collections of partially ordered event sequences with a potentially large amount of mixed-type attributes [3]. Complex dependencies between (sets of) attributes may exist within the same or across separate events. This makes event logs inherently more complex than the data in many other domains, which makes finding anomalies

© The Author(s), under exclusive license to Springer Nature Switzerland AG 2024
A. Marrella et al. (Eds.): BPM 2024 Forum, LNBIP 526, pp. 20–36, 2024.
https://doi.org/10.1007/978-3-031-70418-5_2

in process executions a challenging task [17]. This is further exacerbated by the recent paradigm shift from conventional, case-centric event logs to object-centric ones [2], which replaces the well-defined case-level event sequences with a single log-level sequence that is partially ordered along individual objects [4].

Concretely, there are two main challenges that need to be addressed when finding anomalies in object-centric event logs. First, the introduction of multiple object relations per event makes object-centric event logs inherently multivariate, which necessitates a multi-perspective approach to anomaly detection. However, most existing research focuses on anomalies that pertain to a single process perspective (i.e., event attribute), typically the control-flow [17]. While a few papers explicitly discuss multiple process perspectives and the anomalies that may arise from interactions between them [7, 22, 24], their findings and conclusions have not been applied to object-centric data. Second, object-centric event data comes with new, object-related anomalies, which previous work [6, 21] has only selectively addressed. It is thus currently not known which potential anomalies exist in object-centric event logs and how they can be detected.

In this paper, we address these challenges and adapt anomaly detection to an object-centric context. Specifically, we make the following contributions:

(a) We develop a typology which categorizes the event-level anomalies that can occur in object-centric event data into six different types (Sect. 4).
(b) We discuss how object-related anomalies can be detected in an unsupervised setting: How to group the events and which features to include to distinguish normal from abnormal events (Sect. 5).
(c) We provide two novel object-centric datasets with labelled instances of each anomaly type (Subsect. 6.1).
(d) Using these datasets, we empirically evaluate the extent to which an existing anomaly detection technique can detect each anomaly type (Sect. 6).

This paper is primarily conceptual: By developing a taxonomy, algorithmic adaptations, and labelled datasets, we provide a foundation and a benchmark for others to develop new object-centric anomaly detection approaches in the future.

2   Background and Related Work

2.1   Anomaly Detection

Statistical anomaly detection is an extensively studied problem in data mining literature [11]. It can be conceptualized as a supervised task, where the detection approach is based on observations explicitly categorized as normal or abnormal. Alternatively, it can be set up as an unsupervised task when these labels are not present. For unsupervised anomaly detection, we can further distinguish between semi-supervised approaches, where a clean, anomaly-free dataset is available, and 'true' unsupervised approaches, where no such curated dataset exists.

As an automated technique to detect irregularities in process executions without an explicit process model, anomaly detection has also received considerable attention in the process mining domain (e.g., [7, 8, 14, 16, 19–22, 25]). For a comprehensive overview, we refer to a recent literature review [17] as well as the related research streams on fraud detection [15], noise detection [18], and concept drift detection [9]. Because anomaly labels are unlikely to be available in real-life process mining applications, almost all existing approaches for process anomaly detection are unsupervised ones. In this paper, we also focus on unsupervised process anomaly detection for the same reason.

The majority of contemporary anomaly detection methods, supervised or unsupervised, leverage machine learning techniques [23], also in process mining [14,16,19,22]. Notable exceptions are approaches that emphasize explainability, for example by leveraging association rule mining [8] and likelihood graphs [7] or transition systems [24]. In addition, the majority of anomaly detection approaches require a case-centric event log. The two exceptions are Berti et al. [6], who focus on object-level instead of event-level anomaly detection, and Niro & Werner [21], who detect control-flow and attribute anomalies in object-centric event logs, but do not consider anomalies directly related to objects.

An inherent limitation of statistical anomaly detection is that it can only detect uncommon patterns in process executions, but not determine whether these patterns are actually problematic. Differentiating between rare but harmless and actual erroneous execution patterns usually requires a domain expert.

2.2 Object-Centric Process Mining

A central assumption of traditional process mining is that each event log has a unique case notion that can be used to group events into sequences, each of which corresponds to one execution of the underlying process [3]. Object-centric process mining [2] weakens this assumption, instead postulating that events be associated with one or multiple *objects*. Each object has a *lifecycle* that consists of all events associated with it. Object interactions take place in so-called *synchronization* events that are associated with multiple objects. An object-centric event log can be transformed into a case-centric one by choosing one of the objects as a case notion and thus *flattening* the data.

The object-centric paradigm avoids three problems that often occur when extracting process execution data from an information system and aggregating it into a case-centric event log: (i) the duplication of activities (convergence) [2], (ii) misleading sub-sequences of events (divergence) [2], and (iii) the disappearance of events not associated with the case notion (deficiency) [5]. Object-centric event logs are thus more informative and closer to the source systems than case-centric ones. In turn, they are less structured and for many applications require researchers to develop new or adapt existing process mining techniques.

In this paper, we use a minimal notion of an object-centric event log, only assuming that we know event-object associations and object types. We do not assume any explicit knowledge about attribute affiliations with or relations between objects (unlike, for example, the proposed OCEL 2.0 storage format [1]).

3   Running Example

To illustrate the concepts discussed in the following sections, we use a synthetic object-centric event log as a running example. The underlying process is a variant of an order management process referenced in many other publications on object-centric process mining (e.g., [4,5]). Our version proceeds as follows: A customer *places an order*, consisting of several items. For premium customers, the order's price is immediately reduced by 10% (*apply discount*). Then, the customer receives an *order confirmation*. The items are handled next. *Out-of-stock items* are *reordered* and then *picked*, whereas in-stock items are *picked* directly. At regular intervals, picked items are packaged and dispatched for delivery (*create package, send package*). If a *delivery fails*, it is attempted again. At any point after order confirmation, the customer may *pay the order*. The process concludes once all items are successfully delivered in their respective packages (*package delivered*) and the order has been paid.

Table 1 shows an event log with three exemplary executions of this process. They begin with order placement and handle a single item in this order, up to the delivery of the package with the item to the customer. The first execution (e1 to e10) is typical, without anomalies. The other two (e11 to e22, e23 to e25) contain various exemplary anomalies, which we will discuss in the next section.

Table 1. Running example log with one normal (e1–e10) and two anomalous process executions. Packages are omitted because they do not pertain to any anomalies.

eID	Activity	Timestamp	Price	Status	Orders	Items	Customer
1	place order	Wed 14:00	1,000	premium	o1	i1, i2	c1
2	apply discount	Wed 14:05	900	premium	o1	–	c1
3	confirm order	Wed 14:10	900	premium	o1	i1, i2	c1
4	item out of stock	Wed 18:00	500	–	–	i1	–
5	reorder item	Wed 19:00	500	–	–	i1	–
6	pay order	Thu 15:30	900	premium	o1	–	c1
7	pick item	Thu 17:00	500	–	–	i1	–
8	create package	Fri 09:00	500	–	–	i1, i2	–
9	send package	Thu 13:30	500	–	–	–	–
10	package delivered	Mon 10:00	500	–	–	–	–
11	place order	Wed 14:00	50,000	standard	o2	i3, i4	–
12	apply discount	Wed 14:05	45,000	standard	o2	–	–
13	confirm order	Wed 14:10	45,000	standard	o2	i3, i4, i5	–
14	item out of stock	Wed 18:00	200	–	–	i3	–
15	pay order	Thu 16:00	45,000	standard	o2	–	–
16	reorder item	Fri 08:00	200	–	–	i3	–
17	pick item	Fri 17:00	200	–	–	i3, i4	–
18	create package	Sat 09:00	200	–	–	i3, i4	–
19	send package	Sat 13:00	200	–	–	–	–
20	confirm order	Sat 15:00	45,000	standard	o2	i3, i4	–
21	package delivered	Sun 10:00	200	–	–	–	–
22	<rare activity>	Sun 15:00	–	–	–	–	–
23	place order	Mon 14:00	800	premium	o3	i6	c2
24	confirm order	Mon 14:10	800	premium	o3	i6	c2
25	pay order	Tue 12:00	800	premium	o3	–	c3
...							

4   A Typology for Anomalies

In this section, we introduce a typology of the event-level anomalies that can occur in object-centric event data. For this typology, we rely on the fact that event data can be conceptualized as two-dimensional (cf. Table 1). In the first dimension, it consists of (partially ordered) events. In the second dimension, each event is further characterized by a set of attributes. Drawing on this conceptualization, we categorize process anomalies along these two fundamental dimensions:

- **Cardinality of involved events.** An *atomic* anomaly concerns only a single event, whereas a *collective* anomaly refers to at least two events. In a tabular event log, collective anomalies extend across more than one row.
- **Cardinality of involved attributes.** A *univariate* anomaly refers to a single attribute, whereas a *multivariate* anomaly results from a combination of two or more attributes. In a tabular event log, multivariate anomalies are those that involve more than one column.

We further distinguish multivariate anomalies into those that involve "standard" attributes and those that involve object identifiers. We do this because object associations (as separate entities) are conceptually distinct from event attributes (as mere values). Hence, they differ considerably at the data level and require a different detection approach (which we will discuss in detail later on). This results in a total of six anomaly types, summarized in Fig. 1.

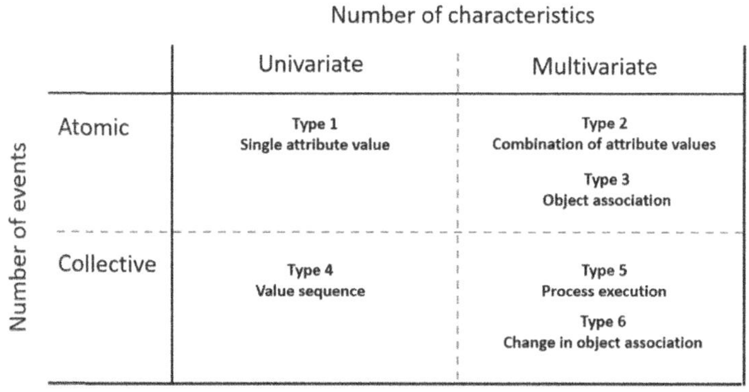

Number of characteristics

		Univariate	Multivariate
	Atomic	**Type 1** Single attribute value	**Type 2** Combination of attribute values **Type 3** Object association
	Collective	**Type 4** Value sequence	**Type 5** Process execution **Type 6** Change in object association

(Left vertical axis label: Number of events)

Fig. 1. Categorization of anomaly types

The distinction between atomic/collective and univariate/multivariate anomalies originally stems from data mining literature [11,13]. In principle, these dimensions have already been applied in previous publications on process anomaly detection [7,17,22,24], though only selectively and with different terminology. Building on this work, we provide a comprehensive typology of process anomalies. In the following, we describe each anomaly type in detail and illustrate them with examples from the running example in Table 1.

Type 1: Anomalous single attribute value. A type 1 anomaly is a classic outlier data point. It manifests as a globally abnormal singular attribute value. Examples are the price of an order that is several magnitudes higher than usual (e11–15), or an activity that is only executed a single time (e22).

Type 2: Anomalous combination of attribute values. A type 2 anomaly is characterized by a combination of two or more attribute values within a single event, where each individual value by itself is not unusual, but their collective presence is. Consider, for example, the *apply discount* activity, which is executed when a customer with membership status *premium* places an order, but is not executed for *standard* customers. Hence, event e12 with activity *apply discount* and membership status *standard* would be considered anomalous.

Type 3: Anomalous object association. A type 3 anomaly occurs when an event is associated with at least one object in an unusual way. For example, if a *pick item* event is usually associated with exactly one item, an association with zero or more than one items will be an anomaly (e17). In synchronization events, this anomaly type also (implicitly) encompasses unusual relations between objects. For example, a *place order* event associated with an order but not with a customer (e11) implies a missing relation between these two object instances.

Type 4: Anomalous value sequence. A type 4 anomaly refers to an attribute value that is unusual in the context of at least one previous value of the same attribute within an ordered collection of events, e.g., an object's lifecycle. The archetypal example of this type are control-flow anomalies, e.g., activities that are missing, executed out of order, or (unnecessarily) repeated. In the running example, this happens in e20, when order o2 is confirmed a second time.

Type 5: Anomalous process execution A type 5 anomaly arises from dependencies between the values of at least two different attributes in separate events, essentially combining types 2 and 4. Consider the *item out of stock* event, which is typically followed by *reorder item* within two hours. If there is an item for which these two events are more than one day apart (e16), this constitutes an anomaly, even though the combination of activities and timestamps within each event is not unusual and the value sequences of both activity and timestamp are normal when considered in isolation. Another example is an order lifecycle without *apply discount*. Depending on whether the associated customer has status *standard* or *premium*, this can either be normal or abnormal (e24).

Type 6: Anomalous change in object association Whereas type 3 relates to static object interactions, type 6 covers anomalies that arise from dynamic interactions, i.e., interactions that extend across more than one event. Examples are an order that is paid for by a different customer than the one who placed it (e25), or an additional item being added to the order in the *confirm order* event (e13). In both cases, the object associations are only anomalous when viewed in the context of the associations observed in previous events.

Of these six anomaly types, existing research has addressed types 1, 2, and 4 [17]. Some proposed techniques [7,8,21,22,24] could in principle be used to detect type 5 anomalies, but have not yet been evaluated on datasets that contain them. Type 3 and type 6 anomalies have not been discussed in previous research.

5   Detecting Anomalies in an Object-Centric Context

In this section, we discuss how event-level anomaly detection needs to be adapted to the object-centric domain so that both (conventional) attribute anomalies and (novel) object anomalies can be identified. This requires solving two problems:

(1) To be able to detect collective anomalies, we must determine how to group events into appropriate *collections* and how to model inter-event relationships within collections. In case-centric event logs, this is straightforward – each case is a separate collection, modelled as an event sequence – but in an object-centric context, it requires some deliberation (Subsect. 5.1).

(2) We must determine how to *encode* event data, i.e., which features are required for an anomaly detection approach to effectively differentiate normal from abnormal events. Previous work has already covered features and feature encoding strategies for event characteristics in case-centric data [17], but it is unclear whether these can be applied in an object-centric context and whether they suffice to detect anomaly types 3 and 6 (Subsect. 5.2).

In the remainder of this section, we assume a basic setting in which an anomaly detection approach is used to build a model of normal process behavior from an event log. The approach then assigns scores to individual events that reflect how well they fit into the model. Events that receive scores beyond some threshold are classified as anomalous by the system. All existing approaches for unsupervised process anomaly detection can be interpreted under this high-level framework. For example, the model of normal process behavior could be a set of constraints mined from the event log and the score the number of constraints violated. Alternatively, the model of normal behavior could be encoded in the parameters of a machine learning model and the score derived from its outputs. The discussion in this section is thus approach-agnostic.

5.1   Grouping Events Into Collections

A suitable grouping notion needs to strike a balance between three requirements:

– *Completeness:* A collection contains all events that are related to each other and has an appropriate structure to model any dependencies between them, so that no anomalies remain undetected.
– *Avoid Deficiencies:* All events should be part of at least one collection.
– *Minimize Divergence:* Grouping the events should not introduce misleading event relations to avoid an incorrect model of process behavior. At least, a collection should not contain events that are completely independent.

Individual object lifecycles do not cover dependencies across object types and are thus incomplete. Flattening the event log from the perspective of one object type introduces deficiencies (Subsect. 2.2). The full event log does not minimize divergence as it contains independent process executions, i.e., unrelated events. Instead, we need to find an appropriate combination of related object lifecycles.

Process Executions. To find a suitable grouping notion, we can draw on the concept of process executions [5]. A process execution is an object-centric generalization of the case concept and can be derived from an object graph consisting of object nodes with an undirected edge between them if they share at least one event. Figure 2 shows part of an object graph from the running example, with one customer and one order containing two items shipped in the same package. A process execution can be defined as a collection of all events that are associated with the objects in a particular subgraph of the object graph.

Selecting an appropriate subgraph requires domain and process knowledge [5]. In the running example, the grouping notion that best meets the requirements is *order – item – package*, i.e., one collection contains all events related to a single combination of order, item, and package. Since no events are only associated with the customer, this object can be omitted from the grouping. From Fig. 2, we thus extract two subgraphs: o1, i1, p1 and o1,

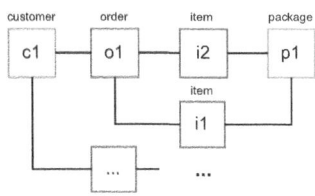

Fig. 2. Object graph excerpt with one customer and a two-item order

i2, p1. With this grouping notion, each collection consists of events that correspond to one process execution, from order placement to package delivery, viewed separately for each item. A collection therefore includes all related events. It also minimizes divergence by ensuring that no unrelated objects appear within the same collection (e.g., two orders that are processed in parallel).

Event Modelling. Next, we need to model the event data per collection. The object-centric standard event model is an event graph that encodes local directly-follows relations between events along each object type [12]. Figure 3 shows an exemplary graph of the events from one order, two items, and one package. Event relations are modelled by directed edges, e.g., *create package* directly follows *pick item* and eventually follows *confirm order* from the perspective of i1.

The problem with event graphs is that they omit dependencies across time if event-associated objects never share a synchronization event. Consider a package that is sent only after the order has been paid. Detecting an anomaly here would require to condition *send package* on the previous occurrence of *pay order*. This is not possible in Fig. 3, because the graph does not model precedence between these two activities. Due to this limitation of event graphs, we argue that for

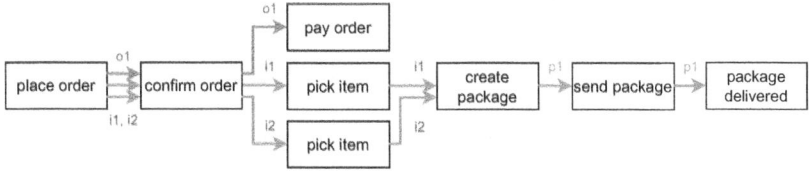

Fig. 3. An event graph with color-coded local directly-follows relations among one order, two items, and one package (customer object omitted for visual clarity)

the purpose of anomaly detection, dependencies within a collection should be modelled *sequentially*, with time as the ordering relation. This preserves global temporal precedence, independent of object types, which is crucial for collective anomalies. If we retain the object association of each event as a characteristic, it also preserves object-level precedence, though less explicit than an event graph.

A manually defined grouping notion and sequence modelling of course very closely resemble a case notion. The only difference is that we propose to flatten the event log using a combination of object types, instead of a single one. This grouping notion does not introduce deficiencies because each event is associated with at least one of the three object types order, item, and package.

Handling Convergence. The duplication of events cannot be avoided when flattening an object-centric event log. We therefore discuss the impact it has on the anomaly detection task and how it can be handled. The main issue is that a detection approach may overemphasize convergent events in its model of normal behavior. Consider an (extreme) example of an order associated with one million items. For our grouping notion, each order-level event would be duplicated a million times, effectively drowning out the patterns in other orders. This is especially problematic if the large order itself is an anomaly. Therefore, a detection approach needs to include a mechanism to limit the impact of convergent events. The nature of this mechanism depends on the detection approach.

5.2   Features for Object Anomalies

With the chosen flattening strategy, existing features and feature encoding schemes for case-level sequences can be applied to object-centric data, so no adaptation is required for normal attributes. In this section, we therefore only discuss whether new object-related features are needed to detect anomalies of types 3 and 6.

For simplicity, we assume that our anomaly detection approach can aggregate previous events in some way, so that it is sufficient to construct features only from the characteristics of the event itself. For example, for event e, we only need to construct a feature that encodes the object characteristics of this event o_e, but not one that encodes the characteristics of the previous event o_{e-1}.

Atomic Object Anomalies (Type 3). Event associations with objects can, in principle, be encoded as multi-valued, categorical attributes. However, this alone is not sufficient to detect all object-related anomalies, because they can arise not only on an individual object, but also on an object *type* level. For type 3 anomalies, we hence consider two main scenarios.

Object Identity Anomalies. Here, the concrete object determines the anomaly. For example, the price of an order may be globally normal, but unusually high compared to other orders placed by the same customer. These anomalies (and their collective counterparts) can be detected by encoding object associations as standard categorical attributes. For some object types, these attributes may have a very high cardinality, which is undesirable in practice because it increases problem sparsity. However, this problem may also occur in case-centric data.

Object Type Anomalies Here, an object anomaly relates only to the object type. Consider a *place order* event, which synchronizes one customer, one order, and any number of items. If two customers are associated with this event, it will constitute a type 3 anomaly. This is the case regardless of the concrete identities of these customers; the anomaly manifests as an unusual number of associations between the event and the object type customer. Consequently, it cannot be detected if the customer is simply encoded as a categorical attribute.

To cover these object type anomalies, we formalize the feature x_1: *number of typed object associations* as follows: Let E be the set of all events, and T be the set of all object types. For each event $e \in E$ and each object type $t \in T$, let $A(e, t)$ be the set of associations that event e has with objects of type t. Feature x_1 for a given event and object type is defined as $x_1(e, t) = |A(e, t)|$.

Collective Anomalies (Type 6). When determining how to detect collective object anomalies, there are two additional scenarios to consider.

Change in Cardinality. An event is associated with more or fewer objects of some type than a previous event when it would normally be associated with the same number (or vice versa). This is essentially the collective version of the object type anomalies discussed above. The subject of the anomaly is the number of object associations, so feature x_1 is sufficient to detect it. For example, when *place order* is associated with 5 items and the subsequent *confirm order* event with only 4, we see that $x_1 = 4$ is abnormal, given the previous value $x_1 = 5$.

Change in Identity. An event is associated with the same number of objects as a previous one, but the concrete object is different, when it would normally be the same (or vice versa). This relates to the identity of objects, but unlike in the atomic case, it is not sufficient to encode the involved objects as categorical attributes, because it is the change in association itself that produces an anomaly. Consider an order that is placed by one customer and paid by another. Similar to the object type anomalies above, it is irrelevant which customer objects are associated with the events: As long as they are not the same, *pay order* is an anomaly. To detect it, we require another feature, formalized as x_2: *prior occurrence indicator*: Let $P_n = \{e_1, e_2, \ldots, e_{n-1}\}$ be the prefix of event e_n, i.e., the sequence of events in the same collection that have occurred before it. Let $O(e, t)$ be the set of objects of type t associated with an event e and $S(e, t) = \bigcup_{e' < e} O(e', t)$ the set of objects of type t that are observed in the prefix. x_2 is defined as:

$$x_2(e, t) = \begin{cases} 1 & \text{if } O(e, t) \subseteq S(e, t) \\ 0 & \text{otherwise} \end{cases}$$

Per object type, x_2 indicates if all objects associated with the current event were also observed in previous events in the collection. If an unexpected change occurs (e.g., we observe a new customer object in a *pay order* order event that is normally associated with the same customer as previous events), it takes value 1; a normal process execution has value 0. If an expected change does not occur (e.g., we do not observe a new item object in an *update order* event that should add an item) it takes value 0; a normal execution has value 1.

6   Evaluation

In this section, we investigate whether these adaptations suffice to detect anomalies in object-centric event logs. We extend a detection approach by the grouping notion and features from Sect. 5 and apply it to two object-centric datasets with anomalies from Sect. 4. Code and data can be found in our repository[1].

6.1   Datasets

Evaluating anomaly detection requires event logs with labelled anomalies, which rules out most publicly available logs. Previous work [7,8,21,22,24] relied on anomaly mutator algorithms to modify events in existing logs ex-post. This is difficult for more complex anomalies, such as types 5 and 6, because injecting those correctly requires understanding the dependencies in the process.

For this paper, we therefore developed two novel log generators to simulate synthetic object-centric event logs for two different processes: (1) the order management process from our running example and (2) a recruiting process. The generators can insert 12 concrete anomalies (two per type, named *a* and *b*) into the logs during simulation. Table 2 lists the anomalies in the order management log (many are examples in the paper). Each anomaly occurs independently, with

Table 2. Anomalies included in the order management logs

ID	Normal Process Execution	Anomaly
1a	Order price is below 6,000	Order price is between 60,000 and 70,000
1b	Events have timestamps from 2019–2020	Events have timestamps from 2010
2a	A delivery attempt (*delivery failed* or *package delivered*) can only be made Monday through Saturday	A delivery attempt is made on Sunday
2b	*Apply discount* is executed only for orders placed by premium customers	*Apply discount* is executed for an order placed by a standard customer
3a	An order is associated with one customer	An order is associated with no customer
3b	An order contains 1–5 items	An order contains 30–40 items
4a	If an item is out of stock, it is reordered and then picked	An out of stock item is directly picked (*reorder item* is skipped)
4b	An order is confirmed only once	An order is confirmed once at the start and then again at a later time
5a	*Reorder item* is executed between 1 and 2 h after *item out of stock*	*Reorder item* is executed between 24 and 48 h after *item out of stock*
5b	*Apply discount* is executed for any orders placed by premium customers	*Apply discount* is not executed for an order placed by a premium customer
6a	An order is paid by the same customer that placed it	An order is paid by a different customer than the one who placed it
6b	*Confirm order* is associated with the same number of items as *place order*	*Confirm order* is associated with one more item than *place order*

[1] https://gitlab.uni-mannheim.de/jpmac/oc-anomaly.

a probability between 1% and 5%, except for 2a, which occurs for 10 orders (plus related items and packages). We label all events affected by an anomaly. For 1a, e.g., all events of the affected order are labeled (*place order*, *apply discount*, *confirm order*, *pay order*), whereas for 2b, only *apply discount* is labeled.

The different probabilities of the anomalies shall prevent excessive label imbalance; otherwise, anomalies that pertain to multiple events per order would appear much more frequently in the log than those that only affect one event.

The log simulation scripts are available in our repository and can be used to reproduce the results reported in the next section, as well as reused and modified to create other object-centric event logs with anomalies. For brevity, we only report the results for the order management process logs in this paper; the results for the recruiting process logs are similar and included in the repository.

6.2 Detection Approach

This evaluation investigates whether, given the proposed adaptations, existing approaches are capable of detecting anomalies in object-centric event data. Therefore, we select a state-of-the-art anomaly detection approach, adapt it according to Sect. 5, and evaluate it on our datasets. The selected approach is BINet [22], because it is designed for multi-perspective detection, performs favorably in benchmarks on case-centric data [19], and scales well to a large number of events and attributes. Its core idea is to train an autoencoder model in a self-supervised manner. The model separately predicts the value of each event feature, using as input an aggregated representation of the previous events in the collection and the other features in the current event (Fig. 4). The previous events are aggregated by a recurrent neural network, then concatenated with the feature values of the current event (except the one to be predicted) and passed through a feed-forward neural network for prediction. Each prediction is then passed through a scoring function that compares it to the true value observed in the log and returns an anomaly score. An event is considered anomalous if any of its feature anomaly scores exceeds a feature-specific threshold.

Setting appropriate thresholds is specific to the problem and the data, so it usually requires domain knowledge. In this paper, we manually select the thresholds to maximize precision and recall. This would not be possible in a realistic unsupervised setting without knowledge of anomaly labels, but it allows us to observe an upper bound for the performance that an anomaly detection approach

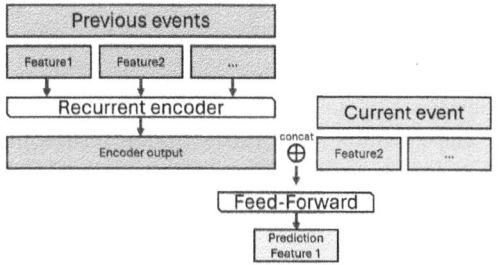

Fig. 4. Architecture of the BINet approach used in the evaluation (adapted from [22])

may achieve with ideal thresholds; this is equivalent to the "best" setting used in other papers [19, 22].

A comprehensive documentation of the implemented detection approach (including features and parameter values) can be found in our repository.

6.3 Results

For the order management process, we apply the detection approach 25 times, on five separately simulated logs with five anomaly detection iterations each, to account for variability in both log generation and training procedure. Runtime is relatively low: For a log with 10,000 orders and about 110,000 events, one iteration (10 epochs) requires roughly 5 min of training time on an Nvidia 3080 GPU. Table 3 shows global precision and recall measures for all anomalous events in each of the 5 logs, averaged across the five iterations. We calculate these measures on the original object-centric event log, not the flattened version. For this, we consider an event as anomalous if the anomaly scores of any of its duplicates exceed the anomaly thresholds.

Table 3. Global precision, recall, and F1 score values (plus standard deviations).

Log	Recall	Precision	F1
1	0.96 ± 0	0.78 ± 0.10	0.86 ± 0.07
2	0.96 ± 0	0.75 ± 0.11	0.84 ± 0.07
3	0.96 ± 0	0.79 ± 0.12	0.86 ± 0.08
4	0.95 ± 0	0.79 ± 0.11	0.86 ± 0.07
5	0.97 ± 0	0.79 ± 0.11	0.87 ± 0.07

We observe consistently high values for recall and precision, with some variation in the latter. Figure 5 shows detailed recall values for the 12 anomaly types (averaged across all logs and iterations). 11 anomalies are detected with very high reliability. For anomaly 5a, however, roughly half the occurrences remain undetected. This is likely because the two attributes involved in this anomaly (activity and time) have rather complex interactions, which the model cannot fit well enough. We can also see that there is some recall variability in four anomalies (2a, 4a, 4b, 5a), whereas recall for the others is very consistent.

These results show that existing techniques for anomaly detection can effectively detect a wide variety of anomalies in object-centric event logs. By extension, they also validate the adaptations we developed in Sect. 5.

6.4 Discussion on Robustness

Although the results indicate a certain robustness of the evaluated approach, there are scenarios in which it does not produce good results. Here, we briefly discuss these scenarios and highlight a fundamental challenge in unsupervised anomaly detection: Ensuring that a model has the right amount of fit.

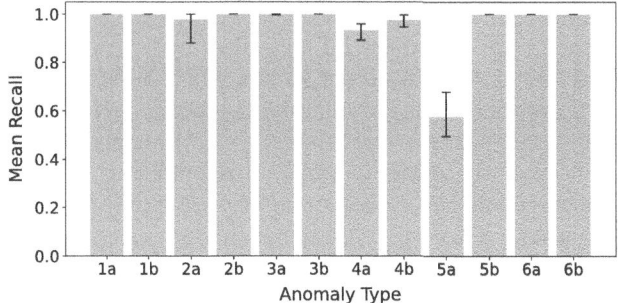

Fig. 5. Recall per anomaly, averaged across 5 iterations each for 5 log versions. Error bars indicate minimal and maximal values.

Both collective and multivariate anomalies are contextual in nature: The observed process behavior is normal in one context, but abnormal in another. In anomaly 2a, the delivery attempt is abnormal when the weekday is Sunday, and normal otherwise. An effective anomaly detection system must be nuanced enough to differentiate these cases so that it labels one as an anomaly, but not the other. If not, it underfits (or overgeneralizes) the observed dependencies.

However, a model can also fit the data *too* well and hence assign low anomaly scores to anomalous events. We demonstrate this by slightly changing how anomalies 4a and 4b in Table 2 are inserted during log simulation. For 4a, we change the execution time of *pick item* so it occurs considerably later, when previously it would have the timestamp of the missing *reorder item*. For 4b, we let the redundant second *confirm order* always be executed right after the first, when previously it could be executed within a large time interval. With these changes, a perfectly fit model would no longer assign high anomaly scores to the affected events, because the combination of relative execution time and object associations now reliably indicates which activity occurs. Figure 6 illustrates this with a likelihood graph: When an item-level event (one with at least one item association) is executed more than two hours after *item out of stock*, its activity is always *pick item*. A perfectly fit model would therefore not classify this event as an anomaly, because it has the expected activity for this very specific context.

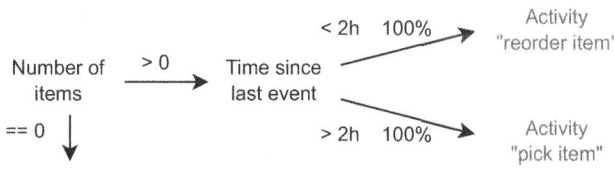

Fig. 6. A likelihood graph for the activity of the event that follows *item out of stock*

To confirm that this also occurs in practice, with models that are not perfectly fit, we repeat our experiment on logs generated with the slightly altered anomalies. Figure 7 shows the recall per anomaly in this setting (again averaged over 5 logs with 5 iterations each). Evidently, the model considerably overfits the data, resulting in a significantly lower percentage of anomaly 4a and 4b events being recognized, while the other anomalies are unaffected. This effect has not been observed in previous work, likely due to few significant attribute dependencies in the used logs. However, it is a challenge in unsupervised multivariate anomaly detection and should be addressed by future research.

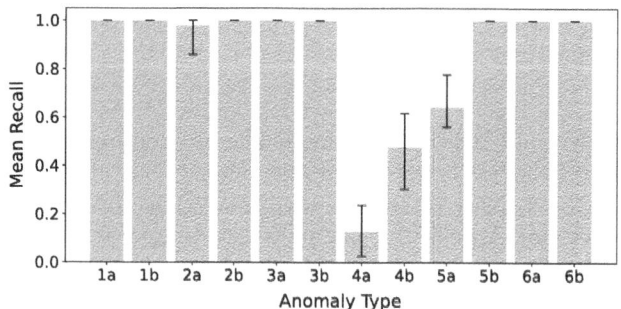

Fig. 7. Recall per anomaly for a log version with which overfitting occurs.

7   Conclusion

In this paper, we make two conceptual contributions to anomaly detection research. The first is a typology of the event-level anomalies that may occur in object-centric event data. The second is a detailed discussion of how anomaly detection can be approached in an object-centric setting. From a practical perspective, we also introduce two novel event logs with a greater range of labelled anomalies than those previously available. Using these datasets, we show that existing anomaly detection techniques can be successfully applied in an object-centric setting, but also point out scenarios in which they are not effective.

There are several limitations to our work. The first concerns the typology in Sect. 4, which is intentionally limited to event-level, statistical anomalies. It does not cover object-level or process-level anomalies [6] unrelated to individual events, e.g., an unusual amount of objects processed at the same time. It also does not cover semantic anomalies [25], which can only be detected with the help of a log-external source of semantic understanding. The typology also remains relatively high-level, and each of the anomaly types could be broken down further, e.g., by differentiating between categorical and numerical attributes.

The anomaly detection strategy in Sect. 5 is effective and can be applied with existing detection techniques without major conceptual innovation. However, we do not argue that this is the only, or even necessarily the best way to detect anomalies in object-centric event logs. Future work may conceive better methods to derive suitable collections and model the event data, as well as more sophisticated features and detection techniques.

We also perform limited experiments on two event logs. The anomalies in these logs are representative of the introduced types, but still arbitrary. We do not know to which extent our findings can be generalized beyond these datasets and anomalies. Additionally, we only evaluate one detection approach, since we are primarily interested in exploring feasibility. A comprehensive evaluation and comparison of multiple detection techniques is still required.

Finally, this paper focuses solely on anomaly *detection*. We do not discuss explainability or prescriptive aspects, which are essential for real-life applications and should also be considered in future work.

References

1. OCEL (object-centric event log) 2.0 specification (2023). https://www.ocel-standard.org/2.0/ocel20_specification.pdf
2. Aalst, W.M.P.: Object-centric process mining: dealing with divergence and convergence in event data. In: Ölveczky, P.C., Salaün, G. (eds.) SEFM 2019. LNCS, vol. 11724, pp. 3–25. Springer, Cham (2019). https://doi.org/10.1007/978-3-030-30446-1_1
3. van der Aalst, W.M.P.: Foundations of process discovery. In: van der Aalst, W.M.P., Carmona, J. (eds.) Process Mining Handbook. LNBIP, vol. 448, pp. 37–75. Springer, Cham (2022). https://doi.org/10.1007/978-3-031-08848-3_2
4. van der Aalst, W.M., Berti, A.: Discovering object-centric Petri nets. Fundam. Informaticae **175**, 1–40 (2020)
5. Adams, J.N., Schuster, D., Schmitz, S., Schuh, G., van der Aalst, W.M.: Defining cases and variants for object-centric event data. In: International Conference on Process Mining, pp. 128–135. IEEE, New York (2022)
6. Berti, A., Herforth, J., Qafari, M., Van der Aalst, W.M.: Graph-based feature extraction on object-centric event logs. Int. J. Data Sci. Anal. (2023)
7. Böhmer, K., Rinderle-Ma, S.: Multi-perspective anomaly detection in business process execution events. In: Debruyne, C., et al. (eds.) OTM 2016. LNCS, vol. 10033, pp. 80–98. Springer, Cham (2016). https://doi.org/10.1007/978-3-319-48472-3_5
8. Böhmer, K., Rinderle-Ma, S.: Association rules for anomaly detection and root cause analysis in process executions. In: Krogstie, J., Reijers, H.A. (eds.) CAiSE 2018. LNCS, vol. 10816, pp. 3–18. Springer, Cham (2018). https://doi.org/10.1007/978-3-319-91563-0_1
9. Bose, R.P.J.C., van der Aalst, W.M.P., Žliobaitė, I., Pechenizkiy, M.: Handling concept drift in process mining. In: Mouratidis, H., Rolland, C. (eds.) CAiSE 2011. LNCS, vol. 6741, pp. 391–405. Springer, Heidelberg (2011). https://doi.org/10.1007/978-3-642-21640-4_30
10. Carmona, J., van Dongen, B., Weidlich, M.: Conformance checking: foundations, milestones and challenges. In: van der Aalst, W.M.P., Carmona, J. (eds.) Process Mining Handbook. LNBIP, vol. 448, pp. 155–190. Springer, Cham (2022). https://doi.org/10.1007/978-3-031-08848-3_5
11. Chandola, V., Banerjee, A., Kumar, V.: Anomaly detection: a survey. ACM Comput. Surv. **41**(3) (2009)

12. Fahland, D.: Process mining over multiple behavioral dimensions with event knowledge graphs. In: van der Aalst, W.M.P., Carmona, J. (eds.) Process Mining Handbook. LNBIP, vol. 448, pp. 274–319. Springer, Cham (2022). https://doi.org/10.1007/978-3-031-08848-3_9

13. Foorthuis, R.: On the nature and types of anomalies: a review of deviations in data. Int. J. Data Sci. Anal. **12**, 1–35 (2021)

14. Huo, S., Völzer, H., Reddy, P., Agarwal, P., Isahagian, V., Muthusamy, V.: Graph autoencoders for business process anomaly detection. In: Polyvyanyy, A., Wynn, M.T., Van Looy, A., Reichert, M. (eds.) BPM 2021. LNCS, vol. 12875, pp. 417–433. Springer, Cham (2021). https://doi.org/10.1007/978-3-030-85469-0_26

15. Jans, M., van der Werf, J.M., Lybaert, N., Vanhoof, K.: A business process mining application for internal transaction fraud mitigation. Expert Syst. Appl. **38**(10), 13351–13359 (2011)

16. Junior, S.B., Ceravolo, P., Damiani, E., Omori, N.J., Tavares, G.M.: Anomaly detection on event logs with a scarcity of labels. In: International Conference on Process Mining, pp. 161–168. IEEE, New York (2020)

17. Ko, J., Comuzzi, M.: A systematic review of anomaly detection for business process event logs. Bus. Inf. Syst. Eng. **65**, 441–462 (2023)

18. Koschmider, A., Kaczmarek, K., Krause, M., van Zelst, S.J.: Demystifying noise and outliers in event logs: review and future directions. In: Marrella, A., Weber, B. (eds.) BPM 2021. LNBIP, vol. 436, pp. 123–135. Springer, Cham (2022). https://doi.org/10.1007/978-3-030-94343-1_10

19. Lahann, J., Pfeiffer, P., Fettke, P.: LSTM-based anomaly detection of process instances: benchmark and tweaks. In: Montali, M., Senderovich, A., Weidlich, M. (eds.) ICPM 2022. LNBIP, vol. 468, pp. 229–241. Springer, Cham (2023). https://doi.org/10.1007/978-3-031-27815-0_17

20. de Lima Bezerra, F., Wainer, J.: Algorithms for anomaly detection of traces in logs of process aware information systems. Inf. Syst. **38**, 33–44 (2013)

21. Niro, A., Werner, M.: Detecting anomalous events in object-centric business processes via graph neural networks. In: De Smedt, J., Soffer, P. (eds.) ICPM 2023. LNBIP, vol. 503, pp. 179–190. Springer, Cham (2024). https://doi.org/10.1007/978-3-031-56107-8_14

22. Nolle, T., Luettgen, S., Seeliger, A., Mühlhäuser, M.: Binet: multi-perspective business process anomaly classification. Inf. Syst. **103**, 101458 (2022)

23. Pang, G., Shen, C., Cao, L., Hengel, A.V.D.: Deep learning for anomaly detection: a review. ACM Comput. Surv. **54**(2) (2021)

24. Pauwels, S., Calders, T.: An anomaly detection technique for business processes based on extended dynamic Bayesian networks. In: ACM/SIGAPP Symposium on Applied Computing, pp. 494–501. ACM, New York (2019)

25. van der Aa, H., Rebmann, A., Leopold, H.: Natural language-based detection of semantic execution anomalies in event logs. Inf. Syst. **102**, 101824 (2021)

Object-Centric Event Logs:
Characteristics, Comparative Analysis
and Road Map

Alexandre Goossens[(✉)], Johannes De Smedt, and Jan Vanthienen

Leuven Institute for Research on Information Systems (LIRIS), KU Leuven,
Naamsestraat 69, 3000 Leuven, Belgium
{alexandre.goossens,johannes.desmedt,jan.vanthienen}@kuleuven.be

Abstract. Process mining aims to comprehend and enhance business processes by analyzing event logs. Recently, object-centric process mining has gained traction by considering multiple objects interacting with each other in a process. This object-centric approach offers advantages over traditional methods by avoiding dimension reduction issues. However, in contrast to traditional process mining where a standard event log format was quickly agreed upon with XES providing a common platform for further research and industry, various object-centric logging formats have been proposed, each addressing specific challenges such as object relations or dynamic attribute changes. This makes that interoperability of object-centric algorithms remains a challenge, hindering reproducibility and generalizability in research. Additionally, the object-centric process storage paradigm aligns well with a wide range of object-oriented databases storing process data.

This paper introduces a characteristics framework from three perspectives originating from studying seminal works in process mining (what should be analyzed), object-centric process modeling (how it should be modeled), and database storage (how it should be stored) perspectives in order to compare and evaluate object-centric log formats. By identifying commonalities and discrepancies among these formats, the study delves into unresolved issues and proposes a road map with potential solutions for further improvements. Ultimately, this research contributes to advancing object-centric process mining by facilitating a deeper understanding of event log formats and promoting consistency and compatibility across methodologies.

Keywords: Object-centric processes · Object-Centric Event log · Process Mining

A. Marrella et al. (Eds.): BPM 2024 Forum, LNBIP 526, pp. 37–54, 2024.
https://doi.org/10.1007/978-3-031-70418-5_3

1 Introduction

The general aim of process mining is to understand, discover and enhance business processes by analyzing event logs recording the different steps occurring in a process. Traditionally, these processes are analyzed through the lens of a single object such as a customer or a product. However, not all processes necessarily fit in this single point of view paradigm as was already indicated in 2001 with the introduction of proclets to model a process from multiple perspectives [2]. The idea that processes can be analyzed from multiple angles has recently gained interest and is more commonly known as object-centric process mining [9]. Compared to traditional process mining, object-centric process mining does not suffer from typical dimension reduction issues such as convergence issues (incorrect multiplication of activities which do not match their actual occurrence) and divergence issues (ambiguity in which activities belong to which objects).

Since the introduction of object-centric process mining, various object-centric logging formats have been proposed with each of them addressing relevant, but different, object-centric process issues such as the eXtensible Object-Centric (XOC) logs [3] and Object-Centric Behavioral Constraint (OCBC) models [27] addressing the need of object relations in object-centric processes as well as the Artefact-Centric Event Log (ACEL) proposal [29]. However, these proposals had scalability issues which were addressed with the Object-Centric Event log (OCEL) 1.0 format [15] at the expense of storing object relations. The Data-aware Object-Centric Event Log (DOCEL) format in turn addressed the issue of changing attribute values in a process. This latter issue was also addressed with the recent OCEL 2.0 format which also includes object relations [8]. Next to that, there is a proposal to store object-centric processes in so-called Event Knowledge Graphs which come with their own set of challenges [14]. All these proposals are happening whilst the IEEE-group is also proposing their own standard object-centric process event log format called Object-Centric Event Data (OCED)[1] which is used as a guideline by the different object-centric event log proposals. Even though there have been other proposals such as MXML [13] prior to the introduction of the eXtensible Event Stream (XES) log format [20], the field of process mining greatly benefited from having the XES standard as it provided a common ground to conduct further research [1,6]. Compared to the XES standard, however, there is currently no consensus on how to store object-centric processes in logs. This suggest that while there are different object-centric process mining algorithms these do not necessarily support other object-centric event log formats limiting the reproducability and generalizability of object-centric process mining research such as in [4,17]. Hence, the object-centric process mining field would benefit from a study investigating what the different object-centric event log formats share and do not share with one another and in doing so hopefully find common ground to have an 'Object-Centric XES'.

This study aims to address this gap by analyzing the different object-centric event log formats using a framework providing an overview of the different issues that an object-centric event log format could or should address. The contributions are as follows:

---

[1] https://www.tf-pm.org/resources/oced-standard.

1. An object-centric event log format characteristics framework is built by investigating three different fields crucial to support a process mining exercise:
   - Databases: given that a wide range of databases are inherently object-centric instead of process-centric making the extraction of object-centric event logs more natural from such databases [28].
   - Process Mining: to discover and analyze the stored processes.
   - Object-centric Process Modeling: representing object-centric processes.
2. All current object-centric event log formats are compared and analyzed using the developed characteristics framework looking for similarities and dissimilarities between event log formats.
3. A road map is offered including solutions for unresolved issues not agreed upon in various object-centric event log formats or left unaddressed altogether.

Based on the previous contributions, this paper can be seen as an object-centric addendum to the process mining manifesto for object-centric event log formats.

The remainder of the paper is structured as follows: Sect. 2 outlines the framework's characteristics and its derivation. Next, using the constructed framework, Sect. 3 compares the different object-centric event log formats and on how they do or do not address the different issues followed by a lengthy discussion on the insights from this comparison in Sect. 4. Next, Sect. 5 proposes further refinements for object-centric event logs. Finally, Sect. 6 discusses the limitations and future work of this study with Sect. 7 concluding this paper.

2   Framework Dimensions and Characteristics

In the next section, object-centric event log format characteristics are derived by investigating three relevant research fields for object-centric processes. The analyzed research fields stem from the process mining manifesto [36] where process mining is explained as using event logs (storage), process models (modelling) to get insights into processes using analytics(process mining). Similarly, the most recent systematic object-centric process mining literature review [9] also classified object-centric analysis techniques among the dimensions 'Log Type', 'Model Type' and 'Technique Type', and 'Quality'.

Therefore, for this study, the following research fields are deemed relevant to derive object-centric event log format characteristics:

1. Process Mining: Given that object-centric process mining is a continuation of process mining we start from its initial conception and investigate this field for characteristics for object-centric event logs.
2. Object-centric Process Modeling: Given that object-centric process mining aims at mining object-centric process models and/or object models, it is necessary to investigate how these models are structured and how objects are intertwined with processes in this field.

3. Databases: Object-centric event log formats are inherently linked to the storage of data. Therefore investigating how different types of databases such as object-oriented databases or temporal databases deal with this is relevant.

For each research field, crucial works were identified such as manifestos, systematic literature reviews and first-time proposals to start the analysis and then backward and forward snowballing search was used to further identify other relevant works which resulted in the identification of the most relevant issues and their implications for object-centric processes.

Framework Dimensions. Object-centric process mining is interested in concepts of process mining, object-centric process modelling and conceptual modeling, focusing on how multiple object instances interact during process execution. It integrates the **Event-to-Event (E2E) Dimension** from traditional process mining, which studies event sequences' relationships, and the **Object-to-Object (O2O) Dimension** from conceptual modeling, which explores object relationships. Note that in this context, objects are considered to have an object type, object identity and have attributes with values. This is in line with how object-oriented databases [5] and OCED consider objects. Additionally, it considers the interplay between events and objects, known as the **E2O** dimension. These dimensions are pivotal in understanding object-centric analysis and storage. A recent study highlights these dimensions' significance, suggesting they can manifest individually or concurrently in an analysis or event log [9]. Apart from **E2E, O2O and E2O**, one other dimension emerges from the analysis: the **Data Quality Dimension**, offering characteristics on data quality.

2.1   Process Mining Research Field Perspective

To gather characteristics from the process mining perspective, two primary sources are identified: the process mining manifesto [36] and the latest object-centric process mining overview study [9].

Process Manifesto: Process Mining Types. The process mining manifesto outlines three types of process mining: process discovery, conformance checking, and enhancement (prediction and prescription). It is clear that any object-centric event log format must support these three types of process mining as well.

Process Manifesto: Process Log Perspectives. Additionally, the process mining manifesto identifies the following perspectives a traditional process log should have:

1. Control-flow: It is important to know how all the activities are related to one another.
2. Organizational perspective: Who does what in an organization?

3. Data-perspective: Associate data with events. Enhance the log with accurate information for each event.
4. Time-perspective: Assign timestamps to events for accurate time-based analysis of processes.

These perspectives need to be integrated in some way to fit into an object-centric process mining context so as to contain at least as much information as traditional process logs.

Process Manifesto: Event Data Criteria. The process mining manifesto also identifies the following criteria for event data and the event log format itself:

1. Trustworthy: The information stored in the event log can be trusted i.e. activities are not duplicated and all cases contain their respective activities.
2. Complete: There is no crucial information missing due to a transformation i.e. convergence or divergence issues.
3. Well-defined semantics: No ambiguity exists in the data format or attributes.

Object-Centric Process Mining. The main finding of [9] is that no identified log format enables analysis on all object-centric dimensions (**E2E, O2O, and E2O**); instead, some formats support two out of three possibilities [9]. The study did not aim to deeply compare the differences between event log formats. Since then, new formats addressing all three dimensions (**E2E, O2O, and E2O**) have emerged, such as OCEL 2.0 and Event Knowledge Graphs (EKGs) [8,14].

Additionally, the study [9] categorizes different object-centric process mining algorithms into these three analysis dimensions. It is not unexpected that most algorithms utilize event logs supporting **E2O and O2O** dimensions, as this aligns with the primary objectives of object-centric process analysis and the availability of these logs initially. Intriguingly, the algorithms predominantly yield process models emphasizing **E2E**, while the other two dimensions (**E2O and O2O**) are less frequently discovered.

Resulting Framework Characteristics. From process mining research, it is essential to note that an object-centric event log format must comply with the requirements outlined in the process mining manifesto. The following **E2E** characteristics are derived from this:

– *C1: An event must have a unique event ID*
– *C2: An event must have a timestamp*
– *C3: An event must have an activity*
– *C4: An event can have other optional attributes*

These characteristics correspond to those in the XES format [20] and the process mining manifesto [36]. Furthermore, the **O2O** characteristics are derived from the **E2E** characteristics adhering to the same principles as events in the XES format.

- *C5: An object must have an object type*
- *C6: An object must have a unique object ID*
- *C7: An object can have optional object attributes*

2.2 Object-Centric Process Modeling Perspective

Object-centric conceptual modeling integrated with processes is predominantly found in two domains, each approaching the issue from a distinct perspective. On one side, the research field of data-aware process modeling treats processes as objects interacting with one another, characterized by well-defined behavior [19, 25, 33]. Conversely, process mining begins with a process model where objects interact with one another. Both will be analyzed to identify common points to derive key characteristics.

1. **Object relations:** In data-aware process modeling, an explicit object model defining object relations is defined prior to modeling the object behavior in a process [25, 26, 33]. In object-centric process modeling, such an object model is not necessarily explicitly defined like with proclets [2] or Colored Petri Nets (CPN) [16, 24]. A recent Petri Net-based approach models the process of multiple individual object instances for read-only data [37]. Conversely, OCBC models explicitly define an object model alongside a declarative process model [3], similar to Guard-Stage Milestone (GSM) [22]. Generally, within the Unified Modeling Language (UML), object relations precede modeling UML behavior diagrams [30]. The idea behind all these approaches is that objects and their relations impact a business process. Storing object relations within an event log can reveal insights into process execution. For instance, why a customer may only have one order at a time or why multiple orders can be packaged together for delivery.
2. **Object-Event relations:** In all mentioned approaches, objects are linked to events either through individual object behavior models resembling state machines [25, 26, 30, 33] or directly integrated into the process model [2, 3, 16, 24, 37]. Regardless of how objects and events are connected, it is essential for both to be associated with each other. This information should also be stored in event logs; otherwise, obtaining a complete view of an object-centric process becomes impossible, as knowing when and which objects interact in a process is crucial. Notably, some approaches require at least two events per object (one for creation and one for deletion), such as PhilHarmonic Flows [26], Merode [33], and UML state machine diagrams [30]. Consequently, every event in these approaches must relate to at least one object. These assumptions have significant implications; no object is considered relevant for a process if it is not involved in an event, and likewise, no event is deemed part of a process without involving at least one object. This means that system events which do not impact any objects are not considered relevant for these approaches such as memory management events. Certain object-centric event log formats incorporate this requirement like OCEL 1.0 [15] and DOCEL [18].

3. **Object information can be updated:** In data-aware process modeling, events can update objects [26,33], a concept also found in UML sequence and activity diagrams. Although the approach in [37] primarily focuses on read-only immutable data, it acknowledges the existence of processes involving mutable data. Similarly, in process mining, the XES standard permits updating case information. This principle should be extended to object-centric processes, as a comprehensive view of object evolution can elucidate routing decisions or final outcomes within a process.

4. **Individual behavior of an object must be recoverable:** Both data-aware process modeling and object-centric process modeling approaches aim to understand how individual object instances undergo a process to analyze exceptions or sub-processes. Additionally, the XES standard explicitly supports this analysis by linking every process execution to an individual trace. Therefore, for a comprehensive understanding of how multiple objects undergo a process simultaneously, an object-centric event log format must allow the discovery of the individual sequence of events involving each object.

Resulting Framework Characteristics. Based on this analysis, the following characteristics for the **O2O** dimensions are added:

- *C8: Object attributes can change values and those changes are traceable*
- *C9: An object instance can be related to other object instances*
- *C10: Supports association classes between object relations*

The first characteristic addresses the need to track changes in object information over time within a process, as processes often interact with both read-only and read-write data. The last characteristic emphasizes the importance of capturing object relationships, a fundamental aspect of conceptual modeling where objects are interconnected. This idea also comes back in object-centric process modeling where object relations are also important. The last characteristic proposes including association classes referred to as relation qualifiers between events and objects or between two objects, endorsed by studies in the object-centric process field [9], OCEL 2.0 [8], and OCED.

For the **E2O** dimension, the following characteristics are derived:

- *C13: An event can be related to 0..N objects*
- *C14: An object can be related to 0..N events*
- *C15: All activities in which an individual object instance participates must be recoverable*

The first two characteristics indicate the minimum cardinalities there are between events and objects, with a stricter formulation chosen for its significant impact on what can be stored in an object-centric event log. The last characteristic ensures the recoverability of the complete process of a single object instance, aligning with concepts proposed in PhilHarmonic Flows [26] and the process mining manifesto's definition of a trace [36]. Note that C15 does not necessarily mean that the notion and knowledge of case ids is not relevant anymore within an object-centric context. Instead, it is perfectly possible to still use the case ids as object ids for that relevant object type but also to have additional object types from which the process can be analyzed.

2.3 Databases Perspective

The databases research field is relevant as it addresses storage issues directly within databases, insights from which may be applicable to object-centric event logs. This section primarily discusses object-oriented databases, as they share similarities with object-centric process storage [5, 10]. Additionally, it touches upon temporal [32] and moving databases [38], as they handle changing information over time, a crucial aspect of process mining. Knowledge graphs [21, 23] are also considered, as certain proposals have utilized them as a foundation for storing object-centric processes [14].

Regarding object-oriented databases, the following relevant requirements are found from the object-oriented databases manifesto [5]:

1. **Complex objects:** The manifesto emphasizes the need for databases to support complex objects and attributes such as lists, sets, and tuples, reflecting the complexity often found in objects. This notion also applies to object-centric processes.
2. **Object Identity:** It is a requirement for object-oriented databases that each object instance must have a unique object ID otherwise it is difficult to uniquely trace the object instance over time [5, 10].
3. **Object types and classes:** In conceptual modeling, objects can be categorized by type or class, each with its own set of attributes or operations [5, 10].
4. **Inheritance:** Object-oriented databases incorporate the concept of inheritance, where an object type inherits attributes and behavior from another object type, including multiple inheritance. However, this aspect has not yet been explored in object-centric process mining.
5. **Updating Object Information:** Object-oriented databases may create new object versions for each update, necessitating a direct link to the previous object version [5, 10]. Temporal databases, introduced in 1986, distinguish between transaction time and valid time [32]. Transaction time denotes when a change occurred in the database, while valid time reflects the time when the change occurred in reality and is accurate [31, 32]. An object-centric event log should only contain accurate and unambiguous information, implying it should only include valid time information. The interplay between transaction time and valid time should be left for the databases themselves. Lastly, moving databases [38] address storing objects with rapidly changing values, distinguishing between static and dynamic attributes based on how frequently their values change over time. The variety of database types addressing this issue highlight the importance of accurately tracking changing attributes values over time in an object-centric process.

6. **Schema Evolution:** In [35], the significance of supporting changes in a database schema for third-generation databases was highlighted, also mentioned as optional in the object-oriented database manifesto [5]. This observation can extend to object-centric processes, where the object model itself may change during a process, potentially affecting the process execution, similar to process mining's identification of concept drift. In object-centric process mining, an *object drift* could induce concept drift later on. It might therefore be interesting to investigate how to support this in object-centric event log formats as well.

7. **Knowledge Graphs:** Whilst not discussed in the object-oriented databases manifesto [5], knowledge graphs store entities as nodes, connected by relations [21,23]. They typically use tuples in the form (tail, relation, head, sometimes (timestamp)), where tail and head are nodes. Knowledge graphs are highly extensible, accommodating various node types. Consequently, they can support updates in object information, complex objects, and object type inheritance. In an object-centric context, this could involve distinct node types for object types, events, and attributes. Therefore, knowledge graphs seem to be able to accommodate for almost the above requirements.

The object-oriented database manifesto outlines additional mandatory features such as **computational completeness**, which ensures any computational function can be expressed in the database language, and **database recovery**. While crucial for databases, these features are irrelevant for an object-centric event log format.

Resulting Framework Characteristics. From the databases research field, there is a strong interest in the **O2O** dimension confirming the importance of the object-object characteristics [5]. It mainly confirms that it is considered good practice to store object information in that manner. Since object-oriented databases also support object inheritance, further exploration of this concept in object-centric process mining could be valuable. The following **O2O** dimensions are found from the database research field:

– *C11: Object relations can change over time*
– *C12: Supports object type inheritance*

The first characteristic, highlighted in database studies [35] and data-aware process modeling [26,33], is exclusively featured in the ACEL object-centric event log format proposal [29]. The last characteristic, allowing for object type inheritance, was underscored in database proposals, considering its close relation to data models [5,10], though currently, none of the object-centric event log formats supports this characteristic.

Next, we identify the following characteristics in the **Data Quality** Dimension, drawn from the process mining manifesto and the databases research field:

- *C16: Data changes must be unambiguous*
- *C17: Data is uniquely identifiable*
- *C18: Data should be minimally duplicated*
- *C19: Data storage should be maximally scalable*

The four data quality characteristics address crucial aspects of data storage: accuracy, uniqueness, consistency, and scalability. An object-centric event log format should enable traceable value changes, ensure unique identification of objects and events over time, minimize data duplication, and ensure scalability of the event log itself.

3 Comparison of the Different Object-Centric Event Log Formats

A framework for object-centric event log characteristics is developed from the previous analysis, consisting of four dimensions: **E2E, O2O, E2O and Data Quality**. The first three dimensions align with the classification of object-centric process techniques proposed in [9], while the **Data Quality Dimension** is crucial considering the event log's role in data storage. These characteristics represent common insights from different research fields, each with its specific focus on object-centric processes [7,25] or storage [5] and should be viewed as characteristics rather than requirements.

Table 1 compares various object-centric event log formats on the characteristics. A ✓ indicates support for the characteristic, while a X indicates lack of support, with clarifications provided if necessary. The comparison includes only object-centric event log formats and the OCED standard, encompassing:

- **XOC** [28]: XOC stores both object-centric data and the complete object model alongside the event that triggered a change in the object model.
- **OCEL 1.0** [15]: OCEL 1.0 was developed after XOC to address scalability issues. It consists of an events table and an objects table. However, it does not support dynamic attributes (attributes changing values over time) [18] or object-object relations [29].
- **ACEL** [29]: ACEL was developed after OCEL 1.0 and mainly addresses the need to store object-object relationships within an object-centric process.
- **DOCEL** [18]: DOCEL was developed after OCEL 1.0 and supports dynamic attributes with the use of foreign keys.
- **OCEL 2.0** [8]: The successor of OCEL 1.0 incorporates dynamic attributes using timestamps and includes object relations in the objects table.
- **EKG** [14]: Based on knowledge graphs, this storage formats creates node types for events, attributes, timestamps and object types. These are then connected to one another with the use of relations.
- **OCED**: It is essential to note that OCED is currently a metamodel, not an object-centric event log format. However, it is included in the comparison because it is developed by the IEEE group responsible for XES, and several event log formats, such as OCEL 2.0 and EKG drew inspiration from OCED.

The comparison therefore does not include the XES format [20] (which covers all the **E2E** characteristics but not the **O2O or E2O** dimensions) or other interesting event log formats which solve specific issues such as IoT [11].

Table 1. Characteristics of object-centric Event Log Formats

Characteristics	XOC [28]	OCEL 1.0 [15]	OCEL 2.0 [8]	DOCEL [18]	ACEL [29]	EKG [14]	OCED
E2E							
C1: An event must have a unique event ID	✓	✓	✓	✓	✓	✓	✓
C2: An event must have a timestamp	✓	✓	✓	✓	✓	✓	✓
C3: An event must have an activity	✓	✓	✓	✓	✓	✓	✓
C4: An event can have other optional attributes	✓	✓	✓	✓	✓	✓	✓
O2O							
C5: An object must have object type	✓	✓	✓	✓	✓	✓	✓
C6: An object must have a unique objectID	✓	✓	✓	✓	✓	✓	✓
C7: An object can have optional object attributes	✓	✓	✓	✓	✓	✓	✓
C8: An object can change values and those are traceable	X (not traceable)	X (not traceable)	✓	✓	✓	✓	✓
C9: An object instance can be related to other object instances	✓	X	✓	X	✓	✓	✓
C10: Supports association classes between object relations	X	X	✓ (relation qualifier)	X	X	X	✓ (relation qualifier)
C11: Object relations can change over time (schema evolution)	X	X	X	X	✓	X	X
C12: Support object type inheritance	X	X	X	X	X	X	X
E2O							
C13: An event can be related to 0..N objects	✓	X	✓	X	✓	✓	✓
C14: An object can be related to 0..N events	✓	X	✓	X	✓	✓	✓
C15: All activities in which an individual object instance participates must be recoverable	✓	✓	✓	✓	✓	✓	✓
Data Quality Dimension							
C16: Data changes must be unambiguous	✓	X	X	✓	✓	✓	/
C17: Data is uniquely identifiable	✓	X	X	✓	✓	✓	/
C18: Data should be minimally duplicated	X	✓	✓	✓	✓	✓	/
C19: Data storage should be maximally scalable	X	✓	✓	✓	✓	✓	/

Regarding the **E2E** dimension, all the object-centric event logs formats agree on the 4 characteristics. For the **O2O** dimension, the different formats all incorporate characteristics $C5$ to $C7$. For characteristic $C8$, the idea is that an object attribute can be changed and that it can be traced back. A key-point here is that tracking attribute changes can be done with timestamps (OCEL 2.0), foreign keys (DOCEL or ACEL), or even object versions (XOC and OCEL 1.0) but in case object versions are used it is important to know which object was the previous version which neither XOC or OCEL 1.0 support. The next **O2O** characteristic ($C9$) deals with storing the object relations an object instance has.

This characteristic is not supported by OCEL 1.0 and DOCEL. Regarding characteristic *C10*, only the OCEL 2.0 supports association classes between object relations called relation qualifiers by OCED. For *C11*, ACEL is the only standard supporting schema evolution whilst no event log format supports object type inheritance (*C12*).

Regarding the **E2O** dimension, the first two characteristics (*C13 and C14*) deal with a more relaxed view of what can be stored in an object-centric event log namely also events without an object or objects without an event. This idea is only dismissed by two formats namely OCEL 1.0 and DOCEL which require each event to be linked to at least one object and each object to be linked to at least one event. For *C15*, all event log formats allow to recover all activities in which a specific object instance participates in.

Moving to the **Data Quality** dimension, *C16* states that data changes must be unambiguous however for OCEL 1.0 and OCEL 2.0 this is not guaranteed. Given that in OCEL 1.0 attribute changes have to be stored with the events, it is unclear to which object instance these changes apply [18]. For OCEL 2.0, attribute value changes are given a timestamp but in case two events happen simultaneously using the same object instance it is unclear which event is responsible for this value change. This is also why OCEL 1.0 and OCEL 2.0 do not follow characteristic *C17*. For *C18*, only XOC does not minimally duplicate information given that for each event that updates an object relation, the whole object model is returned which in turn makes the event log format not very scalable and hence does not support characteristic *C19* neither [15]. Regarding *C19*, two additional event log formats need to be discussed. OCEL 2.0 allows to store object relations but given that object models can be very large this might get very quickly very large. The same observation goes for ACEL where changing attribute values and relationship changes are stored together with the events themselves which might also not scale very well. From the previous observations, it is clear that storing object-object relationships is a challenge when it comes to scalability but it comes with the advantage of providing a more holistic overview of an object-centric process.

4 Discussion

Attribute Value Changes. A first interesting insight is the fact that different object-centric event log formats allow for attribute value changes but that they differ in their implementation:

- **OCEL 2.0:** As explained in the previous section, OCEL 2.0 allows to link a timestamp to an attribute value. This makes it possible to know when an attribute value was changed. However, it does not guarantee that the event responsible for this change can be known. In case an object instance is involved in two events at the same time, it is not possible to know which event is responsible for the attribute value change.
- **DOCEL:** In order to store changing object attribute values, DOCEL stores these in a different table where it makes use of two foreign keys namely the

object ID to indicate to which object instance the changed value belongs to and the event ID to indicate which event is responsible for the change.

- **ACEL:** In order to store changing attribute values, ACEL stores a subtable containing the objectID, attribute and the new value, together in the event row responsible for the change. Even though this solves the issue, ACEL also stores in a similar manner the changing object-relationships making the overall event table not adhering to the normalization principles limiting a scalable and easy analysis [12].
- **EKG:** For EKGs it is possible to link an attribute value to an event or a timestamp. Both are possible.

Regarding attribute value changes it seems that all the possible ways to store these have been covered (including using object versions by XOC and OCEL 1.0). Whether a certain way of storing these dynamic attributes is better mainly depends on what exactly is the purpose of the final analysis. In case an analysis focuses on both control-flow and the dynamic attributes, it is important to know that certain event log formats do not allow to analyze dynamic attribute changes.

Object Relations. Moving on to object-object relationships, it is clear from XOC that storing complete object models with each event quickly causes scalability issues [15,18]. However even with OCEL 2.0 and ACEL which also support the object-object relations, scalability can not necessarily be guaranteed. The main reason for this is that there is no clear policy on which object-object relationships should be stored. First of all, it is unclear if and how the XOC and OCEL 2.0 proposals store transitive object-object relationships as this can cause duplication. This makes it complicated to correctly derive an object model from the stored relationships given that it is unclear which object relations are directly or indirectly connected. Related to that is the storage of N:M relationships because storing the object-object relationships on both sides causes duplication of information. One possible way is to reify a N:M relationship to a 1:N relationship [34]. Practically, this would mean that all the object instances on the N-side of the relationship store the object-instance which is on the 1-side of the relationship. This would ensure scalability as all the object-object relationships can be traced back correctly whilst transitive relationships do not need to be stored anymore given that these can be traced back anyhow. This proposal however does not support schema evolution, but this could also be done with the use of dynamic attributes. In turn, it would allow to analyze the impact a schema evolution has on the object-centric process. A final aspect worth mentioning is that none of the event log formats allow to indicate whether an object type is a subclass of another object type. Adding this might however provide new insights as it might be possible to discover what the typical process execution is for a superclass whilst a subclass might have a specific process execution linked to them. Of course, it is possible to analyze this by object type but is currently not possible to know whether an object type A is a superclass of object type B.

In short, it seems that with the current object-centric event log formats, the **O2O** dimension still has the most to gain given, currently, that if an event log format supports it, scalability is not always guaranteed. Next to that, certain intricacies of object modeling such as schema evolution, relation qualifiers and (especially) object type inheritance are not fully supported yet even though these might provide interesting insights for an object-centric process analysis.

Event-Object Relations. A third insight is that both OCEL 1.0 and DOCEL require that each event has at least 1 object connected to it and each object is at least involved in one event. Regarding the requirement that 1 event must have at least 1 object, this stems from the fact in XES every event must be assigned a case. Therefore a similar assumption is taken in OCEL 1.0 and DOCEL. But there is also the possibility that an event is relevant for a process but not for a specific set of objects. Secondly, regarding the requirement that an object must be involved in at least 1 event, it might be that redundant process information should not be stored in an object-centric event log. But of course, it is possible that even though certain object instances are not directly involved in an event these might be important through their relations with objects which are directly involved in the process.

Final Remarks. Even though ACEL cover the most characteristics, it unfortunately is not practical for analysis as it is not sufficiently normalized [12]. Next, EKG also cover a lot of characteristics, but unfortunately do not have a wide support on the analysis side as is concluded in [9] given that EKG can not directly use algorithms based on tabular data given that EKG are knowledge graphs in essence. In third place, there is OCEL 2.0 which sees more applications from the research community but still lacks support for specific characteristics such as in the **O2O** dimension. Another interesting insight is that none of the proposed event log formats deal with streaming object-centric event data. Given that most of these formats have rigid table and column structures, it seems that these formats would not function well in the beginning of streaming where it is unclear which data and data types will be streamed unless the whole data model is known upfront. Addressing this aspect in future event log formats might be an interesting research avenue where there is more robustness incorporated against novel data types.

5   Road Map for Refining Object-Centric Event Logs

According to the recent object-centric overview paper [9], OCEL 1.0 is currently the object-centric event log format with the widest support for database extraction and object-centric process analysis. Given that the latest OCEL 2.0 iteration included object relations and dynamic relations on top of what OCEL 1.0 supports, it is more than probable that OCEL 2.0 will be the first object-centric event log format researchers will go to for an object-centric analysis. Hence, we propose a set of important but feasible solutions to further improve OCEL 2.0:

- **Traceable Dynamic Attributes:** Instead of using timestamps to trace attribute changes, it might be better to also support the use of event ids as foreign keys linking both the object instance and event responsible for the change in cases where the events are known.This way, it is unambiguous which event is responsible for a change where with the timestamp it might be that two events happened simultaneously.
- **Scalable Object Relations:** In order to ensure scalable object relations storage, it would be beneficial, if possible, to reify all N:M object relations to 1:N relations. Hence only a limited set of object relations need to be stored whilst still being able to discover the complete object model without storing the transitive object relations which might cause duplication.
- **Dynamic Object Relations:** Finally, in order to support characteristic *C18* for changing object relations, a quick solution would be to consider object relations as dynamic attributes which can also change over time. This way OCEL 2.0 would also support schema evolution.

By incorporating the previous suggestions, OCEL 2.0 would support all dimensions namely **E2E,O2O and E2O** whilst keeping everything scalable and traceable. On top of that, it would also support two additional relevant characteristics namely supporting schema evolution and relation qualifiers.

6   Limitations and Future Work

Even though the framework was constructed based on insights from multiple research fields and validated by all the authors of this study, the framework would of course benefit from a wider validation from the object-centric process mining community. Another limitation is that the conclusions of this study are based on the current insights from the investigated research fields. Given that, especially, the object-centric process mining field is evolving rapidly, future studies might make that certain characteristics of the framework should be updated or adapted. In the future, investigating how an object-centric event log format can store more information whilst still remaining scalable is certainly worth investigating. Especially, incorporating more **O2O** characteristics such as schema evolution and object type inheritance whilst ensuring scalability is relevant.

7   Conclusion

This study began by noting the various object-centric event log formats available for analyzing object-centric processes. It introduces a framework for object-centric event log characteristics, derived from analyzing three pertinent research areas: process mining, object-centric process modeling, and databases. Through a comparison with existing event log formats, several key findings emerge. Firstly, it is clear that an object-centric event log format should support the tracking of attribute value changes but this is currently done in different ways each with their advantages and disadvantages. Secondly, storing object-object relations is not an easy feat as it quickly generates scalability issues. Related to that is that schema evolution is a well-researched topic in databases and conceptual modeling but is not always supported in object-centric event log formats (except one). Lastly, object type inheritance is also highlighted as important in the databases and conceptual modeling fields but is also not supported by any object-centric event log format. With these insights, a set of feasible solutions to further improve OCEL 2.0 is proposed.

References

1. van der Aalst, W.M.P.: Process mining: a 360 degree overview. In: van der Aalst, W.M.P., Carmona, J. (eds.) Process Mining Handbook. LNBIP, vol. 448, pp. 3–34. Springer, Cham (2022). https://doi.org/10.1007/978-3-031-08848-3_1
2. van der Aalst, W.M., Barthelmess, P., Ellis, C.A., Wainer, J.: Proclets: a framework for lightweight interacting workflow processes. Int. J. Coop. Inf. Syst. **10**(04), 443–481 (2001)
3. van der Aalst, W.M., Li, G., Montali, M.: Object-centric behavioral constraints. arXiv preprint arXiv:1703.05740 (2017)
4. Adams, J.N., van der Aalst, W.M.P.: OC π: object-centric process insights. In: Bernardinello, L., Petrucci, L. (eds.) PETRI NETS 2022. LNCS, vol. 13288, pp. 139–150. Springer, Cham (2022). https://doi.org/10.1007/978-3-031-06653-5_8
5. Atkinson, M., Dewitt, D., Maier, D., Bancilhon, F., Dittrich, K., Zdonik, S.: The object-oriented database system manifesto. In: Deductive and Object-Oriented Databases, pp. 223–240. Elsevier (1990)
6. Augusto, A., et al.: Automated discovery of process models from event logs: review and benchmark. IEEE Trans. Knowl. Data Eng. **31**(4), 686–705 (2018)
7. Berti, A., van der Aalst, W.: Extracting multiple viewpoint models from relational databases. In: Ceravolo, P., van Keulen, M., Gómez-López, M.T. (eds.) SIMPDA 2018-2019. LNBIP, vol. 379, pp. 24–51. Springer, Cham (2020). https://doi.org/10.1007/978-3-030-46633-6_2
8. Berti, A., et al.: OCEL (object-centric event log) 2.0 specification (2023)
9. Berti, A., Montali, M., van der Aalst, W.M.: Advancements and challenges in object-centric process mining: a systematic literature review. arXiv e-prints pp. arXiv–2311 (2023)
10. Bertino, E., Martino, L.: Object-oriented database management systems: concepts and issues. Computer **24**(4), 33–47 (1991)
11. Bertrand, Y., Veneruso, S., Leotta, F., Mecella, M., Serral, E.: NICE: the native IoT-centric event log model for process mining. In: De Smedt, J., Soffer, P. (eds.) ICPM 2023. LNBIP, vol. 503, pp. 32–44. Springer, Cham (2024). https://doi.org/10.1007/978-3-031-56107-8_3
12. Codd, E.F.: A relational model of large shared data banks (1970). In: Ideas That Created the Future: Classic Papers of Computer Science. The MIT Press (2021)
13. van Dongen, B.F., Van der Aalst, W.M.: A meta model for process mining data. EMOI-INTEROP **160**, 30 (2005)
14. Fahland, D.: Process mining over multiple behavioral dimensions with event knowledge graphs. In: van der Aalst, W.M.P., Carmona, J. (eds.) Process Mining Handbook. LNBIP, vol. 448, pp. 274–319. Springer, Cham (2022). https://doi.org/10.1007/978-3-031-08848-3_9

15. Ghahfarokhi, A.F., Park, G., Berti, A., van der Aalst, W.: OCEL standard. Process and Data Science Group, RWTH Aachen University, techreport **1** (2020)
16. Ghilardi, S., Gianola, A., Montali, M., Rivkin, A.: Petri net-based object-centric processes with read-only data. Inf. Syst. **107**, 102011 (2022)
17. Goossens, A., De Smedt, J., Vanthienen, J.: Extracting process-aware decision models from object-centric process data. arXiv preprint arXiv:2401.14847 (2024)
18. Goossens, A., De Smedt, J., Vanthienen, J., van der Aalst, W.M.P.: Enhancing data-awareness of object-centric event logs. In: Montali, M., Senderovich, A., Weidlich, M. (eds.) ICPM 2022. LNBIP, vol. 468, pp. 18–30. Springer, Cham (2023). https://doi.org/10.1007/978-3-031-27815-0_2
19. Goossens, A., Verbruggen, C., Snoeck, M., De Smedt, J., Vanthienen, J.: Aligning object-centric event logs with data-centric conceptual models. In: van der Aa, H., Bork, D., Proper, H.A., Schmidt, R. (eds.) BPMDS EMMSAD 2023. LNBIP, pp. 44–59. Springer, Cham (2023). https://doi.org/10.1007/978-3-031-34241-7_4
20. Günther, C.W., Verbeek, H.M.W.: XES standard definition. IEEE Std (2014)
21. Hogan, A., et al.: Knowledge graphs. ACM Comput. Surv. (CSUR) **54**(4), 1–37 (2021)
22. Hull, R., et al.: Business artifacts with guard-stage-milestone lifecycles: managing artifact interactions with conditions and events. In: Proceedings of the 5th ACM DEBS, pp. 51–62 (2011)
23. Ji, S., Pan, S., Cambria, E., Marttinen, P., Yu, P.S.: A survey on knowledge graphs: representation, acquisition, and applications. IEEE Trans. Neural Netw. Learn. Syst. **33**(2), 494–514 (2022)
24. Kleijn, J., Koutny, M., Pietkiewicz-Koutny, M.: Regions of Petri nets with a/sync connections. Theor. Comput. Sci. **454**, 189–198 (2012)
25. Künzle, V.: Object-aware process management. Ph.D. thesis, University of Ulm (2013)
26. Künzle, V., Reichert, M.: PHILharmonicFlows: towards a framework for object-aware process management. J. Softw. Maint. Evol. Res. Pract. **23**(4), 205–244 (2011)
27. Li, G., de Carvalho, R.M., van der Aalst, W.M.P.: Automatic discovery of object-centric behavioral constraint models. In: Abramowicz, W. (ed.) BIS 2017. LNBIP, vol. 288, pp. 43–58. Springer, Cham (2017). https://doi.org/10.1007/978-3-319-59336-4_4
28. Li, G., de Murillas, E.G.L., de Carvalho, R.M., van der Aalst, W.M.P.: Extracting object-centric event logs to support process mining on databases. In: Mendling, J., Mouratidis, H. (eds.) CAiSE 2018. LNBIP, vol. 317, pp. 182–199. Springer, Cham (2018). https://doi.org/10.1007/978-3-319-92901-9_16
29. M'baba, L.M., Assy, N., Sellami, M., Gaaloul, W., Nanne, M.F.: Process mining for artifact-centric blockchain applications. Simul. Model. Pract. Theory **127**, 102779 (2023)
30. OMG: UML: Unified Modeling Language 2.5.1 (2017). https://www.omg.org/spec/UML/2.5.1/About-UML/. Accessed 23 June 2022
31. Pelekis, N., Theodoulidis, B., Kopanakis, I., Theodoridis, Y.: Literature review of spatio-temporal database models. Knowl. Eng. Rev. **19**(3), 235–274 (2004)
32. Snodgrass, R., et al.: Temporal databases. Computer **19**(09), 35–42 (1986)
33. Snoeck, M.: Enterprise information systems engineering. The MERODE Approach (2014)
34. Snoeck, M., Dedene, G.: Existence dependency: the key to semantic integrity between structural and behavioral aspects of object types. IEEE Trans. Softw. Eng. **24**(4), 233–251 (1998)

35. Stonebraker, M., et al.: Third-generation database system manifesto. ACM SIG-MOD Rec. **19**(3), 31–44 (1990)
36. van der Aalst, W., et al.: Process mining manifesto. In: Daniel, F., Barkaoui, K., Dustdar, S. (eds.) BPM 2011. LNBIP, vol. 99, pp. 169–194. Springer, Heidelberg (2012). https://doi.org/10.1007/978-3-642-28108-2_19
37. van der Werf, J.M.E.M., Rivkin, A., Polyvyanyy, A., Montali, M.: Data and process resonance. In: Bernardinello, L., Petrucci, L. (eds.) PETRI NETS 2022. LNCS, vol. 13288, pp. 369–392. Springer, Cham (2022). https://doi.org/10.1007/978-3-031-06653-5_19
38. Wolfson, O., Xu, B., Chamberlain, S., Jiang, L.: Moving objects databases: issues and solutions. In: Proceedings of the Tenth International Conference on Scientific and Statistical Database Management (Cat. No. 98TB100243), pp. 111–122. IEEE (1998)

Explainable DMN

Carl Corea[1,2]([✉]), Timotheus Kampik[2,3], and Marco Montali[4]

[1] Institute for IS Research, University of Koblenz, Koblenz, Germany
ccorea@uni-koblenz.de
[2] SAP Signavio, Walldorf, Germany
timotheus.kampik@sap.com
[3] Umeå University, Umeå, Sweden
[4] Free University of Bozen-Bolzano, Bolzano, Italy
montali@inf.unibz.it

Abstract. We investigate means for the *explainability* of Decision Model and Notation (DMN) models. These are especially relevant for industrial settings, where the scale and complexity of DMN models can otherwise quickly make it unfeasible for companies to understand and maintain their decision logic. To this aim, we present a formal approach for measuring the *impact* of decision inputs on the decision output. In particular, we show how the decision logic of a DMN model can be transformed into a coalitional game based on (Datalog) queries over the decision tables, which allows one to apply the game-theoretic underpinning of Shapley values for measuring impact. Intuitively, the inputs of the decision act as the players of a coalitional game, and the payoff is the impact of an input/player on the decision output. The motivation of this work stems from real-life settings where means for understanding decision models are crucial, e.g., models of industrial complexity and domains such as fraud management. We implement our approach and evaluate it with real-life DMN models from the SAP-SAM dataset.

Keywords: DMN · Explainability · Shapley Values

1 Introduction

While DMN is meant to provide an intuitive means for modeling business rules, this does not scale well if models get more complex or are maintained by multiple modellers [2,5]. In practice, DMN is typically applied in large and complex organizations. In such settings, DMN models can easily have a double- or triple digit amount of *decision inputs* (e.g., think of inputs from a customer case such as Name, Age, Credit Score, or other system variables). In turn, it can become very challenging for organizations to understand the true functioning of the decision logic on an object-level, in particular, which of these decision inputs actually

This research is supported in part by SAP Signavio.
C. Corea—Visiting Researcher.

A. Marrella et al. (Eds.): BPM 2024 Forum, LNBIP 526, pp. 55–71, 2024.
https://doi.org/10.1007/978-3-031-70418-5_4

impact the decision *output*. This also makes it difficult to ensure the decision logic is compliant on a meta-level, e.g., the decision on a loan application may need to be independent of—and not discriminate against—certain data inputs from the application such as the gender of the applicant.

Hence, means for *explainability* of DMN models are needed, especially to support modelers in understanding which inputs have an impact on the decision. To develop such capabilities, the aim of this work is to create a formal framework for measuring the impact of decision inputs to the decision output. To this end, we show how the decision logic of a DMN model can be transformed into a coalitional game based on queries over the decision tables. This provides the basis to measure impact through the application of the game-theoretic underpinning of *Shapley values* [15]. Intuitively, decision inputs act as players of a coalitional game, and the payoff is the impact of that input on the decision output.

There are some existing approaches [2,13] that try to solve the problem of "DMN explainability" by layering Machine Learning (ML) on top of DMN. They train an ML model to reproduce the DMN logic, and then use standard XAI methods to reason about impact through the ML model. However, this essentially takes the (whitebox) DMN model and turns it into a blackbox ML model, transferring the explainability problem to a subsymbolic domain. As discussed in [2], there are many limitations to this ML-based approach. In particular, it cannot be assured that i) the explanations remain faithful to the original, symbolic DMN model, and ii) the approach is scalable to real-world applications where thousands of ML models would need to be trained. In this work, we natively assess explainability on the symbolic DMN model itself, thus overcoming by design the shortcomings arising when lifting the approach to ML models.

In the broad explainability spectrum, we are interested in providing insights on the *impact* that decision inputs have on the output of the decision model. In this light, our contributions are as follows:

- We introduce and formalize *impact* in the context of DMN, focusing on S-FEEL (graphical) decision tables. We then present a novel Shapley value-based approach to measure the impact of decision inputs, relying on queries over DMN tables. We refer to such values as *Shapley DMN values*.
- We show that the resulting framework guarantees desirable explainability properties. First and foremost, we tackle the *efficiency* principle, which provides an interpretation to our Shapley DMN values in terms of how they explain the marginal contribution towards an overall payoff. In addition, we show that our approach is semantic in spirit or, as we call it, is *representation independent*, in the sense that it provides the same treatment to syntactically different tables that convey the same decision logic. This means that representation-level changes, e.g., merging two individual numbers 1 and 2 into a compact interval [1..2] in a rule, does not affect the outcome.
- To show feasibility, we implement our approach and evaluate it with real-life DMN models from the SAP-SAM dataset (containing 5668 DMN models), as well as synthetic data. Our experiments show that Shapley DMN values can be computed sufficiently fast for real-world applications.

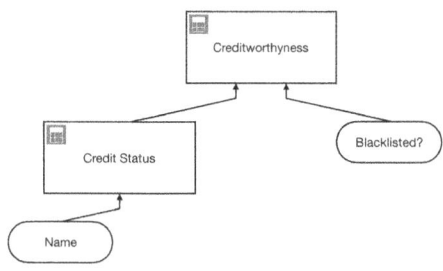

Creditworthiness		
Input	**Input**	**Output**
Status	*Blacklisted*	*Creditworthy*
Gold	True	True
Silver	True	False
-	False	False

Customer Status	
Input	**Output**
Name	*Status*
A	Gold
B	Silver

(a) Exemplary DRD with sub-decision

(b) Corresponding tables (U)

Fig. 1. DMN model for determining the creditworthiness of a customer.

Based on preliminaries in Sect. 2, we present our approach and the foundational results in Sect. 3. We then provide the evaluation in Sect. 4. Finally, we conclude with a discussion and future outlook in Sect. 5.

2 Preliminaries

We provide necessary preliminaries on DMN and Shaply values.

2.1 DMN

We focus on DMN decision models that rely on (graphical) *decision tables*, as well as a *decision requirements diagram* (DRD). An example is shown in Fig. 1.

The DRD is an acyclic graph that specifies the elements of the decision. In Fig. 1, the square elements represent decisions (corresponding to the decision tables) and the round elements are decision inputs. Decisions can be chained, so that an output of one table provides the value for an input of another table; e.g., the output **Status** of the *Customer Status* table maps to the **Status** input of the *Creditworthiness* table. For the formal semantics of DRDs, see [3].

A decision table comes with (input and output) attributes, each of which can be seen as a variable with values belonging to a specific datatype. We consider DMN tables that employ the S-FEEL expression language [3,6].

Definition 1. *An* (S-FEEL) condition *is inductively defined as follows:*

$$\langle condition \rangle ::= \langle disjCond \rangle \mid \text{``not(''} \langle disjCond \rangle \text{'')''}$$
$$\langle disjCond \rangle ::= \langle atomicCond \rangle \mid \langle atomicCond \rangle \text{'',''} \langle disjCond \rangle$$
$$\langle atomicCond \rangle ::= \text{``-''} \mid k \mid \langle compOp \rangle \ k \mid \langle openInt \rangle \ k_1 \text{'',''} k_2 \ \langle closeInt \rangle$$
$$\langle compOp \rangle ::= \text{``<''} \mid \text{``≤''} \mid \text{``>''} \mid \text{``≥''}$$
$$\langle openInt \rangle ::= \text{``[''} \mid \text{``(''}$$
$$\langle closeInt \rangle ::= \text{``]''} \mid \text{``)''}$$

where quotes are used for reserved keywords, and k, k_1, k_2 *are constants.*

The intuition of a condition φ is that it confines which values an attribute must carry, e.g., a value k. A special case is the wildcard "-", which represents a "don't care" condition, matching any value.[1] With conditions at hand, we can now define decision tables. We limit ourselves to the constitutive elements in our treatment; in particular, we omit facets and defaults. Also, we focus on tables with a *unique hit policy*; this is w.l.o.g., as other policies can be "uniqueified" [1].

Definition 2. *A (unique-hit) decision table is a tuple $\langle N, I, O, R \rangle$, where: (i) N is the table name; (ii) I and O are disjoint, finite ordered sets of input and output attributes; (iii) R is an ordered set $\langle r_1, \ldots, r_p \rangle$ of rules. Each rule r_j is a pair $\langle \mathsf{If}_j, \mathsf{Then}_j \rangle$, where If_j is an input entry function that maps each input attribute $\mathbf{a} \in I$ to a condition, and Then_j is an output entry function that associates each output attribute $\mathbf{b} \in O$ to a value in the domain of the datatype of $\mathbf{b}$. If_j must be defined so that, for every combination of values for input attributes, no two rules can match.*

A *DMN model* is then defined as a set of decision tables and a DRD.

2.2   Queries over DMN Tables

Similarly to the idea proposed in [3], the approach in this work relies on performing queries over DMN decision tables. For this, we define the structure of a decision table as a relational schema, supporting Datalog-like queries.

Let Const be a finite set of constants, which denote in our setting, S-FEEL conditions as per Definition 1. Then, define $\vec{c} = (c_1, \ldots, c_n)$ as a tuple over Const. A relation r is a set of tuples over Const, each with the arity $\mathsf{len}(r)$. Let $\mathbf{S}$ be a schema of relational symbols R (each with arity $\mathsf{len}(R)$). In our setting, the relational symbols R correspond to the table names. Then, a DMN database[2] $\mathbf{D}$ is defined as a finite set of entries of the form $R(c_1, \ldots, c_n)$, denoting that $(c_1, \ldots, c_n) \in \mathsf{Const}^n$ is in $R^{\mathbf{D}}$. We denote $\mathcal{D}(\mathbf{S})$ as the universe of DMN databases over $\mathbf{S}$. We also call *DMN fact* a constant used in some condition of a decision table of interest. For example, constants "1..10" and "-" are two facts of (the first rule in) table Status of Fig. 2.

Relative to $\mathbf{D}$, a relational query q is a function $q : \mathcal{D}(\mathbf{S}) \to 2^{\mathsf{Const}^n}$ that maps a DMN database $\mathbf{D}$ to a relation $q(D)$. We say that every tuple $\vec{c} \in q(D)$ is an *answer* to q w.r.t. $\mathbf{D}$. As customary, we denote a query q as a rule of the form

$$q(\vec{x}) :- R_1(\vec{t_1}), \ldots, R_n(\vec{t_n})$$

where $R_1, \ldots, R_n$ are relational symbols over $\mathbf{S}$, and $t_1, \ldots, t_n, \vec{x}$ are tuples containing variables or constants. A tuple $\vec{c}$ is an answer to q (w.r.t. $\mathbf{D}$) if there exists a homomorphism from the query to the tuple s.t. (i) every constant a in $\vec{x}$ maps to the corresponding constant a in $\vec{c}$, and (ii) every variable X in $\vec{x}$ maps to a constant a in $\vec{c}$ such that the query holds true (i.e., via grounding).

---

[1] It is not clear whether "any value" also means that the condition evaluates to true if no value is provided. We will come back to this point later in the paper.

[2] We use term "DMN database" for the set of relational tables of the DMN model.

Example 1. Consider the following DMN table D_1.

Status	
Input	**Output**
Name	Status
A	*Gold*
B	*Silver*

Now, consider the following query q_1, with $q_1(x, y) :- \text{STATUS}(x, y)$. Then we have that (w.r.t. D_1) q_1 has the following answer:

$$\{(A, Gold), (B, Silver)\}$$

As we will show, the impact of individual DMN facts can be made measurable by interpreting DMN tables as a coalitional game: the decision inputs act as the players of the game and the "payoff" is the impact they have on the decision output. To quantify this notion of "payoff", we build on so-called *numerical queries*, which are functions $q_{num} : \mathcal{D}(\mathbf{S}) \to \mathbb{R}$ that assign to a DMN database a numerical value. In [11], concrete instantiations of numerical queries are presented. They are referred to as aggregate-relational queries.

Definition 3 (Aggregate-Relational Query). *Let D be a DMN database over a schema $\mathbf{S}$ and q a relational query. Then, an aggregate-relational query is an aggregation function $\gamma : 2^{\mathsf{Const}^n} \to \mathbb{R}$ that takes the result of a query $q(D)$ and maps this relation to a numerical value $\gamma(q(D))$.*

In the remainder, we denote q^λ as the aggregate-relational query λ w.r.t. q. The two aggregate-relational queries from [11] we consider are as follows:

- $\mathsf{count}(q(D)) = |q(D)|$;
- $\mathsf{distinct}_i(q(D)) = |\{\vec{c}[i] \mid \vec{c} \in q(D)\}|$;

(i.e., the number of query results, resp., distinct entries over the ith column).

Example 2. Recall the DMN database D_1 and query q_1 from Example 1. Then we have that $\mathsf{count}(q_1(D_1)) = \mathsf{count}(\{(A, Gold), (B, Silver)\}) = 2$. If both tuples in $q_1(D_1)$ contained the return value Gold, $\mathsf{distinct}_2(q_1(D_1))$ would return 1.

Remark 1. For any aggregate-relational query q^λ, we assume $q^\lambda(\emptyset) = 0$.

2.3 Shapley Values

In this work, we aim to measure the impact of a decision input by transforming a decision model into a transferable utility game (TU game) in coalitional form. A *TU game in coalitional form* is a pair $g = (N, v)$, where $N = \{1, \ldots, n\}$ is a set of n players, and $v : 2^N \to \mathbb{R}$ is a characteristic function assigning a "payoff" $v(S)$ to every subset $S \subseteq N$, with $v(\emptyset) = 0$. In our setting, the "payoff" is the impact that decision inputs have on the decision.

Let $\mathcal{G}^n$ be the universe of all games with n players. Then, a *value* is a function $\varphi : \mathcal{G}^n \to \mathbb{R}^n$ which assigns to each game $g \in \mathcal{G}^n$ an n−dimensional vector $\in \mathbb{R}^n$ s.t. $\varphi_i(g) = v(\{i\})$ is the payoff of player i in the game g. A very important value in this context is the Shapley value [15], which determines the payoff by considering the *marginal contribution* of a player i w.r.t. a coalition S, given by $v(S) - v(S \setminus \{i\})$. The Shapley value φ_i of a game $g = (N, v)$ is then defined via

$$\varphi_i(g) = \sum_{S \subseteq N} \frac{(|S| - 1)!(|N| - |S|)!}{|N|!} (v(S) - v(S \setminus \{i\})) \tag{1}$$

In our DMN setting, *the facts of the decision tables (i.e., the constants appearing in their conditions), are the players*. Here, an important distinction has to be made between so-called *endogeneous* and *exogeneous* "players" of a DMN database. Consider the DMN model in Fig. 1(a). The input to the Credit Status decision is the case-dependent data element via the data-input node Name. This is referred to as an *endogeneous entry*, and the facts appearing in that attribute for the different rules will then be endogenous players. In the consequent Creditworthiness decision, the status attributes depends on other (endogeneous) entries, due to the chaining of the decision tables, and thus it is considered there as a so-called *exogeneous entry* (defined in the Credit Status decision). Hence, its facts will be exogenous players. For a DMN database D, we denote this distinction as $D = D_x \cup D_n$ (where D_x are facts to be considered exogenous, and D_n facts to be considered endogenous).

Consider now some DMN rule, and the fact of an input attribute **a** appearing therein, under the hypothesis that **a** is classified as an endogenous entry. To check whether that fact indeed participates as an endogenous player, a special consideration must be given to *wildcards*. Unfortunately, wildcards have an ambiguous interpretation in the DMN standard: the current standard[3] states that *"if the input entry is '-' [...] that particular input is irrelevant"* (p.61). In this light, one possible interpretation is that "−" is syntactic sugar for the condition that defines the entire facet/domain of **a**. Under this mild interpretation, "−" means that every value that can be assigned to **a** matches the condition. This means that "−" behaves like a standard condition, and thus it plays as an endogenous player, being **a** and endogenous entry. A second interpretation is that "−" indicates that **a** should not be considered at all when taking the decision. Under this strong interpretation, "−" means that the rule is applied as if the attribute does not exist, and so it must be completely ignored, that is, treated as an exogenous player.

In this work, we adopt this second interpretation, for the very simple reason that it implicitly covers also the first one: whenever we want to handle "−" under the mild interpretation, it is enough to explicitly substitute it with the explicit condition matching the facet/domain of the attribute. As a result, all attributes with wildcards in a rule will be treated as exogenous players in the context of that rule (and hence, they will not have any impact).

---

[3] https://www.omg.org/spec/DMN/1.4/PDF.

Status		
Input	**Input**	**Output**
Age	*City*	*Status*
1..10	–	Gold
11..18	C	Silver

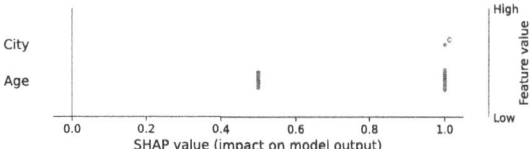

Fig. 2. Exemplary DMN model D_1.

Fig. 3. Beeswarm plot for the impact of `Age` and `City` in D_1 (w.r.t. data values).

3   Measuring Decision Impact with Shapley Values

We now present how to measure the impact of decision inputs. To illustrate our notion of impact, we begin with a motivational example, shown in Fig. 2.

The table contains two rules, based on the input columns `Age` and `City`. However, the *importance* of these columns differs between the two rules:

1. For the first rule, there is a wildcard in the `City` column. So the `City` column has no impact on the decision if the age is between 1 and 10 (included).
2. For the second rule, given the `Age` is between 11..18, we also need the matching `City` input. So `Age` and `City` both "evenly" contribute towards the outcome of the decision in this case.

Thus, in regard to the "players" from D_1, the constant 1..10 should be considered to have a higher impact than 11..18. So concretely, with "impact" we mean to convey that 1..10 (alone) can trigger 1 rule, but 11..18 is only "half" of what is needed. In our approach, the impact of a DMN fact is consequently quantified by measuring its contribution towards the result of a numerical query q^λ such as count [11]. This also allows us to understand how the concrete input *values* (e.g., from a customer form) impact the decision, as illustrated in Fig. 3. Figure 3 conveys that a lower age value (blue) will have a high(er) impact on the decision logic than for higher age values – specifically: the impact of any value between 1..10 is higher but decreases from age 11 onward. Identifying such break-points could be particularly useful in domains such as fraud management: for example, an expert can make sure to verify the `Age` on the application is correct if it is near the break-point, as here, even a small change in the age may have a large effect on the decision output.

Practically useful interpretations of Shapley DMN values on *model*-level are, for example: *(i)* if the impact of an attribute or fact is disproportionally high or low given its business relevance as assessed by a domain expert, this may hint at undesirable behaviors, where the technical specification may not reflect the actual business requirements; *(ii)* if the impact of an attribute or fact changes when a DMN model is adjusted, this may—in some cases—point to unintended side effects of the changes; *(iii)* if the impact of an attribute, fact, or rule is 0 this means that the DMN model can be simplified.

3.1 Shapley DMN Values

We now show how to systematize what we have described so far, by introducing the notion of *Shapley DMN values* to quantify the impact of DMN facts. As motivated above, for a numerical query q^λ, the impact of a DMN fact should be quantified as its contribution towards the query result. We begin by stating some desirable behaviors that we expect from any such form of impact measure.

In general, the impact of all DMN facts together should be *distributive*, in the sense that it explains the overall result of the numerical query. For example, if the count measure has counted n rules, the impact of all DMN facts should "add up to", or, "explain", n. This ensures that the impact measure can be interpreted as the marginal contribution of an individual fact towards the overall result of the numerical query. This also ensures that DMN facts with an equal impact are assigned an equal value. Furthermore, from the discussion in Sect. 2.3, it follows that exogeneous facts should not have any impact, and hence have a marginal contribution of 0.

As we will show, the provided intuitions exactly correspond to classical properties of Shapley values. Here, every player (in our case: DMN fact) is assigned a payoff value based exactly on the contribution towards the (decision) output. Technically, given a set N of players, the TU games $g = (N, v)$ and $g' = (N, v')$, and two players x_i, x_j from N, the properties of Shapley values are as follows.

Efficiency: $\sum_{i \in N} \varphi_i(g) = v(N)$
Symmetry: If $v(S \cup \{x_i\}) = v(S \cup \{x_j\})$ for all $S \subseteq N \setminus \{x_i, x_j\}$, then $\varphi_{x_i}(g) = \varphi_{x_j}(g)$
Additivity: $\varphi_i(g + g') = \varphi_i(g) + \varphi_i(g')$[4]
Dummy: If x_i is s.t. for all $S \subseteq N : v(S \cup \{x_i\}) = v(S)$, then $\varphi_{x_i}(g) = 0$

Efficiency states that the payoff of the grand coalition should be the sum of the individual players' payoffs. In our setting, this is in line with the above motivation that we want to measure the impact of an individual DMN fact towards the result of the numerical query. *Symmetry* states that two players that contribute equally should be assigned an equal payoff value (also in line, see above). It also follows via *Additivity* that the impact measures are additive, e.g., the impact of a compound results from the aggregation of the individual impacts. Finally, *Dummy* states that a player with no impact should be assigned a payoff of 0. Following Sect. 2.3, this applies to all exogeneous facts, that is, facts from input attributes that are exogenous entries, or wildcards.

As the properties of Shapley values conceptually match the desired behavior of an impact measure, we propose to measure impact by viewing q^λ as a TU game over a DMN database D. For this, we need some further notation.

Definition 4 ($\overline{F}$-Difference). *Let a DMN database $D = D_n \cup D_x$, $F \subseteq D$ a set of facts F, and $F_n = F \cap D_n$ be those facts in F that are also endogenous. Then, the $\overline{F}$-difference of D is defined as $\overline{F}(D) = D \setminus \{a \in D_n : a \notin F_n\}$.*

---

[4] with $\varphi_i(g + g') = \varphi_i((N, v + v'))$ for any $g = (N, v), g' = (N, v')$.

In other words, the $\overline{F}$-difference of D is a set $D' \subseteq D$ s.t. all endogeneous facts but those in F are removed from D.

Example 3. We recall the DMN table D_1 from Fig. 2. Consider a set of facts $F = \{B, Gold, Silver\}$. Then, the $\overline{F}$-difference over D_1 is the set $D_1' \subseteq D_1$ from which all endogeneous facts except $B, Gold, Silver$ have been removed (so A gets removed). This is illustrated in the following table.

Status	
Input	**Output**
Name	Status
-removed-	Gold
B	Silver

The $\overline{F}$-difference is used to filter exogenous facts before applying queries. E.g., in the above example we have $q_1(D_1') = \{(B, Silver)\}$, hence $\mathsf{count}(q_1(D_1')) = 1$.

We are now ready to define Shapley values for DMN models.

Definition 5 (Shapley DMN values). *Let D be a DMN database over a schema $\mathbf{S}$, q^λ a aggregate-relational query λ w.r.t. q, and f a fact in D. The Shapley DMN value $\varphi_f(D, q^\lambda)$ is the value $\varphi_i(N, v)$ as in Eq. (1), where:*

1. $N = D_n$
2. $v(F) = q^\lambda(\overline{F}(D))$, for all $F \subseteq N$
3. $i = f$

In other words, for a DMN database D and an aggregate-relational query q^λ, the payoff of a coalition of players F is defined via q^λ over the $\overline{F}$-difference of D. A simple corollary from the definition of the $\overline{F}$-difference is that all exogenous players receive a payoff of 0 as expected.

Example 4. Consider the DMN models D_1, D_2, and D_3, shown next (ranges are to be understood inclusive)

Status	
Input	**Output**
Name	Status
A	Gold
B	Silver
D_1	

Status		
Input	**Input**	**Output**
Age	City	Status
1..10	-	Gold
11..18	C	Silver
D_2		

Status		
Input	**Input**	**Output**
Age	City	Status
1..10	B	Gold
1..10	C	Gold
11..18	B	Silver
11..18	C	Gold
D_3		

We recall the query $q_1(x, y) :- \text{STATUS}(x, y)$, and define the query q_2 via $q_2(x, y, z) :- \text{STATUS}(x, y, z)$. Then we have

$q_1(D_1) = \{(A, Gold), (B, Silver)\}$

$q_2(D_2) = \{(1..10, -, Gold), (11..18, C, Silver)\}$

$q_2(D_3) = \{(1..10, B, Gold), (1..10, C, Silver), (11..18, B, Silver), (11..18, C, Gold)\}$

The Shapley DMN values for the examples are as follows. For D_1, we have:

$$\varphi_A(D_1, q_1^{count}) = 1 \qquad\qquad \varphi_B(D_1, q_1^{count}) = 1$$

This is intuitive, as both facts A and B have an equal impact on the possible decision outcome. The shown Shapley DMN values (which are w.r.t. count) can be interpreted such that both A and B trigger 1 (of 2) rules.

For D_2, we have that

$$\varphi_{1..10}(D_2, q_2^{count}) = 1 \quad \varphi_{11..18}(D_2, q_2^{count}) = 0.5 \quad \varphi_C(D_2, q_2^{count}) = 0.5$$

As discussed in the motivational example, note that the fact 1..10 has a higher impact due to the wildcard in the city column. Thus, the shown Shapley DMN values (which are w.r.t. count) capture the desired behavior that 1..10 (alone) triggers one rule and the other two facts both trigger 0.5 rules.

For D_3, notice that there are differences in the number of yield distinct outputs. For Age, if 11..18 is removed, only Gold can still be reached, whereas if 1..10 is removed, both Silver and Gold are reachable. In this sense, removing 11..18 has a higher impact on the number of reachable outcomes. Thus, the shown Shapley DMN values (w.r.t. distinct) can be interpreted such that 11..18 contributes towards 2/3 of all distinct outputs. Similar considerations apply to B vs C.

$$\varphi_{1..10}(D_3, q_3^{distinct}) = 1/3 \qquad\qquad \varphi_B(D_3, q_3^{distinct}) = 2/3$$
$$\varphi_{11..18}(D_3, q_3^{distinct}) = 2/3 \qquad\qquad \varphi_C(D_3, q_3^{distinct}) = 1/3$$

3.2   Logical Properties

General Properties. We now turn to logical properties of the proposed approach. As a first result, the proposed Shapley value for DMN satisfies the desirable properties stated at the beginning of Sect. 3.1.

Proposition 1. *Let D be a DMN database over a schema S, q^λ a numerical query λ w.r.t. q, and f a fact in D. Then, the Shapley DMN value $\varphi_f(D, q^\lambda)$ satisfies Efficiency, Symmetry, Additivity and Dummy.*

Proof. As per Definition 5, the Shapley DMN value is defined as the Shapley value of a TU game in coalitional form. Consequently, it follows from the well-established properties of the Shapley value [15] that the Efficiency, Symmetry, Additivity, and Dummy properties are satisfied if for the characteristic function v, it holds that $v(\emptyset) = 0$. This clearly is the case, as we have earlier (in Remark 1) stipulated that $q^\lambda(\emptyset) = 0$. Also, if $F = \emptyset$, $\overline{F}(D)$ will only contain exogeneous players, which, by definition, have a payoff of 0.   □

Notably, these desirable properties only hold for the proposed Shapley DMN values, but will not hold when using simple, *naïve* approaches, such as the simple marginal contribution. This is immediate by considering DMN model D_3 and distinct, under the assumption that the output of the first column is *Silver*.

Relationship to DRD. An important aspect for the DMN setting is how the impact of individual players (here: facts) commutes from tables to the DRD-level. Let D be a DMN database over $\mathbf{S}$, for every $R \in \mathbf{S}$ with arity n, we denote $C_1^R, ..., C_n^R$ as the columns corresponding to the DMN table defined via R. Furthermore, with $C_i^{R\downarrow}$, we denote a tuple of all *distinct* facts of the ith column C_i^R, i.e., all distinct constants c in $R^{\mathbf{D}}$ with index i. This creates a projection of distinct facts per column. We denote $\varphi(C_i^{R\downarrow}, q^\lambda)$ as the vector containing all Shapley DMN values (w.r.t. q, λ) for all facts in C_i^R.

*Example 5. We recall the DMN model D_2 and the corresponding Shapley DMN values shown in Example 4. Then we have for the 1st column (**Age**) that $C_1 = (1..10, 11..18)$ and $\varphi(C_1^\downarrow, q^\lambda) = \begin{bmatrix} 1 \\ 0.5 \end{bmatrix}$.*

Via *Additivity*, this yields *column impacts*, characterized as the the sums of all Shapley values of facts in the column of an attribute of interest.

*Example 6. The DMN model D_2 and the corresponding Shapley DMN values from Example 4 yield the following column impact values via **sum**$(\varphi(C_i^\downarrow, q^\lambda))$:*

$$\text{Importance Name: } 1 + 0.5 = 1.5 \qquad \text{Importance City: } 0.5$$

These (column) Shapley values capture the desired behavior that the Name attribute triggers 1.5 (of 2) rules and the City attribute triggers 0.5 rules (cf. Example 4). The Name attribute can therefore be interpreted to have a higher impact. Especially in large-scale models, this is useful for visualizing the impact of decision inputs on the DRD level, as shown in Fig. 4.

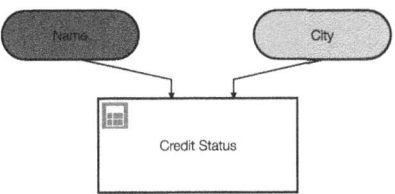

Fig. 4. Visualization of data input node importance (DRD-level) for D_2.

In the extreme case where a column has only wildcards, its importance would be 0: the attribute can be removed without altering the decision logic.

The described behavior of how the impact on q^λ commutes from the individual players in $\mathbf{D}$ to a column- (or DRD-) level can be captured as follows, where the column importances are equal to the impact of all players in that column.

Distribution. $\sum_{R \in \mathbf{S}} \sum_{i=1..\text{len}(R)} \text{sum}(\varphi(C_i^{R\downarrow}, q^\lambda)) = \sum_{f \in \mathbf{D}} \varphi_f(\mathbf{D}, q^\lambda)$

Proposition 2. *Let $\mathbf{D}$ be a DMN database over a schema $\mathbf{S}$, q^λ a numerical query λ w.r.t. q, and f a fact in D. Then, the Shapley DMN value $\varphi_f(D, q^\lambda)$ satisfies Distribution.*

Proof. Recall that $\mathbf{D}$ consists of a finite set of entries $R(c_1, ..., c_n)$. Then we have

$$\sum_{f \in \mathbf{D}} \varphi_f(\mathbf{D}, q^\lambda) = \sum_{R \in S} \sum_{i=1..\text{len}(R)} \sum_{x \in C_i^{R\downarrow}} \varphi_x(\mathbf{D}, q^\lambda)$$

$$= \sum_{R \in S} \sum_{c \in R(c_1,...,c_n)} \varphi_c(\mathbf{D}, q^\lambda) = \sum_{c \in \mathbf{D}} \varphi_c(\mathbf{D}, q^\lambda) \qquad \square$$

Independence of Representation. An important aspect of DMN modeling is that "how" the rules are syntactically written depends on the modeler. In addition, syntactically different rules may be semantically equivalent, that is, capture the same decision logic [8]. In this context, it is important to understand whether Shapley DMN values depend on the syntactic choices taken by the user, or only relate to the semantics of the decision logic. In short, our approach is independent of the syntactic representation, which we will show in the following.

The DMN standard contains many aspects where the same semantics can be represented in different ways, and we cannot cover all of these here. In the following, we focus on the representation of *intervals*, i.e., representing an interval as [1..2] or two separate rules 1 and 2. This relates to the problem described above as the modeler may decide to write a single rule where an attribute $\mathbf{a}$ has condition [1..2], or two distinct rules where all the other attributes keep the same conditions, while $\mathbf{a}$ uses 1 in the first, and 2 in the second - a clear case of semantic equivalence. An example is shown in Fig. 5.

Customer	
Input	**Output**
Age	*Status*
[1..2]	Gold

(a) Table with compact interval

Customer	
Input	**Output**
Age	*Status*
1	Gold
2	Gold

(b) Same table with expaned rules

Fig. 5. Semantically equivalent DMN models with different representations.

The sensitivity of the Shapley values to the concrete representation in our case depends on the specific aggregate-relational query. For example, moving from (a) to (b) in Fig. 5 changes the value of count as a new row is introduced, but distinct remains the same. This is described by the following property for DMN databases D, with q^λ being an aggregate-relational query λ over q.

Representation Independence [Interval]. For a fact f in $R(\vec{c})$ of the form $[x_1, ..., x_n]$, let $R' = R(\vec{c}) \setminus \{f\}$ and $D' = (D \setminus \{R(\vec{c})\}) \cup \{R' \cup \{x_1\}, ..., R' \cup \{x_n\}\}$. Then $\lambda(q(D)) = \lambda(q(D'))$.

Representation Independence [Interval] states that the value of λ should remain the same even if an interval is "expanded" from a compact representation into individual rows. As stated above, this holds, e.g., for distinct. So inherently, with

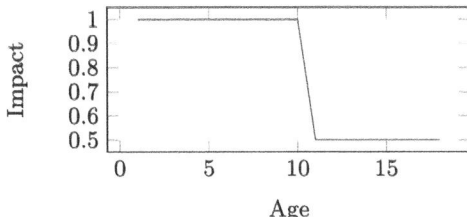

Fig. 6. Lineplot of impact over `Age` value (breaking-point at 11).

our approach we are working at the *behavioral* level, i.e., the Shapley DMN values depend on the semantics of the decision logic captured in the diagram, and is not influenced by the syntactic representation of that logic. In Fig. 5, inspecting the individual values of the interval [1..2], we indeed see that regardless of the numerical query, the impact *per* individual point is semantically the same for both representations: the impact for values 1 and 2 is the same in both tables.

We thus get the following result (the proof is straightforward).

Proposition 3. *Let D be a DMN database over a schema S, q^λ a numerical query λ w.r.t. q, and f a fact in D. Then $\varphi_f(D, q^\lambda)$ is insensitive to the representation of intervals if λ satisfies* Representation Independence [Interval]

3.3 Further Use Cases

The presented framework also offers support for further explainability use-cases.

First, next to the input column in general, the concrete *value* for this column (e.g., stemming from a customer form) can also heavily impact the decision outcome. For example, in a sales process for buying alcohol, changing the age on the order form from 17 to 18 may greatly impact the outcome of the decision.[5] Such a form of drill-down analysis thus allows to identify "break-points" of the impact of a decision input. Consider, e.g., D_2 from Example 4. For numerical ranges (here: `Age`), we can plot the importance per range value as in Fig. 6.

This visualization allows to check the difference of impact for (discrete) neighboring points. For example, here, there is a break-point at $10/11$. The case worker should therefore inspect the case more closely if `Age` is set to 10 or 11, to make sure that there was no fraud attempt. Furthermore, the shown distribution of Shapley DMN values can be used to compute a general assessment of the inputs' sensitivity to value changes with "fluctuation" measures from the literature, such as entropy or elasticity—the idea being that a distribution where the impact never changes is less problematic. For instance, entropy[6] provides an assessment of how the impact values are spread, i.e., if entropy is low, the impact does not change much even if the data values are changed. Likewise, elasticity measures

---

[5] We refer to the legal drinking age of 18 years (as in most European countries).

[6] For a distribution of impact values I, the entropy of these values is defined as $-\sum_{v \in I} \frac{f_v^I}{|I|} \ln \frac{f_v^I}{|I|}$, where f_v^I is the frequency of v in I.

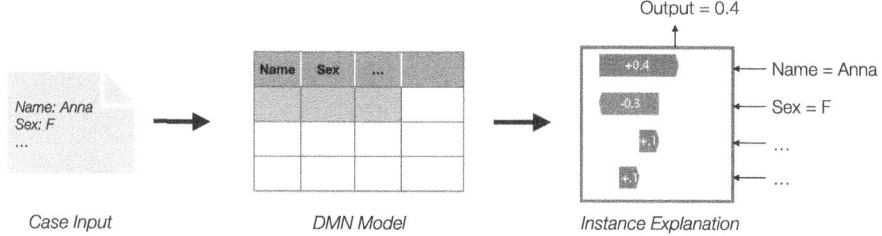

Fig. 7. Instance-level decision explainability with Shapley DMN values.

the change of impact over the change in the data value, e.g., *"will changing the input value by a small amount greatly change the impact on the decision?"*.

As a further use-case, the presented framework can be applied for explaining decisions on concrete instances, an example of which is shown in Fig. 7. For a given case-input, the corresponding values from the case can be matched to the respective rules they trigger (this can directly be done via our query-based approach by entering case facts in the query). Then, the Shapley DMN values from the corresponding DMN facts can be matched to explain the impact of the specific case values/variables on the case decision. This provides a comprehensive understanding of input contributions to the decision outcome, enhancing interpretability and trust in the decision logic and decision-making process.

4   Proof of Concept

An open question for our framework is whether the computation of DMN Shapley values can be feasibly obtained for real DMN decision models. To answer this question and investigate feasibility, we implemented our approach and evaluated it against real-life and synthetic datasets. Our implementation takes as input a DMN model and returns the Shapley DMN values for all DMN facts, using count for aggregation. For iterating through sets of players, we applied a *contraction* algorithm design [14]. The contraction pattern starts with the grand coalition, and then tries to decompose it into smaller subsets of players. However, in our setting, for any player set with a payoff of 0, any subset will also have a payoff of 0 because the payoffs are non-negative. Hence, our implementation does not further decompose any such player sets. This very effectively reduces the space complexity w.r.t. combinations that have to be considered (see below discussion).

Corollary 1. *Let D be a DMN database over a schema $\boldsymbol{S}$, q^λ a numerical query λ w.r.t. q, F a set of facts in D, and $\varphi_F(D, q^\lambda)$ be the sum of payoffs for all players $f \in F$. If $\varphi_F(D, q^\lambda) = 0$, then $\varphi_{F'}(D, q^\lambda) = 0$ for any $F' \subseteq F$.*

We continue with our evaluation results. All experiments were run on an M2 processor with 16GB RAM. The implementation code can be found online[7].

---

[7] https://anonymous.4open.science/r/shapleyDMN-F483/.

Description	Value	SD
# of analyzed models	5 668	-
Avg. # of decision tables	1.96	2.92
Avg. # of decision inputs	3.96	3.63
Avg. # of rules per table	5.62	7.24

Fig. 8. Statistics of SAP-SAM dataset.

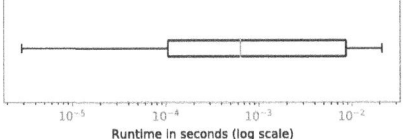

Fig. 9. Runtimes for computing Shapley DMN values for SAP-SAM (s).

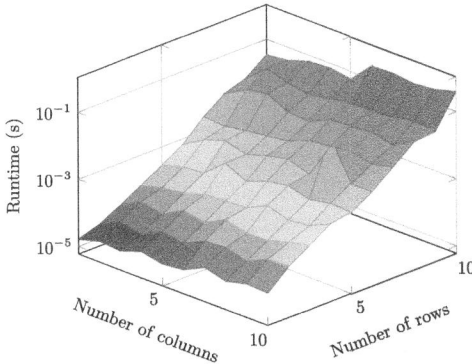

Fig. 10. Runtimes for the analysis of synthetic decision tables.

Real-Life Data (SAP-SAM). The SAP-SAM dataset is a large repository of process, decision, and other business models, created over the course of around ten years (from 2011 to 2021) for academic purposes [10]. From this dataset, we tested all DMN models (excluding some auto-generated duplicates). Figure 8 shows general properties of the analyzed dataset (N=5668 DMN models). The runtimes for computing the Shapley DMN values are reported in Fig. 9, with virtually all models being computable in seconds (the largest model took 29.3 s).

Synthetic Data. To evaluate feasibility with more comparable models, we generated synthetic DMN tables. As parameters, we selected the number of rows from $\{1, 2, ..., 10\}$ and the number of columns from $\{1, 2, ..., 10\}$. For every parameter combination ($10 \times 10 = 100$), we created a DMN table with the respective settings, where every table cell contained a unique string value.[8] We then computed the Shapley DMN values for these models - cf. Fig. 10. For the largest model (10×10), the runtime was <1 s. As can be seen, the slope is much steeper when the number of rows increases (vs. number of columns). This makes sense, as via the query-approach, if one endogeneous column is removed, the entire row can be disregarded. In this context, the applied algorithm design seems suitable.

All in all, our evaluation witnesses feasibility over both datasets, which we consider as a strong concrete basis for the effective application of our approach.

---

[8] The dataset is provided in the code repository.

5   Discussion and Conclusion

The basic idea of DMN is that it *is* explainable – the decision logic is explicitly modelled by means of business rules. However, this does not seem to scale if size and complexity increase [2,9]. This work inserts an additional explainability layer over DMN models, giving modelers an intuitive quantification of how decision input attributes impact on output attributes. This enables modelers to check sanity and robustness of business decision logic in settings that are highly complex and/or trigger frequent updates on how decisions are taken.

For the explainability use-case, there have been some works [2] trying to layer ML models on top of DMN. In Sect. 1, we already identified possible limitations of such an ML approach, the most prominent beeing that it takes a whitebox DMN model and turns it into a blackbox model. Besides this general problem, a concrete problem of the ML-based approach is that the utilized XAI techniques in [2] (SHAP) are geared towards numerical outputs. It is hence challenging to apply such techniques for categorial outputs [12]. In our approach, any output can be considered as we work on numerical queries, at the same time also guaranteeing faithfulness to the decision logic through logical properties.

Some limitations to our approach, which we intend to tackle in consequent work, are as follows. *First*, computing Shapley values can be computationally expensive [11]. Our investigation indicates that their computation is feasible on real-life DMN models, but our approach may still incur high computational cost when it is applied to (many) large DMN models, especially if their tables contain many rules. Yet, our approach is, in this regard, still superior to the aforementioned approaches [2,13] that require the training of an ML model *in addition* to the computation of feature attributions using Shapley-based explainability methods. In future works, one could rely on sampling-based approaches that give approximate guarantees with respect to principle satisfaction, or existing methods for computing Shapley values for facts w.r.t. queries (cf. [4,7]). *Second*, in domains where combinatorics of different columns play a prominent role, the current player-based view might be too narrow. E.g., for fraud management, one may need to assess whether cheating on a combination of columns affects the outcome, in a setting where manipulating the values in just one would not suffice. Our framework can be easily extended to consider combinations of columns by merging different players, and exploiting the *Additivity* property. The computation of break-points can then be implemented by considering multi-dimensional optimization techniques, which we aim to integrate in future works. *Third*, we have only focused on DMN tables with a *unique* hit policy, assuming that, if the original policy is different, one first "uniqueifies" the table [1]. It would be interesting to study how variants of our framework can be defined to natively deal with other hit policies, in particular for what concerns the definition of endogenous players. For example, a hit policy like *first* has an inherent defeasible semantics that can be used to define Shapley DMN values differently, considering that if a rule having higher priority matches, all the lower priority rules become "unreachable" and hence irrelevant. Here, we believe there is a strong tie between the (in)sensitivity of Shapley DMN values to hit policy changes and

DMN soundness [5,6] – an example is that the values are insensitive if the DMN table has no overlaps [1]. In future works, we aim to investigate this relationship of DMN soundness and Shapley DMN values. *Finally*, in this approach we focused on measuring the *impact* of decision inputs. In this sense, our methods allow to explain the impact of certain inputs/input values on the outcome. Arguably, the term "explainability" can also refer to many other concepts, such as causality, which we do not focus on in this work. In future works, we aim to extend our approach to also be capable of explaining other aspects or dimensions of decision models.

References

1. Batoulis, K., Weske, M.: Disambiguation of DMN decision tables. In: Abramowicz, W., Paschke, A. (eds.) BIS 2018. LNBIP, vol. 320, pp. 236–249. Springer, Cham (2018). https://doi.org/10.1007/978-3-319-93931-5_17
2. Bork, D., Ali, S.J., Dinev, G.M.: AI-enhanced hybrid decision management. Bus. Inf. Syst. Eng. **65**(2), 179–199 (2023)
3. Calvanese, D., Montali, M., Dumas, M., Maggi, F.M.: Semantic DMN: formalizing and reasoning about decisions in the presence of background knowledge. Theory Pract. Logic Program. **19**(4), 536–573 (2019)
4. Chen, H., Covert, I.C., Lundberg, S.M., Lee, S.I.: Algorithms to estimate Shapley value feature attributions. Nat. Mach. Intell. **5**(6), 590–601 (2023)
5. Corea, C., Kampik, T., Delfmann, P.: Empirical evidence of DMN errors in the wild - an SAP Signavio case study. In: De Weerdt, J., Pufahl, L. (eds.) BPM 2023. LNBIP, vol. 492, pp. 326–336. Springer, Cham (2024). https://doi.org/10.1007/978-3-031-50974-2_25
6. De Leoni, M., Felli, P., Montali, M.: Integrating BPMN and DMN: modeling and analysis. J. Data Semant. **10**(1–2), 165–188 (2021)
7. Deutch, D., Frost, N., Kimelfeld, B., Monet, M.: Computing the Shapley value of facts in query answering. In: Proceedings of the 2022 International Conference on Management of Data, pp. 1570–1583 (2022)
8. Figl, K., Mendling, J., Tokdemir, G., Vanthienen, J.: What we know and what we do not know about DMN. EMISAJ **13** (2018)
9. Hasić, F., Vanthienen, J.: Complexity metrics for DMN decision models. Comput. Stand. Interfaces **65**, 15–37 (2019)
10. Kampik, T., Warmuth, C., Sola, D., et. al.: Sap Signavio academic models (2022). https://doi.org/10.5281/zenodo.6964945
11. Livshits, E., Bertossi, L., Kimelfeld, B., Sebag, M.: The Shapley value of tuples in query answering. Logical Methods Comput. Sci. **17** (2021)
12. Lundberg, S.M., Lee, S.I.: A unified approach to interpreting model predictions. In: Advances in Neural Information Processing Systems, vol. 30 (2017)
13. Park, G., Küsters, A., Tews, M., Pitsch, C., Schneider, J., van der Aalst, W.M.P.: Explainable predictive decision mining for operational support. In: Troya, J., et al. (eds.) ICSOC 2022. LNCS, vol. 13821, pp. 66–79. Springer, Cham (2023). https://doi.org/10.1007/978-3-031-26507-5_6
14. Peppas, P.: Belief revision. Found. Artif. Intell. **3**, 317–359 (2008)
15. Shapley, L.S.: A Value for N-Person Games. RAND Corporation (1952)

Trace vs. Time: Entropy Analysis and Event Predictability of Traceless Event Sequencing

Peter Pfeiffer[1,2(✉)] [ID] and Peter Fettke[1,2] [ID]

[1] German Research Center for Artificial Intelligence (DFKI),
Saarbrücken, Germany
{peter.pfeiffer,peter.fettke}@dfki.de
[2] Saarland Informatics Campus, Saarbrücken, Germany

Abstract. Process mining offers powerful techniques to analyze real-world event data, aiming to improve processes. Typically, the data is stored and examined in event logs as traces, where each trace contains the sequence of events pertaining to a specific process case. A case can, e.g., represent the management of a customer request or the sequence of events from ordering to delivering a product to a customer in online retail businesses. While this approach allows to analyze and gain insights from complex event data, it also isolates events that in reality are correlated, potentially concealing important process behavior. In this paper, we motivate and conceptualize the approach to describe the observations generated by the underlying system as a single event sequence that is ordered as being executed. We study and compare how much the event order and trace notion affect the entropy rates of different real-life processes. Further, we investigate how predictable next activities in event sequences are. Our study indicates that ordering the events as executed does not necessarily increase the entropy rates of the process. We discuss these findings and their implications for future research.

Keywords: Event Data · Entropy · Prediction

1 Introduction

Process mining techniques analyze event data that is collected by actions taking place in the real world. Often, humans and machines are involved to achieve a certain objective, forming a socio-technical system [9]. Each action is tracked and recorded by information systems. Examples include incident management systems where some actions are triggered by humans (clients and workers) while others are controlled by algorithms; production facilities like car manufacturing where actions are taken by humans and machines; retail businesses where customers order goods that are picked and packed by machines; to travel departments involving many interactions between humans. When using process mining such data is analyzed from traces in event logs. The event logs contains and

© The Author(s), under exclusive license to Springer Nature Switzerland AG 2024
A. Marrella et al. (Eds.): BPM 2024 Forum, LNBIP 526, pp. 72–89, 2024.
https://doi.org/10.1007/978-3-031-70418-5_5

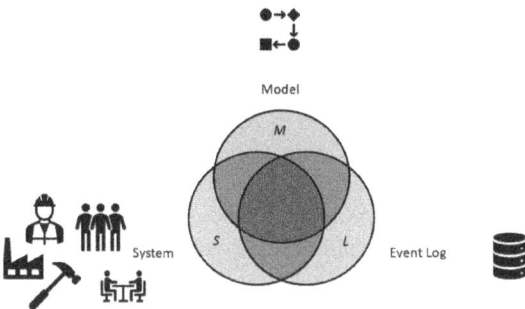

Fig. 1. Venn diagram showing the behavior of the system S, captured in the log L and represented by a model M(adapted from [6]).

Table 1. Example Event Log L_0 with three cases.

Case ID	Activity	Timestamp	Index
1	Start	2024-01-01 08:00:01	1
1	Collect Data	2024-01-01 09:01:02	2
1	Inspect	2024-01-03 09:05:03	5
1	Approve	2024-01-09 09:47:04	13
1	Send	2024-01-10 09:30:05	16
1	Finish	2024-01-11 10:00:06	19
2	Start	2024-01-02 08:04:07	3
2	Collect Data	2024-01-03 16:00:08	8
2	Inspect	2024-01-04 12:00:09	11
2	Inspect	2024-01-04 16:00:09	12
2	Approve	2024-01-09 10:01:10	14
2	Send	2024-01-10 10:10:01	17
2	Finish	2024-01-11 12:00:02	20
3	Start	2024-01-03 08:02:03	4
3	Collect Data	2024-01-03 12:00:04	6
3	Inspect	2024-01-03 15:00:05	7
3	Inspect	2024-01-04 07:00:05	9
3	Inspect	2024-01-04 09:00:05	10
3	Approve	2024-01-09 11:20:06	15
3	Send	2024-01-10 17:00:07	18
3	Finish	2024-01-12 15:00:08	21

describes the behavior of the process as a collection of traces. Each trace contains the behavior of a certain process case as a sequence of events, ordered as recorded by the information system.

If we think of the real world as a socio-technical system, whose behavior is recorded and stored in the event log, the event log represents the actions happening in this system in a case-centric way. Table 1 shows one event log L with three traces where each trace shows the sequence of events of an incident management case from start to end. At any point in time, there is not only one incident handling case ongoing, but multiple. Events happening in a period will not necessarily be assigned to one case but can happen for multiple process cases. The actual execution order of the system is given by *Index* in Table 1.

As many processes require interactions between humans, machines, and sensors in such a socio-technical system, there are very likely arbitrary dependencies between actions carried out. For instance caused by shared resources, bottlenecks, synchronization of products and goods before delivering, shifts, or weekend breaks, just to name a few reasons. If events are stored as traces, such dependencies might be scattered across various traces, depending on the trace notion chosen [2], even if being executed shortly after each other. When analyzing the events trace by trace, this behavior might remain concealed. We suspect that dependencies between events in event logs on event level exist. Although inspecting events in a trace-less way looks very unstructured initially, such patterns, if they exist, could aid process analysis and prediction.

We embed our analysis in the conceptualization where the event log L contains and represents the behavior of the unknown system S which we want to analyze [6]. One core task in process mining is to discover a model M, that represents the behavior of S from the event log L. Figure 1 shows the system, the log, and the model as three sets containing and representing process behavior in different ways. In this work, we will define entropy estimators [19], building on the idea of observing the behavior of the system S as a stochastic process [7] (instead of a set as done in [6]), to estimate and analyze its entropy rates directly on the events as being executed. For this, we assume the outcome of S to be individual events and study it using random variables of activity *Ngrams* [10]. The results help to investigate how much useful information can be found in the original event sequences, how much harder analyzing S without information about process cases is, and how predictive approaches are affected by it.

In the next section, we motivate the analysis of event sequences instead of traces. In Sect. 3 we conceptualize the system S as a stochastic process before studying and comparing its entropy rates and predictability in Sect. 4. Section 5 discusses the implications and limitations of the experiments. Afterwards, related work is summarised in Sect. 6 and the paper concludes in Sect. 7.

2   Motivation

We motivate the idea of this paper using a very simple, artificial example, inspired by [19], before going to a more complex one from real data. Assume a system S with three activities A, B, and C. The activities always happen in this order and describe one process case from start to end, captured as one trace in the event log. Thus, all symbols are identically distributed, but not independent. We can write the sequence of events as ordered by cases as

$$ABC\ ABC\ ABC \tag{1}$$

Process mining techniques process and analyze such data from the event log trace by trace, i.e., as ABC, three times in succession. In reality, multiple process cases are ongoing at the same time in the system S. Therefore, the order in which the events happened in reality is likely different. S can be described as a sequence of events in the order they have happened, e.g., as

$$ABAACBBCC \tag{2}$$

The case-ordered sequence (1) found in the event log L is more structured than (2). After having seen $ABC\ ABC$ we can be very certain that the next element will be A, followed by B and so on. Within cases the next activity is always clear. The probability that the next activity will be from another case is $\frac{1}{3}$ while the probability that the next event is from the same case is $\frac{2}{3}$. In contrast, the second sequence, ordered as events were executed in reality, is less structured. The probability for any sequence of length 3 is less than $\frac{1}{3}$. After seeing the first 6 elements $ABAACB$ there is a chance for the next element to be B or C. At any point, it is way more surprising what the next activity will be. The probability that the next event will be from another case is higher than $\frac{1}{3}$ while the probability that it will be from the same case is less than $\frac{2}{3}$.

Fig. 2. Seven process cases of the BPIC 13 Incidents event log (excerpt), layered by trace and colored by activity. The red vertical line is for orientation. (Color figure online)

The same observations can be made for real-world processes and events. The behavior of a system can appear unorganized if the events are ordered as executed in a single sequence. However, since humans are involved, there might be a certain structure, that is not directly visible but describes the behavior of such a system. We assume that some of this behavior becomes present as *Ngram* patterns when inspecting long sequences of events in the order as executed. Such patterns, if present, make the process structured and less surprising.

We can find such patterns in real processes. Figure 2 shows 7 cases from the data provided by the Business Process Intelligence Challenge 2013 (BPIC13 Incidents) where some sequences of events were removed for better displayability. This recording contains a total of 7.554 cases from an incident management process from Volvo with a total of 65.533 events collected by the system called *VINST*. Each case describes the handling of an incident that involves activities from multiple organization units, exemplified in the event log presented in Table 1. Each layer in Fig. 2 shows the sequence of events from one case, similar to a dotted-chart [20]. Events are ordered by their timestamp and spaced according to the duration between events on the y-axis, indicating their temporal relation. Different colors indicate different activities. As one can see, the workload changes with time, indicated by a higher or lower frequency of events.

When investigating the events by case, one can see very similar behavior. At some point, an incident case starts, information is collected, incidents are investigated, and accepted, information is sent to the customer and, finally, the incident is closed. Some incidents require more investigation than other cases, indicated by more or less blue-colored events. The behavior in each case is very similar which makes it look very structured. For many tasks in process mining, focusing on events per case reduces the complexity to a level that allows solving them in a traceable way, e.g., for discovery and conformance checking [1].

However, we can also find patterns that span across multiple cases. Some events in different cases directly follow each other. When looking over all cases, there seems to be additional behavior, besides the behavior within cases, that all cases follow. For instance, the red-colored events in Fig. 2 are *Accept* activities that are executed in a short period. The three red events immediately before and the two after the red vertical line are one week apart and were done by *Earl* in the Netherlands. This indicates that *Earl* accepts incidents once a week. Such behavior, known as batching, is common if the responsible person for approval accepts all cases at once. Similarly, the red events in the end are also *Accept*

Fig. 3. The same events from Fig. 2 as a single sequence colored by activity. (Color figure online)

activities, this time performed by *Juan*, before closing the incident. This shows that events can be correlated, e.g., through time or by participants in the process. When processing events trace by trace, as stored in the event log L, such behavior can remain unseen or be observed differently as happened in reality.

To emphasize the temporal relation and correlations between events across cases we think of S to produce a single sequence of events, which is eventually grouped into traces in the event log. Figure 3 displays the events shown in Fig. 2 as a single sequence. The per-case patterns discussed before become present visually. In the beginning, a sequence of blue and orange events indicates that cases started. After some time, some red-colored activities happen before red and green events alternate again, indicating the approval activities performed in batches. Other patterns repeat frequently such as three blue events or orange followed by two blue events. This aligns with the observation made from Fig. 2: events in the system S happen through time and not per case. At most points in time, there are multiple cases ongoing. The next event to happen in S does not need to be from the same case. Following this observation, we suspect that there are influences between cases on event level, i.e., dependencies between events as they happen through time.

These observations are also important for process prediction tasks. Predicting the probability distribution of next elements in a sequence has many applications, e.g., in natural language processing or time series analysis. In business process management, such approaches are used to predict the future behavior of an ongoing case, e.g., the activity, timestamp, or resource of the next event. Existing approaches make predictions at some point in time per case and use the sequence of previous events from within this case only. Thereby, events that have happened (concurrently) in other cases are not considered. The amount of information about S contained in one case is limited while patterns describing previous event behavior that has happened in the system S can contain much more information for predictive tasks. For instance, knowing whether and when *Accept* activities in other cases have been executed should make predicting future behavior beyond the red line easier. Similarly, predicting what will happen next in one particular case does not necessarily give information about what will happen next in S. Instead, predicting what event(s) will happen next in S is an interesting task for learning the system behavior and may enable additional predictive tasks like load prediction, case synchronization prediction, or compliance prediction.

Intuitively, one would assume that ordering events by cases makes the sequence less surprising as ordering them as being executed in the system - resulting in a lower entropy and higher predictability. However, certain patterns and information from other cases might make the next event foreseeable - positively affecting its entropy and predictability. To study the mentioned observations quantitatively, we choose to measure and compare the entropy over sequences of events as originally done by Shannon for English text [18,19]. Entropy, as defined by Shannon [19], measures the level of information or surprise of a random variable and is a fundamental concept in information theory. In the following, we formally conceptualize S as a stochastic process which allows studying the entropy of event sequences.

3 Concept

We start by conceptualizing S as an unknown source from which we observe a sequence of events before defining different sequences that differ how events are ordered. We denote a random variable as X and the stochastic process as an indexed sequence of random variables $\langle X_1, X_2, \ldots, X_N \rangle$.

3.1 System S

As the system S is largely unknown to the observer [6] we do not know the behavior of it. S can only be observed by the events it generates through time. Each event indicates that a certain activity in a certain context has been executed with arbitrary dependencies on other events. This randomness in observing S can be characterized by random variables. Through observation, S generates an integer-indexed sequence of random variables $S := \langle X_1, X_2, \ldots, X_N \rangle$ that describes its behavior.

In this paper, we deal with finite sequences of random variables of length N that have been recorded. Infinite sequences, similar to event streams [22], are not considered. We expect the events to be ordered in the stochastic process as they have been executed in S giving the index the interpretation of time. Note that the index is not the timestamp but the order in which the events have been observed, i.e., the events in S are ordered by time but not by trace. We assume an information system that records and indexes all events in S unequivocally.

The outcome of the random variables generated by S are events $e \in \mathcal{E}$ with $\mathcal{E}$ being a finite, discrete set of events. Events e are *tuples* of attribute values $(a_1, \ldots, a_j)$. Typical attribute values are the type of *activity* carried out, the *timestamp* of the event, the *CaseID*, the *resource* performing the activity, objects like products, goods involved, the cost associated with it, etc. For process mining, the attributes *CaseID*, *activity*, and *timestamp* are required to perform analysis. The outcome of the stochastic process is a finite sequence of events $\langle e_1, \ldots, e_N \rangle$.

3.2 Event Log L

In the event log, events generated by S are grouped and stored by their *CaseID* as traces. A trace is a subsequence of random variables from the original sequence of random variables generated by S. Note that the order of events within a trace does not change, i.e., the events within a trace are still ordered by their index given by S. Formally, a trace c_{id} of length m is a sequence of events $\langle e_1, \ldots, e_m \rangle$ where all events e have the same $CaseID = id$. We refer to the number of traces in L as $|L|$. The event log is a subset of all possible sequences over the events $\mathcal{E}$ in S, i.e., $L \subseteq \mathcal{E}^*$.

4   Entropy and Predictability of S

In this section, we will analyze how the entropy rates of S are affected by the event order and trace notion and how predictable next activities in S are.

4.1   Entropy Estimation

For a single discrete random variable X, with alphabet $\mathcal{X}$ and $p(x)$ the probability of $x \in \mathcal{X}$, the entropy $H(X)$ is defined by Shannon [19] as:

$$H(X) = - \sum_{x \in \mathcal{X}} p(x) \, log_2 \, p(x) \tag{3}$$

Stochastic processes generate sequences of events which, as motivated in Sect. 2, can contain correlations between events. As for natural language, where words are correlated, the entropy of S is not computed on single events but on *Ngrams* [10], i.e., sequences of events of length n, to incorporate such dependencies between events. The entropy rate, or per-symbol entropy, accounts for this characteristic [7]. Intuitively, it quantifies the change in entropy if using one more element in the sequence of observed random variables, i.e., if increasing n to $n + 1$. The limit of the entropy rate for $n \to \infty$ quantifies the entropy of the stochastic process (if the limit exists) [7,10].

$$H(X_1, \ldots, X_N) = \lim_{n \to \inf} \frac{1}{n} H(X_1, \ldots, X_n) \tag{4}$$

For stationary processes, the limit $\lim_{n \to \inf} H(X_n | X_1, \ldots, X_{n-1})$ of the conditional entropy is equal to Eq. 4 (if it exists) [7,10,18].

We define two entropy rate estimators for S which differ in the way the *Ngrams* are built and compare them to an existing entropy rate estimator for event logs L. We denote ES as the sequence of events e ordered as executed by S. In contrast, the trace sequence TS is generated from the event log L by concatenating all traces in the order of L. For an event log with $|L|$ traces and the set of $CaseIDs = \{1, \ldots, |L|\}$, the trace sequence is obtained by $TS := \langle c_1 + c_2 + \cdots + c_{|L|} \rangle$ where $+$ signifies concatenation. This gives a trace-wise view of the system S. The entropy rate estimator $H_n^{ES}(S)$ builds *Ngrams* on ES where *Ngrams* can contain events from various cases. $H_n^{TS}(S)$ estimates the entropy by building *Ngrams* on TS where *Ngrams* contain events from multiple cases at the point where traces are concatenated only.

We compare $H_n^{ES}(S)$ and $H_n^{TS}(S)$ to an existing event log entropy rate estimator called *ratio-based k-block entropy rate estimator* (denoted as $rate_k^r$ in [4]

Table 2. Frequency and probability for different *Ngrams* on the example L_0

$act_{1:n}$ (activity *Ngram*)	Frequency *ES*	Frequency *TS*	$p^{ES}(act_{1:n})$	$p^{TS}(act_{1:n})$
"Start"	3	3	$\frac{3}{21} = 0.143$	$\frac{3}{21} = 0.143$
"Inspect, Inspect"	3	3	$\frac{3}{20} = 0.150$	$\frac{3}{20} = 0.150$
"Inspect, Approve"	1	3	$\frac{1}{20} = 0.050$	$\frac{3}{20} = 0.150$
"Finish, Start"	0	2	$\frac{0}{20} = 0.000$	$\frac{0}{20} = 0.150$
"Approve, Approve, Approve"	1	0	$\frac{1}{19} = 0.053$	$\frac{0}{19} = 0.000$

with *k-blocks* being *Ngrams*) which follows a similar idea. However, different from $H_n^{ES}(S)$ and $H_n^{TS}(S)$, it respects the traces, i.e., *Ngrams* are built using events from within the traces in L only, preventing that *Ngrams* contain events from more than one case. The authors claim that it satisfies a number of properties for promising entropy estimators, indicating that $rate_k^r$ converges towards the true entropy of processes [4]. We refer to it as $H_n^L(S)$. Note that $H_n^L(S)$ is case-centric by design as it measures the entropy strictly within traces. $H_n^{TS}(S)$ is implicitly case-aware through the order of events in TS. In contrast, $H_n^{ES}(S)$ contains no information about cases.

For reliable entropy estimation, the probabilities for *Ngrams* have to be approximated as accurately as possible. We resort to the maximum likelihood estimation (MLE) for this [10]. In the sequence of events we count the number of times the *Ngram* has been observed and divide it by the total number of *Ngrams* built. However, the probabilities will become unreliable for event sequences quickly. If using the alphabet $\mathcal{E}$ with events being tuples of attribute values, we would get a count of 1 for each event e generated by S, as each e is unique due to the *CaseID* in combination with the *timestamp* attribute. The probability for each event would tend towards $\frac{1}{N}$ which gives imprecise and unreliable entropy estimations. To circumvent this problem, we focus on activity sequences only, i.e., build and count *Ngrams* in the control-flow of event sequences, similar to [4]. This increases the frequency of how often each *Ngram* is observed and increases the probability towards $\frac{freq(Ngram)}{\#Ngrams}$ with *freq* being the frequency of the *Ngram* and $\#Ngrams$ the number of individual *Ngrams*. In turn, the MLE becomes more precise and the entropy reliable. However, the length of *Ngrams* has to be limited as the number of individual *Ngrams* grows exponentially with n. At some point, the number is too high to get reliable estimations as most *Ngrams* appear only once or twice. Therefore, we additionally use a constraint that limits n to a value that ensures reliable estimations.

We define $p(act_{1:n})$ as the probability of the activity *Ngram* of length n in the sequence of events which is obtained using the maximum likelihood estimation, i.e., counting the number of times $act_{1:n}$ appears in the sequence and dividing it by the number of individual $act_{1:n}$. Table 2 shows the frequency and probability for some activity *Ngrams* in the event log in Table 1 for *ES* and *TS*. Note that the unigram frequency and the entropy for $n = 1$ are the same for *ES* and *TS*. However, for $n > 1$, the frequencies and probabilities can be different. This shows

Table 3. Event Logs used in the experiments.

	$|L|$	# Events N	# Activities K	max length c	H_1^{ES}
BPIC 2013_I	7,554	65,532	5	123	1.35
BPIC 2017	31,509	1,202,266	26	180	3.79
BPIC 2018	43,809	2,514,266	41	2,973	3.57
RFM	150,370	561,470	11	20	2.64

that the case notion affects the probabilities for *Ngrams* although *ES*, *TS* and *L* contain the same activities. We can now introduce the three activity entropy rate estimators $H_n^{ES}(S)$, $H_n^{TS}(S)$ and $H_n^L(S)$ for data generated by S.

$$H_n^{ES}(S) = -\frac{1}{n} \sum_{act_{1:n} \in ES} p^{ES}(act_{1:n}) \, log_2 \, p^{ES}(act_{1:n}) \tag{5}$$

$$H_n^{TS}(S) = -\frac{1}{n} \sum_{act_{1:n} \in TS} p^{TS}(act_{1:n}) \, log_2 \, p^{TS}(act_{1:n}) \tag{6}$$

$$H_n^L(S) = -\frac{1}{n} \sum_{act_{1:n} \in \mathcal{E}^*} p^L(act_{1:n}) \, log_2 \, p^L(act_{1:n}) \tag{7}$$

With increasing n we run into the problem that not enough samples are available to obtain accurate probabilities $p(act_{1:n})$. However, if n is too small the entropy rate is computed using short sequences only, resulting in a very high entropy rate that does not consider pattern over longer sequences. The question is how far we can n let grow before the MLE becomes inaccurate. This is especially critical for the event log entropy estimator where $act_{1:n}$ are built and counted within traces only. If n reaches the length of the longest trace in L, there is only a single $act_{1:n}$ left for probability estimation [4]. For ES and TS, n can grow larger as $act_{1:n}$ can span multiple traces. Nevertheless, the number of individual $act_{1:n}$ can grow very quickly and reach a point where not enough statistical information is left. This problem is discussed in [14] for short sequences and in [4] for the event log entropies.

Different constraints on the maximum size of n can be used, depending on the assumptions made about the underlying stochastic process [14]. A lenient constraint specifies n on the vocabulary size K and the running entropy rate h as $n < \frac{log_2 K}{h}$ [14]. This, however, allows n to grow to very large values. A more strict constraint is to choose n depending on the total sequence length N[1] and the number of non-negligible *Ngrams* that contribute to the entropy [4,14] which is 2^{nh} according to the Shannon-McMillan-Breiman theorem [3].

$$Nh > n2^{nh} log_2(K) \tag{8}$$

---

[1] For the event log entropy estimator N is set to the size of the longest trace in L.

We will use constraint (8) in the experiments for reliable entropy rate estimations. Since the constraint tends to be too strict [4], we will also allow n to grow larger to get an indication of how the rates will continue to behave.

In the experiment, we will investigate how the different entropy rate estimators compare with increasing n. To ensure that event dependencies exist, only real-life logs are used. For BPIC 2017, BPIC 2018, Road Fine Management (RFM), we used all events but ignored the lifecycle transition. For BPIC 2013 we used the incident version (Table 3).

Results. Figure 4 shows the different entropy rate estimates. The rates have solid lines until the constraint (8) applies while dashed lines indicate how the rates will continue after the constraint is violated. For $H_n^L(S)$ we let n continue with dashed lines to the median trace length to prevent having little to no activity *Ngrams* left. For $H_n^{ES}(S)$ and $H_n^{TS}(S)$ we let n grow to larger values as *Ngrams* can always be built. For $n = 1$, the entropy rates are the same for all estimators as single activities have the same MLE in L, TS, and ES.

$H_n^L(S)$ becomes unreliable quickly, especially for logs with mostly short traces, while the others are reliable longer. For event logs with long traces $H_n^{TS}(S)$ seems to behave similarly to $H_n^L(S)$. In all event logs, the case-ordered entropy rates start to decline faster as $H_n^{ES}(S)$. This is expected as the number of distinct activity *Ngrams* $act_{1:n}$ in ES is much higher as in TS or L and many

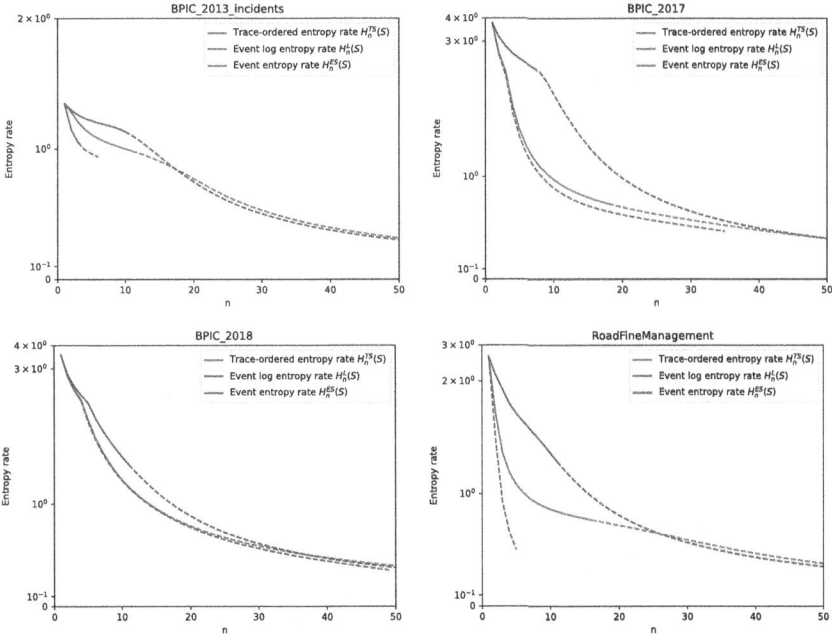

Fig. 4. Comparison of different entropy rate estimators. Dotted lines indicate how the entropy rates continue after constraint (8) is violated.

patterns in ES are longer. If n increases and the *Ngrams* become longer, the entropy rates for most event logs come closer. Given that the constraint used is very strict, the entropy rates are likely reliable a little further than indicated by the solid lines, reducing the gap between $H_n^{ES}(S)$ and $H_n^{TS}(S)$ even further.

4.2 Estimating Next Activities

We now investigate how well prediction models can learn from S. In general, a sequence with low entropy, indicating little "surprisingness" of that sequence, should be easier to learn than a sequence with high entropy. For this, we test how accurately the probability for the next activity can be predicted, i.e., the next step prediction task in process mining. Although this task seems to be trivial, predicting the next element is widely used for learning text, time series, or other sequential data. Note the close link between the probability estimation for an *Ngram* and the probability for the next element given the previous $n-1$ elements in a sequence [10,18], which is given by the chain rule of probability.

$$p(X_1, \ldots, X_n) = \prod_{k=1}^{n} p(X_k \mid X_1, \ldots, X_{k-1}) \tag{9}$$

For instance, the probability for the activity *2gram* $p(act_1, act_2)$ is equal to $p(act_1)p(act_2|act_1)$ which includes the conditional probability of seeing act_2 after act_1 has been observed. Even though a low entropy rate indicates little randomness for a sequence, it does not necessarily translate to a high fidelity in estimating next events. The entropy rates are estimated on the whole log while a prediction model must learn from the train set to perform well on the test set. For this, process prediction models PPM must be capable of learning the statistical distribution of event sequence pattern, i.e., *Ngrams* accurately. While there are many approaches for predicting the next activity in a trace, predicting the probability for next activities in S has not been done so far. We expect the latter to be much more challenging due to a higher entropy and the need to process and learn from longer sequences. We define three prediction approaches PPM that estimate the probability for the next activity act_n based on a history of previous $n-1$ activities $hist_{n-1}$.

$$p(act_n \mid hist_{n-1}) \tag{10}$$

The three PPM differ in the way $hist$ is built, following the same idea as $H_n^{ES}(S)$, $H_n^{TS}(S)$ and $H_n^{L}(S)$. PPM_L is the standard approach for process prediction that respects traces in the event log L. It builds the history from activities within traces and predicts next activities for traces. PPM_{ES} samples the previous $n-1$ activities and act_n from ES while PPM_{TS} samples them from TS. We focus on the control flow only and do not use additional event attributes.

We train and test three different prediction models on the prediction scenarios to validate how feasible the scenarios are. As baseline, we choose a random forest (RF) model consisting of 100 trees with entropy criterion. A 2-layer

Table 4. Performance of different PPM models on the test split.

		PPM_{ES}			PPM_{TS}			PPM_L		
Cross-entropy (CE) in log_2 (lower is better)										
	H_1^{ES}	RF	LSTM	TF	RF	LSTM	TF	RF	LSTM	TF
BPIC 13 I	1.35	1.27	**1.07**	1.22	1.26	**1.04**	1.27	1.10	**0.97**	0.98
BPIC 17	3.79	2.56	**1.61**	1.63	0.75	0.49	**0.47**	0.61	**0.46**	0.47
BPIC 18	3.57	2.35	**2.11**	2.29	2.30	2.23	**2.13**	0.87	**0.57**	0.62
RFM	2.64	1.50	**1.30**	1.33	0.87	**0.64**	0.66	**0.67**	**0.67**	**0.67**
Accuracy (higher is better)										
BPIC 13 I		69.53%	**70.48%**	69.18%	64.94%	**67.53%**	61.48%	68.82%	**69.67%**	65.59%
BPIC 17		51.58%	**63.22%**	62.86%	86.15%	87.63%	**87.90%**	86.80%	**87.98%**	87.81%
BPIC 18		51.50%	**58.27%**	55.61%	59.76%	61.30%	**65.65%**	81.60%	**85.52%**	84.37%
RFM		65.05%	**66.81%**	65.57%	81.50%	**83.05%**	82.90%	80.91%	**80.92%**	**80.92%**

$LSTM$-based PPM is used as the state-of-the-art model in process prediction. Additionally, we trained a transformer-based model (TF) with four attention heads. All models return a probability distribution over the next activities (i.e. Eq. (10)) which allows computing the discrepancy between the predicted and ground truth activity using the cross-entropy (CE) loss function[2]. In addition, we report the accuracy in predicting the next activity. We split the data on a temporal basis by 70/30 into train and test - either in the number of traces $|L|$ (for PPM_L) or by the number of events N (for PPM_{ES} and PPM_{TS}) - such that all training events/traces have timestamps up to date d and all test events/traces a timestamp after d. For all models, we use an input sequence of 128 activities. Shorter sequences need to be padded for PPM_L while this is not required for PPM_{ES} and PPM_{TS}. We train each prediction model 5 times with early stopping and report results for the run with the lowest CE (Table 4).

Results. All prediction models reach a lower cross-entropy as the corresponding entropy rate H_1^{ES}. This indicates that all models learned to estimate conditional probabilities for next activities more accurately than estimating them by their frequency. Overall, the lowest cross-entropies are reached in PPM_L followed by PPM_{TS} and PPM_{ES}. This ranking aligns with the entropy rate analysis where the entropy for $H_n^{ES}(S)$ is higher as $H_n^{TS}(S)$ and lowest for $H_n^L(S)$. For all but the BPIC 18 log, the CE of PPM_{TS} is only a little higher than PPM_L or even slightly lower for the RFM data ($LSTM$ and TF model). Between PPM_{TS} and PPM_{ES} the CE increases noticeably, except for BPIC 13 I and BPIC 18.

In PPM_L, all prediction models perform very similarly while differences are more significant for PPM_{TS} and PPM_{ES}. For the latter task, the difference between the CE of RF and neural models is the biggest for the BPIC 17 and RFM data, which are also the logs in which the difference between the respective entropy rates is the biggest. This indicates that neural approaches are more

---

[2] Note, that we report the CE in log_2 while most tools, including pytorch, use ln.

accurate in such difficult scenarios. The *LSTM* model performs best across all settings. Interestingly, the CE was reduced the least for BPIC 13 I, compared to the entropy rate. For all other logs, we notice a more significant reduction. For instance, for BPIC 13 I the CE of the *RF* model in PPM_{ES} is only a little lower as H_1^{ES} while the performance of all models on this log is most consistent across all tasks. Given that BPIC 13 I is the log with the lowest entropy rates, we expected the *PPM* models to reduce the CE more significantly.

In general, the accuracies are consistent with the CE. However, there are exceptions where a model with higher CE reached a higher accuracy. For instance, on BPIC 13 I between *RF* and *TF*. Similarly, accuracies can be very similar although the CE differs considerably, e.g., for RFM. We suspect this to come from label ambiguity [15], a phenomenon where the same prefixes have different continuation options resulting in an arbitrariness in the prediction labels.

Reproducibility. Code and figures are available in the GitHub repository[3]

5   Discussion of the Results

We will discuss the implications and limitations of the experiments as well as potential applications for process mining tasks.

5.1   Implications and Applications

The results of the entropy rate analysis in Fig. 4 show that the rates with and without case notion can behave similarly and might decrease to similar values. Further, all tested predictive approaches reduce the prediction error considerably indicating that relations between previous and next events can be learned on S.

Entropy Rates. The slopes of entropy rates indicate that the probabilities for activity sequences $p(act_{1:n})$ can behave similarly in TS and ES. Since the activity *Ngrams* $act_{1:n}$ are different when being built on ES, TS, or in L, the similar slopes show that (different) activity *Ngrams* repeat with similar frequency. From these similarities, we conclude that a structure in the event sequence ES, generated by the system, exists. While the case notion structures the events in the log L and TS, e.g., the order of activity executions in traces, there is no such structure in ES. Instead, it must come from another phenomenon. While the entropy rates do not give information about where the structure comes from we suspect, as motivated, that human behavior or other participants are the reason for it. If considering the entropy estimates as reliable a little further, which is reasonable since the constraint used is very strict, the difference between $H_n^{ES}(S)$ and $H_n^{TS}(S)$ becomes smaller and possibly approaches a very similar value. Analyzing events as executed including inter-case behavior can be more practical than expected. It can also help if no *CaseID* or multiple options for it are available, in choosing the attribute that reduces the entropy the most.

---

[3] https://github.com/ppfeiff/EventSequenceEntropy.

Predictions. In the prediction experiments we found the cross-entropy values of PPM_{ES} to be consistently lower than H_1^{ES}, indicating that the models were able to learn from longer sequences. For BPIC 13 I, the cross-entropy of PPM_{ES} is only a little higher than PPM_L. Given that making predictions with PPM_{ES} is supposed to be much harder than for PPM_L and PPM_{TS}, the results are encouraging for learning the behavior of S. While the final task for process mining applications should not necessarily be predicting the next activity happening in S, it shows that PPM_{ES} learn patterns that are decisive for the next activity and might be helpful features (containing inter-case behavior) for solving other process prediction tasks.

Existing, case-centric prediction methods like PPM_L, can be supplemented by the prediction approaches presented in this paper. For instance, by using ES as an additional source of information in *hist* for making predictions for a particular trace. Since making predictions on ES might be feasible, trace predictions could also be obtained from the sequence of next events predicted by PPM_{ES}. For instance, by enumerating the traces currently running in *hist* and add this information to the events which would allow such a model to distinguish events between traces. Instead of predicting the next activities(s) in a trace, the sequence of the next activities in S could be predicted and all traces, that are running currently, be reconstructed from the predicted sequence.

As such models are aware of inter-case behavior, they can also be used for a variety of downstream tasks. For instance, to detect infrequent behavior in S for anomaly detection or predicting other unwanted behavior such as process deviations or compliance issues [17], which are hard to detect if considering single traces only. Further, concept drifts could be analyzed, detected, or even predicted in advance using the sequence of events from S.

5.2   Limitations

Certain limitations have to be considered when interpreting the results. For entropy rate analysis as well as for prediction, we took only the activities into account. Thereby, other attributes that contain important information are not considered, influencing the interpretation of the results. Further, only a limited number of event logs have been analyzed so far.

Entropy Rates. Events in the system S are characterized by more than just the activity. It can be challenged whether estimating the entropy of S by its control flow results in an accurate estimation of its "true" entropy. Two events with the same activity can be regarded as different if executed in a different situation, e.g., at a different time, by other resources, or in a different trace.

The entropy rates after the violating constraint (8), i.e., the dotted lines in Fig. 4, are subject to uncertainty due to the problem of undersampling. Having more data can increase the reliable range. However, it remains open whether there will be enough data to reach convergence (for finding the true entropy of S) and whether the same observations can be made for all event logs.

Constraint (8) assumes the process to be stationary and ergodic [14], which follows from the Shannon-McMillian-Breimann theorem [3]. As already pointed out by previous work on event log entropies [4], event sequences from real-life processes are most likely neither stationary nor ergodic, since processes can change through time. However, its effect on the validity is unclear.

Analyzing the entropy of event sequences ordered as executed can give the same results as the case-ordered sequence, e.g., for serial productions. It can also create questionable correlations for processes with isolated traces. For instance on click data from UI logs where each case captures the interactions from one individual user that usually has no relations to other users.

Predictions. While the prediction results are encouraging they are just a first impression on the feasibility of making predictions on S. Approaches need to be tested on downstream tasks like anomaly detection or remaining time prediction to evaluate their applicability. So far, no additional attributes but only the activity have been considered. While additional attributes in the history of previous events can contain information that makes predicting the next activity more accurate, predicting more attributes than the activity of the next event(s) makes the task much more challenging. One might have to balance carefully which attributes to consider in *hist* and which to predict, depending on the use case. The limit of using 128 activities in *hist* might also limit the performance of *PPM*. If multiple cases are running concurrently, the limit of 128 activities is reached quickly which limits the available information per trace.

6   Related Work

Related work covers entropy estimators for event logs, event streams, stochastic process mining, and research on process behavior spanning across traces. Several entropy estimators have been proposed by [4], specifically tailored for event log data, aiming to classify logs for suitability for declarative or imperative process discovery. All estimators respect traces, i.e., they keep traces separate when building *Ngrams*. The authors argue against building *Ngrams* across traces to avoid "spurious correlations among events" [4]. As motivated and shown, we argue that correlations between events are not per se spurious. Further, the entropy rate estimators $H_n^{TS}(S)$ and $H_n^{ES}(S)$ remain reliable longer as $H_n^L(S)$.

Another line of research deals with event streams as an infinite sequence of events [22]. Some properties of event streams also apply to our work like the assumption of a single-channel stream (single-source in our case) and that events arrive in the order of their execution. However, we are not aware of any work that defines an event stream as a stochastic process. Furthermore, event streams are often used to characterize situations where events are emitted continuously, and analysis, therefore, has to be performed "live". In contrast, we defined S to emit a finite sequence of random variables whose outcomes are events.

There is also research that characterizes and analyzes event data for process mining stochastically, which is often referred to as *stochastic process mining*. To the best of our knowledge, approaches in that line of research assume the outcome of random variables X to be traces, i.e., they describe the log L with stochastic means [5, 12, 13, 16] but not the system S which generated the individual events. In this paper, we assume S to be a stochastic process that generates a sequence of random variables, i.e., the random variables in our case range from single events to *Ngrams* of events and are, at first, independent of the trace-notion. In contrast to existing work, this allows us to describe and measure process behavior between events across traces.

Inter-case behavior, spanning across traces, is also of interest for other tasks in process mining. A framework for cross-case association patterns is presented in [8]. They formulate nine cross-case patterns and propose techniques to detect two in event logs. An approach to detect system-level behavior as temporal event patterns across cases for detecting dynamic bottlenecks is presented in [21]. For several prediction tasks, features that describe inter-case behavior have shown to increase performance, e.g., remaining time prediction [11]. Predicting deviations and compliance in processes requires learning inter-case behavior as well [17]. Object-centric process mining [2] is another way of describing and mining relations in events. Relations in object-centric process mining are described on objects involved across traces which allows deriving descriptive patterns. In contrast, no objects or object relations are required for the analysis performed in this work.

7 Conclusion

In this paper we have motivated and shown that *Ngram* patterns, characterizing the behavior of the underlying system S, can be found in its sequence of events ES. The system S has been conceptualized as being observable as a stochastic process that produces a sequence of random variables. Two ways to measure the entropy rate of S, either in the order as executed or ordered by trace have been proposed. To avoid having insufficient information for reliable entropy estimation, constraints limiting n have been implemented.

The results confirm that analyzing event sequences by traces is the least surprising, as the reliable entropy rates of $H_n^L(S)$ and $H_n^{TS}(S)$ are lower than $H_n^{ES}(S)$. However, the entropy rates $H_n^{ES}(S)$ of S without trace information and ordered as executed behave not too differently from the entropy measured of events ordered by traces in L. This observation is interesting as it shows that activity sequences, i.e., *Ngrams*, repeat regularly in S and give the event sequence ES emitted by S a certain structure. The prediction results confirm that this structure can be learned which motivates using all events that have happened previously in ES as an additional source of information. For tasks where precise estimations of events and inter-case behavior is important, e.g., anomaly detection or prediction tasks, this information can be useful.

It also motivates future work in which we want to investigate what other analyses can benefit from using event sequences as executed. We are especially interested in training models that learn inter-case correlations from ES for better representations of the underlying system that generated the event log. Learning S with a prediction model, instead of L (with information about traces if available), can enable a new generation of predictive approaches that are more accurate and versatile than existing ones. They are by design aware of inter-case relations and can generate sequences of future behavior of S. Further, the higher complexity of ES fits the capacity of neural-network-based $PPMs$ better than learning from L. In the era of large models that can easily process sequences of tens of thousands of tokens, processing very long sequences of events executed in S is technically feasible. However, building such a model will be challenging and requires large amounts of data.

Apart from that, other definitions of *Ngrams* that account for typical process behavior like concurrency might help to lower the entropy and increase predictability further. Such phenomena cause a literal explosion in the number of individual *Ngrams*, leading to high entropies and early violation of the constraint. Also, extending the definition of entropy for processes towards cross-entropy could be a way to formally define generalization for process data.

References

1. van der Aalst, W.M.P.: Process Mining: A 360 Degree Overview. Springer International Publishing, Cham (2022). https://doi.org/10.1007/978-3-031-08848-3_1
2. van der Aalst, W.M.: Object-centric process mining: unraveling the fabric of real processes. Mathematics **11**(12) (2023)
3. Algoet, P.H., Cover, T.M.: A sandwich proof of the Shannon-Mcmillan-Breiman theorem. Ann. Probab. (1988)
4. Back, C.O., Debois, S., Slaats, T.: Entropy as a measure of log variability. J. Data Semant. **8**(2) (2019)
5. Bogdanov, E., Cohen, I., Gal, A.: SKTR: trace recovery from stochastically known logs. In: 5th International Conference on Process Mining (ICPM)
6. Buijs, J.C., van Dongen, B.F., van der Aalst, W.M.: Quality dimensions in process discovery: the importance of fitness, precision, generalization and simplicity. Int. J. Coop. Inf. Syst. **23**(01) (2014)
7. Cover, T., Thomas, J.: Elements of Information Theory. Wiley, Hoboken (2006)
8. Dubinsky, Y., Soffer, P., Hadar, I.: Detecting cross-case associations in an event log: toward a pattern-based detection. Softw. Syst. Model. **22**(6) (2023)
9. Grisold, T., Kremser, W., Mendling, J., Recker, J., Vom Brocke, J., Wurm, B.: Generating impactful situated explanations through digital trace data. J. Inf. Technol. (2023)
10. Jurafsky, D., Martin, J.H.: Speech and Language Processing, vol. 3 (2021)
11. Klijn, E.L., Fahland, D.: Identifying and reducing errors in remaining time prediction due to inter-case dynamics. In: 2020 2nd International Conference on Process Mining (ICPM)
12. Leemans, S.J.J., Mannel, L.L., Sidorova, N.: Significant stochastic dependencies in process models. Inf. Syst. **118** (2023)
13. Leemans, S.J.J., Polyvyanyy, A.: Stochastic-aware precision and recall measures for conformance checking in process mining. Inf. Syst. **115** (2023)
14. Lesne, A., Blanc, J.L., Pezard, L.: Entropy estimation of very short symbolic sequences. Phys. Rev. E **79**(4) (2009)

15. Pfeiffer, P., Lahann, J., Fettke, P.: The label ambiguity problem in process prediction. In: Business Process Management Workshops (2022)
16. Polyvyanyy, A., Moffat, A., García-Bañuelos, L.: An entropic relevance measure for stochastic conformance checking in process mining. In: 2nd International Conference on Process Mining (ICPM). IEEE
17. Rinderle-Ma, S., Winter, K., Benzin, J.V.: Predictive compliance monitoring in process-aware information systems: state of the art, functionalities, research directions. Inf. Syst. **115** (2023)
18. Shannon, C.E.: Prediction and entropy of printed English. Bell Syst. Tech. J. **30**(1) (1951)
19. Shannon, C.E.: A mathematical theory of communication. Bell Syst. Tech. J. **27**(3) (1948)
20. Song, M., van der Aalst, W.M.: Supporting process mining by showing events at a glance. In: Proceedings of the 17th Annual Workshop on Information Technologies and Systems (WITS)
21. Toosinezhad, Z., Fahland, D., Ö, K., Aalst, W.M.P.V.D.: Detecting system-level behavior leading to dynamic bottlenecks. In: 2020 2nd International Conference on Process Mining (ICPM) (2020)
22. van Zelst, S.: Process mining with streaming data. Thesis (2019)

Data-Driven Decision Support for Business Processes: Causal Reasoning and Discovery

Ali J. Alaee[1], Matthias Weidlich[2], and Arik Senderovich[1(✉)]

[1] School of Information Technology,
Faculty of LA&PS, York University, Toronto, Canada
{alialaee,sariks}@yorku.ca
[2] Department of Computer Science,
Humboldt-Universität zu Berlin, Berlin, Germany
matthias.weidlich@hu-berlin.de

Abstract. Various types of decisions influence the execution of a business process, e.g., in terms of control-flow and resource assignments. Data recorded during process execution can be used to identify which decisions are informed by data and by previous decisions, to predict their outcome, and to guide interventions as part of a what-if analysis. The latter requires causal models that explain the factors that influence decisions. Yet, existing causal techniques for business processes are limited: they focus on control-flow decisions only, ignore data variables, and use ad-hoc methods to resolve causal conflicts. In this paper, we fill this gap by introducing a causal decision modeling framework that incorporates variables, which allows us to uncover, for example, confounding effects, and capture resource decisions. Moreover, we provide a process-aware causal discovery algorithm, based on the notion of temporal tiers, that takes process precedence into account, without the need for heuristic conflict resolution between process discovery and causal discovery. We demonstrate the effectiveness of our approach through experiments using synthetically generated data, and show a proof-of-concept implementation on a real-world dataset.

1 Introduction

Decisions play a central role in business processes [6]. They influence the outcome of individual cases and, in the larger context, implement organizational policies. Methods to improve decision-making, therefore, are vital to the success of organizations [20]. With the ample availability of process-related data, the opportunity to develop data-driven decision support has been widely recognized and methods to learn, predict, and prescribe decisions have been developed [3–5,11,13,21,25].

Traditional approaches to data-driven decision support, however, are not grounded in causal models. As a consequence, they cannot support accurate

A. Marrella et al. (Eds.): BPM 2024 Forum, LNBIP 526, pp. 90–106, 2024.
https://doi.org/10.1007/978-3-031-70418-5_6

'what-if?' analysis and also lack means to quantify the uncertainty associated with decisions under various changes to the process. Only recently, first approaches to overcome these limitations have been proposed, such as [19]. However, the existing techniques exhibit two substantial drawbacks: The impact of data variables on decisions is disregarded and the models are limited to the control-flow, neglecting decisions related to other process perspectives, such as resource assignments.

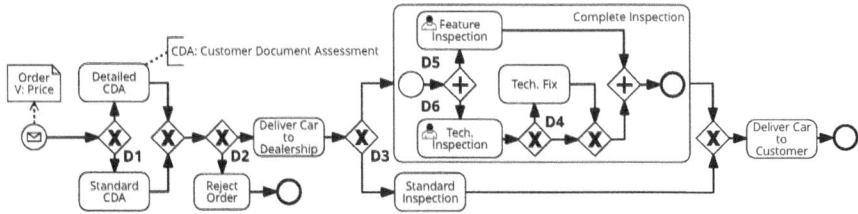

Fig. 1. Running example: a BPMN model of a car purchase and delivery process.

To illustrate the above aspects, Fig. 1 gives a BPMN model of a process at a car dealership. It starts with an order (with price V) and ends with either the rejection of the order or the delivery of the car. The process features several control-flow decisions: D_1 on the type of the customer document assessment; D_2 on the rejection; D_3 on the type of pre-delivery inspection, and D_4 on the need for a technical fix. Also, the process involves decisions on resource assignments, exemplified as D_5 and D_6 for the feature inspection and technical inspection.

For this process, decisions D_3 and D_4 have a large impact on the cycle time, due to the duration of a complete inspection, in particular when involving technical fixes. Support for decision-making shall therefore explore the factors that influence these decisions. Yet, existing techniques neglect contextual variables which could serve as a confounder, mediator, or collider in a respective causal model [19]. For our example, consider the following situation: While the exact conditions are unknown, the decision D_1 on the document assessment influences decision D_2 on the rejection, (e.g., a detailed CDA often reveals customer details that favour rejection), but not the decision D_3 on the inspection (i.e., it is independent of the CDA). In addition, the order price V is a confounder, influencing all three decisions (e.g., a higher price favours a detailed CDA, a rejection, as well as a complete inspection). Existing work would ignore the dependence on V and, hence, identify a spurious causal effect of D_1 on D_3. Similarly, the resource assignment decisions D_5 on D_6 (i.e., the selection of a particular person that assumes the role defined for the BPMN activity) influence decision D_4. However, state-of-the-art methods cannot capture their effect, since the decisions do not materialize as control-flow choices in the data. Both limitations severely bias the estimation of causal effects, which leads to misleading analyses, as we later illustrate empirically.

In this paper, we address the above research gap with a unified framework for data-driven decision support that is based on discovery of a causal model. In particular, we make the following contributions:

(1) We introduce a model to facilitate causal reasoning for process decision support. It goes beyond the state of the art by integrating confounders and contextual variables, which enables the inclusion of various types of decisions.
(2) We propose a novel tier-based causal discovery method that identifies the causal model from event data collected during process executions. This approach seamlessly integrates processes into causal reasoning.

We demonstrate the feasibility of our approach through a proof-of-concept implementation. Moreover, we report on experimental results that illustrate the improvement achieved by our technique over the state of the art.

In the remainder, we first give preliminaries (Sect. 2), before we turn to our causal model for decision support (Sect. 3) and its discovery from event data (Sect. 4). Subsequently, we present experimental results (Sect. 5), review related work (Sect. 6), and conclude the paper (Sect. 7).

2 Preliminaries

We first provide essential background on causal modeling and interventions (Sect. 2.1), before turning to the discovery of causal models (Sect. 2.2).

2.1 Causal Modeling and Reasoning

Causal Models. A causal model captures relations between variables and a joint probability distribution over their values. We recall basic definitions using [23].

Definition 1 (Causal Model). *A causal model is a pair* $(\mathcal{G}, P)$ *with:*

○ $\mathcal{G} = (\mathcal{V}, \mathcal{E})$ *being a directed acyclic graph (DAG) over a set of variables* $\mathcal{V}$. *Every vertex is a variable in* $\mathcal{V}$, *and there exists an edge* $(V_i, V_j) \in \mathcal{E}$, *iff* $V_i \in pa(V_j)$, *with pa being a function that returns the parents of a node* $V \in \mathcal{V}$; *the parents are interpreted as the* direct causes *of* V.
○ P *being a joint probability distribution, such that*

$$P(v_1, \ldots, v_n) = \prod_{i=1}^{n} P(v_i \mid pa(v_i)), \tag{1}$$

which is a standard Markov decomposition of the joint distribution.

The structure of the causal model of our running example from Fig. 1 is illustrated in Fig. 2. The price of the order, variable V influences the choices on the customer document assessment (D_1), on the rejection of the order (D_2), and on the inspection (D_3). The document assessment (D_1) also directly influences the order acceptance (D_2). Moreover, the resource assignment for the technical inspection (D_6) influences the decision on the need for a technical fix (D_4).

We write $P(v)$ to denote the probability that a random variable V takes a value v, i.e., $P(v)$ is defined as $P(V = v)$. For a set of random variables $\mathcal{V}$, aka a configuration, we write $P(\mathcal{V} = v)$ to refer to $P(V_1 = v_1, \ldots, V_n = v_n)$, for all $V_i \in \mathcal{V}$. Moreover, we write $P(v \mid w)$ for $P(V = v \mid W = w)$.

Returning to the example in Fig. 2 and focusing solely on the variables of the main process, we can write the joint probability as:

$$P(d_1, d_2, d_3, v) = P(d_3 \mid v)P(d_2 \mid v, d_1)P(d_1 \mid v)P(v). \tag{2}$$

Using historical data, one can estimate each of the quantities and answer *statistical* (i.e., predictive) queries about the decisions in Fig. 2. However, interventional queries ('what would have been the outcome of decision D_2, if the price changes?') cannot be answered only using historical data and the decomposition in Eq. 2. Instead, one would require the notion of *interventions*.

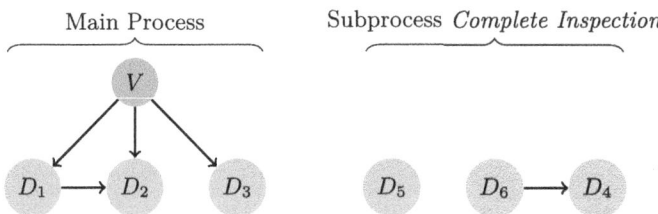

Fig. 2. The DAG of the causal model representing the relations between variables of the example given in Fig. 1; variable V denotes the price of the ordered car.

Interventions on Causal Models. An intervention is a deliberate change of the values of some variables, usually to establish the causal effect of one variable on other variables. It is represented by a *do* operator, with $do(V = v)$ (or $do(\mathcal{V} = v)$ for a set of variables) denoting that the value of variable (or set thereof) has been set to v. We consider interventions as a modification of a causal graph $\mathcal{G}$ to a mutilated graph, $\mathcal{G}'$, such that all incoming edges to the intervened variables are eliminated, noting that the pre- and post-intervention distributions are different. For an intervention $do(V_i = v_i)$, the transformation corresponds to a truncated factorization formula:

$$P(v_1, \ldots, v_n \mid do(V_i = v_i')) = \begin{cases} \prod_{j \neq i} P(v_j \mid pa(v_j)) & \text{if } V_i = v_i' \\ 0 & \text{if } V_i \neq v_i' \end{cases}. \tag{3}$$

Eq. 3 reflects the removal of the term $P(v_i \mid pa(v_i))$ from the product of Eq. 1, since $pa(v_i)$ no longer influences V_i. Returning to Fig. 2 and focusing solely on the variables of the main process, if we intervene on the price V and set it to be 10 (thousand dollars), the probability distribution changes as follows:

$$P(d_1, d_2, d_3, v \mid do(V = 10)) = \& P(d_3 \mid do(V = 10)) \cdot \quad (4)$$
$$P(d_2 \mid do(V = 10), d_1) P(d_1 \mid do(V = 10)).$$

One cannot estimate this causal quantity only using historical data, unless the do-operator can be removed from Eq. 4. For this specific example, there exists a way to remove the do-operator using so-called backdoor adjustment. Intuitively, it conditions on variables to block paths in the causal model, in order to enable conclusions on causal effects from systematic correlations. This way, the impact of the intervention on the price on the three decisions can be estimated.

An important quantity that measures the impact of an intervention is the average treatment effect (ATE), which is defined as:

$$\text{ATE} = \mathbb{E}[Y \mid do(T = 1)] - \mathbb{E}[Y \mid do(T = 0)], \quad (5)$$

where Y is an outcome, and T is a binary treatment (our intervention; it can be easily extended to non-binary treatments). For the example in Eq. 4, we may wish to assess the impact of setting $D_1 = 1$ (choosing the upper branch in the process model) on the expected outcome of decision D_2. Since D_2 is also binary (reject or not), we can write the ATE as:

$$\text{ATE} = \sum_v P(D_2 = 1 | D_1 = 1, V = v) P(V = v) - \quad (6)$$
$$- \sum_v P(D_2 = 1 | D_1 = 0, V = v) P(V = v).$$

The two quantities (and difference) can then be theoretically estimated using historical data, while, in practice, statistical considerations are required to obtain a valid estimate of the probability expressions.

2.2   Constraint-Based Causal Discovery

Given some observational data, causal discovery aims at uncovering the 'true' causal model that generated the data [23]. Approaches for causal discovery can be divided into three categories [16]: (1) constraint-based discovery, (2) scored-based discovery, and (3) learning functional causal models (FCMs).

FCM-based methods are mostly suitable for data with continuous variables, and thus are less suitable for discovery of decisions. Constraint-based methods conduct a series of conditional independence tests between variables in a fully connected causal model to eliminate (or allow) possible causal arcs. Score-based methods are a relaxation of constraint-based discovery, where independence tests are replaced with goodness-of-fit tests: Instead of searching for a graph that contains true independence relations, 'good enough' graphs are discovered.

Without introducing any additional background knowledge, causal discovery can lead to ambiguity in the resulting structure since different causal models may manifest the same dataset. Therefore, in this paper, we employ [12], an approach that allows to choose between constraint-based and score-based methods and combines historical data with background knowledge in the form of temporal constraints, e.g., partial orders over variables, and forbidden edges.

3   A Framework for Causal Decision Support in Processes

In this section, we introduce a data-driven framework for causal decision modeling in business processes. First, we introduce the formal model (Sect. 3.1), followed by an application of the model to flow analysis (Sect. 3.2).

3.1   Modeling Causal Decision Structures

We adopt a causal model to capture the decision structures of a business process as follows. Following Definition 1, a causal model is a pair $(\mathcal{G}, P)$ of a causal graph and a probability distribution. Specifically, the causal graph $\mathcal{G} = (\mathcal{V}, \mathcal{E})$ is defined over the set of all possible business process variables $\mathcal{V}$ that influence the execution of the process, e.g., control-flow decision points, resource assignment decisions, and contextual variables (such as price in our running example). The set of edges, $\mathcal{E}$, represents direct causal effects between those variables.

While we address the task of discovering the graph, $\mathcal{G}$, later (Sect. 4), we note that the estimation of P without interventions is a purely statistical problem that does not require $\mathcal{G}$. Yet, to learn the behavior of P under a *do* intervention, one requires to obtain the causal graph first.

When modeling the decision structure of a business process with a causal graph, we distinguish between decision variables, and contextual variables, as they play different roles in any analysis. Specifically, decisions may disable other decisions, while variables cannot be disabled and cannot disable decisions. Consider decision D_2 in the running example. Choosing to reject the order will disable decision D_3. This difference between decisions and contextual variables is crucial as it calls for a different treatment and representation.

Decision Nodes. Our aim is to model and analyze decisions from a given causal model. To this end, we assume the existence of m decision points that correspond to m variables, $\mathcal{D} = \{D_1, \ldots, D_m\} \subseteq \mathcal{V}$. Examples of decision points include the choice of the next activity from a set of activities when routing a case, choosing a resource to perform an activity from a set of resources, etc. Decision points can be detected given a business process model e.g., via the method proposed in [5] for control-flow decisions. They can also be mined from an event log via standard algorithms for process discovery. Specifically, for control-flow decisions one can apply the approach from [1], while for the extraction of resource profiles to identify possible resource assignments one can use the method in [17].

A decision point $D_i \in \mathcal{D}$ takes values in a corresponding discrete bounded domain $\mathbf{D}_i$. For instance, if D_i is the decision on which resource will perform an activity (e.g., activity 'Feature Inspection' of our example), we can define the domain as $\mathbf{D}_i = \{R_1, \ldots, R_l\}$ with l being the number of resources eligible to perform the task. For control-flow decisions, the domain is given by an enumeration of the branches that are activated once the decision has been taken. In our running example, all control-flow decisions $D_1, \ldots, D_4$ happen to be binary, so that their domain is represented as $\mathbf{D}_i = \{0, 1\}, 1 \leq i \leq 4$.

Decision points may be evaluated more than once throughout the execution of the business process. Reasons for that may be control-flow loops or several resource allocations that happen as part of multiple-instance activities. In the remainder, we shall consider recurrent decisions to be distinct decision points. For example, if a loop may take the process back to a previous decision, we shall consider it to be a new decision point.

Context Nodes. In addition to decision points, we also consider contextual variables that may be present and have an impact on the outcomes. We consider the existence of n contextual variables such that $\mathcal{C} = \{V_1, \ldots, V_n\} = \mathcal{V} - \mathcal{D}$. Contextual variables of a decision can include static case features that do not change during execution (e.g., the age of a patient checking in at a hospital), dynamic case features (e.g., body temperature measures of a patient), and inter-case features (e.g., the number of cases in the system). Each variable V_i takes a value from its domain $\mathbf{V}_i$, as we consider all of them to be discrete variables.

Dynamic contextual variables can potentially change after every event in the log. Thus, multiple instances of variables that correspond to their measurements throughout the process must be incorporated into the causal model as separate nodes. The arrangement of these nodes in terms of causal edges is determined by the causal discovery algorithm explained in Sect. 4. In this case, for each dynamic variable V_x there are i measurements, $V_{x,1}, \ldots, V_{x,i}, 1 \leq i \leq k$ with k being the maximal number of measurements allowed for a variable.

In the dealership example, we focus on a single contextual variable, namely the price of the ordered vehicle. While this variable is naturally continuous, its exact value can be considered to be of minor importance for decision support. We therefore discretize it and assume its domain to be $\{1, 2, 3\}$, representing different price levels of the ordered car.

3.2 Flow Analysis for What-If Interventions via Causal Decisions

In this part, we demonstrate an application of our causal model for decision support when computing key performance indicators (KPIs), such as cycle time, and flow times between two events, under various interventions. This analysis is especially useful for conducting 'what-if?' analysis.

Without interventions, computing the cycle time, for instance, can be performed via exact analysis by computing expectations of flow times using *flow analysis* [29], where a closed-form mathematical expression of the expected values of a KPI can be obtained. To this end, one would not require a causal model. However, if one is interested in an interventional analysis, e.g., quantifying the impact of various decisions on flow times, one must use a causal model in order to compute these KPIs. Specifically, the computational formulae adopted in flow analysis will remain the same, but the expected values that one plugs in are different. In fact, they depend on the estimated P of the causal model, in conjunction with the mutilated causal graph $\mathcal{G}'$ (see Sect. 2).

We illustrate this approach with our running example. Let X_1 be the duration of activity 'Detailed CDA', X_2 the duration of activity 'Standard CDA', and X_3 the duration of activity 'Deliver Car to Dealership'. Our KPI of interest is the expected flow time F between process start until the car is delivered or until the order is rejected. Without an intervention on D_1, the flow time is, given by the following flow analysis expression:

$$\mathbb{E}(F) = P(D_1 = 1)\mathbb{E}(X_1) + P(D_1 = 0)\mathbb{E}(X_2) + P(D_2 = 0)\mathbb{E}(X_3), \qquad (7)$$

which involves statistical quantities that can be easily estimated from data.

Alternatively, if we intervene on D_1, say by setting it to 1 (i.e., choosing detailed CDA) and wish to compute the KPI again (thus measuring the impact of this decision), we get the following:

$$\mathbb{E}(F) = \mathbb{E}(X_1) + P(D_2 = 1 \mid do(D_1 = 1))\mathbb{E}(X_3), \qquad (8)$$

which is a causal expression that cannot be computed without a causal model. However, if we obtain the true causal model (see Fig. 2), we can turn this causal expression into a statistical expression, using the do-calculus rules [23]:

$$\mathbb{E}(F) = \mathbb{E}(X_1) + \sum_v P(D_2 = 1|D_1 = 1, V = v)P(V = v)\mathbb{E}(X_3), \qquad (9)$$

which can again be estimated using historical data. This simple example emphasizes the difference between causal quantities, and statistical quantities when computing the impact of decisions on performance measures.

Flow analysis of pre-interventional KPIs can be performed easily using historical data as it corresponds to the latter. However, post-interventional KPIs can only be computed using a causal model.

4   Process-Aware Causal Discovery

In this section, we introduce a novel process-aware causal discovery algorithm. To achieve this, we combine constraint-based discovery from [12] with temporal constraints that we mine from an event log. These temporal constraints are fed into the constraint-based algorithm as additional background knowledge.

In this work, we consider two specific types of temporal constraints, namely temporal tiers, and forbidden edges. The former type of constraints corresponds to a set of time-ordered tiers indicate the ordering between the variables (decisions and contextual data). Forbidden edge constraints are employed to ensure that variables that always succeed others, yet do not fit into separate tiers, do not parent their predecessors.

Figure 3 presents the steps of our discovery algorithm. Step 1 detects a partial order over decisions, and constructs temporal tiers, such that each tier contains

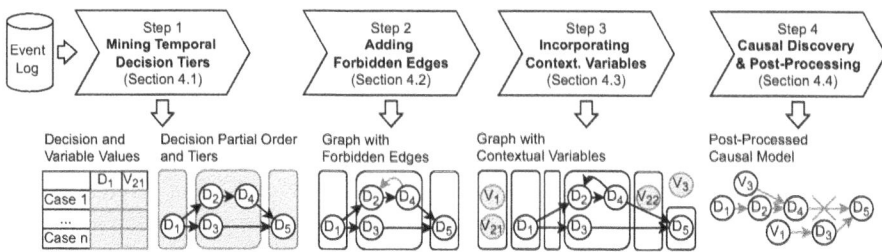

Fig. 3. Overview of our approach to process-aware causal discovery.

a set of unrelated decisions. Step 2 adds forbidden edge constraints that enforce a temporal ordering within tiers. Step 3 adds a subset of contextual variables from the event log into temporal tiers, when the ordering between variables and decisions can be derived. All other contextual variables remain 'floating' without an assigned tier. Lastly, Step 4 employs the approach from [12] to discover a causal graph, and post-processes the result to remove disabling edges, i.e., edges that induce false causal relations between decisions.

4.1   Step 1: Mining Temporal Decision Tiers

Assuming we have a process map and its decisions mined by some process discovery algorithm (e.g., inductive miner), we start by running a partial order discovery algorithm on the event log to obtain the partial order over all activities and decisions. Here, any algorithm to derive a partial order may be used, for instance the one presented by Kourani and van Zelst [18]. Then, we project the partial order on decision points, which yields a set of pairs of ordered decision points, PO, of the following structure:

$$PO = \{(D_1, D_2), \dots, (D_x, D_y), \dots\} \qquad (10)$$

where (D_x, D_y) represents a temporal precedence between decisions D_x and D_y.

Using PO, we encode the temporal ordering among decisions into a disjoint set of tiers. Denote by $T_D = \{t_1, \dots, t_k\}$ the set of tiers, i.e., distinct sets of decisions, $t_i \subseteq \mathcal{D}, 1 \leq i \leq k$ with $t_i \cap t_j = \emptyset, 1 \leq i, j \leq k$ and $i \neq j$, that also respect the partial order:

$$\forall\, t_i, t_j : i < j \iff \forall\, D_x \in t_i\; D_y \in t_j : (D_x, D_y) \in PO. \qquad (11)$$

We construct the tiers T_D by performing a topological sort over the DAG that corresponds to the partial order. The tiers are then elicited from the resulting strongly connected components.

4.2   Step 2: Adding Forbidden Edges

When constructing decision tiers some temporal constraints that may exist within the tier may be neglected. To illustrate this scenario, consider the 'Complete Inspection' subprocess of our example in Fig. 1, which includes two resource decisions, and a control-flow decision. The top concurrent branch represents the 'Feature Inspection' activity, which can be executed by multiple resources (D_5). Meanwhile, the bottom branch consists of the 'Technical Inspection' activity, also operated by multiple resources (D_6). After technical inspection, there is a decision point regarding whether a 'Technical Fix' is required (D_4). The concurrency between D_5 and $\{D_4, D_6\}$ leads to lack of ordering between the decisions when considering tiers. However, from the temporal order in the log, we can learn that D_4 will never precede D_6.

We incorporate such knowledge into our approach by assessing whether there exists an ordering between decisions within the same tier that is obeyed in all recorded process executions. If so, we inject the obtained knowledge by imposing constraints on forbidden edges that would violate the identified ordering. Specifically, we restrict the existence of edges with an opposite direction with respect to time. In our example, the edge (D_4, D_6) is disallowed.

4.3   Step 3: Incorporating Contextual Variables into Tiers

Next, the ordered decision tiers T_D are expanded with contextual variables. Note that it is not required to assign a tier to every variable.

Recall that $\mathcal{C}$ is the set of all contextual nodes in the causal graph. We split this set into two disjoint sets corresponding to static and dynamic variables, which we denote by $\mathcal{C}_s$ and $\mathcal{C}_d$, respectively. To distinguish the two sets, we check the variable values throughout the recorded executions in the event log. If V_x remains unchanged throughout the execution of all cases it is assigned to $\mathcal{C}_s$; otherwise, it is assigned to $\mathcal{C}_d$.

Having established the two sets, we aim to extend our decision tiers T_D to a set of tiers that would also include contextual variables. To this end, we create new variable tiers between two consecutive decision tiers, which is captured by an extended set $T_{D,V}$ of tiers:

$$T_{D,V} = \{\text{tier}(i) \mid i \in \mathbb{N}_0, i \leq 2k\} \text{ with } \text{tier}(i) = \begin{cases} \emptyset & \text{if } i\%2 = 0 \\ t_{\frac{i+1}{2}} & \text{otherwise.} \end{cases} \quad (12)$$

Based on Eq. 12, each decision tier (t_i) corresponds to a new tier in the final tier set, $t_i = \text{tier}(2i - 1) \in T_{D,V}$. In addition, we create a tier between decision tiers (from T_D), and we can now assign variables from both $\mathcal{C}_s$ and $\mathcal{C}_d$ to tiers.

Packing Tiers with Static Variables. Static variables are case attributes for which the value is set upon the start of case executions. Thus, they are placed into a variable tier that precedes all decision tiers. In other words, we allocate all variables in $\mathcal{C}_s$ to tier$(0) \in T_{D,V}$.

Packing Tiers with Dynamic Variables. The dynamic variables are not as straightforward to assign to tiers. Consider a scenario where we wish to assign a dynamic variable that we observe changing z times throughout a case. We let $V_{x,1}, \ldots, V_{x,z}$ be those z measurements. The objective is to categorize each $V_{x,i}, i = 1, \ldots, z$ into a tier in $T_{D,V}$, necessitating an analysis of its temporal ordering as captured in the event log. This entails recording the timestamp $\tau_{V_{x,i}}$ corresponding to the measurement of variable $V_{x,i}$, as well as the timestamps of the decisions. For each decision tier, we ascertain the timestamps of the earliest and latest decision made by bounding the preceding and succeeding activities that appear before and after the decision, respectively. Subsequently, we evaluate $\tau_{V_{x,i}}$ against this temporal ordering. If it falls within the time range defined by the earliest and the latest decision timestamps for a specific t_i, then $V_{x,i}$ is allocated to tier$(2i - 1)$ (a decision tier). Conversely, if it falls between the latest decision timestamp of t_i and the earliest decision timestamp of t_{i+1}, then $V_{x,i}$ is assigned to the variable tier identified by tier$(2i)$.

Note that the tier assignment described above is specific to the variable within a *single* case. Therefore, there may be temporal inconsistencies when assigning variables while considering multiple cases. Heuristically, we can assure that for a majority of cases, the tier of a contextual variable is consistent. A solution to resolve these inconsistencies in a more systematic manner is left to future work.

4.4   Step 4: Constraint-Based Discovery and Post-processing

Having obtained $T_{D,V}$ and the set of forbidden edges, we feed them along with the dataset containing all variables into a causal discovery algorithm that supports input constraints. We refer to it as *tier table*. Note that the method is agnostic to the specific causal discovery algorithm. The result is a causal graph, that in conjunction with P (that one estimates from the data), can be used to analyze and support decisions. To create a dataset suitable for causal discovery, we set up a table with all the selected variables, including decision points and a subset of contextual measurements. Each row in the table corresponds to a case in the event log, with values filled in for each column. For cases where a valid value for a specific decision is absent (as illustrated in the next paragraph), we use a placeholder number to represent null.

In some cases, control-flow decisions disable parts of the process. For example, in Fig. 1, if decision D_2 selects the bottom branch, the respective process execution will not face decisions D_3 and D_4. This yields a deterministic disablement dependency between the values of D_2 and D_3, which in turn produces a spurious edge in the discovery phase between D_2 and D_3. This discovered edge does not appear in Fig. 2.[1] Thus, to obtain an accurate causal graph, we must pre-process the data from Step 1 by adding *null* values to each of the decision domains for decisions that could be disabled at execution time.

Based thereon, we post-process the result of causal discovery that we get in the current step by removing spurious disablement edges. To achieve this, we search for decisions that are always disabled whenever upstream decisions take on specific outcomes (e.g., the rejection of an order). Specifically, we consider a decision D_i that is always mutually exclusive with a decision D_j, which comes later in the ordering relation, i.e., $(D_i, D_j) \in PO$. Then, if $\exists \, o \in D_i : P(D_j =$

---

[1] Note that this is a unique feature of the business processes domain that stems from the dependencies between a given choice and its down-stream decisions.

$null \mid D_i = o) = 1$, we are guaranteed that a deterministic causal edge will appear between D_i and D_j. In that case, We collect (D_i, D_j) into a set E_{null}, and eventually remove E_{null} from the causal graph. Note that multiple edges leading into D_j may be removed, as multiple upstream decisions can disable D_j.

5   Evaluation

In this section, we first test the proposed approach on data generated from our running example (Sect. 5.1). Furthermore, we compare our framework against the method proposed by [19], to highlight the advantages of our causal model. Subsequently, we apply our approach to a real-world dataset, and show its ability to uncover a causal graph, as well as to estimate the ATE for the corresponding decisions (Sect. 5.2). An implementation of our approach is publicly available, along with the experimental setup.[2]

5.1   Synthetic Data for Running Example

First, we assigned conditional probabilities as in Table 1a, in correspondence with the model presented in Fig. 2 and generated a tabular dataset that consists of 10,000 samples across all variables. Then, we aimed to construct an event log that would correspond to the running example. To this end, we utilized PM4Py [7] to generate event data from BPMN models, using the process from the running example (Fig. 1). The choice probabilities in the generative model depended on the previous choices and variables in accordance to the tabular data, and the probabilities in Table 1a. With a synthetic event log in hand, our objective was to execute our process-aware discovery algorithm. To achieve this, we ran the steps in Sect. 4 and obtained the final tier table presented in Table 1b.

For causal discovery, we used the *Bayesys Package*[3], which implements the algorithms in [12]. Our approach accurately produced the true causal graph depicted in Fig. 2. Utilizing this causal graph to estimate the Average Treatment Effect (ATE) of D_1 on D_3 resulted in an estimate of 0, indicating no causal path from the treatment to the outcome. However, the method proposed in [19], which assumes no additional confounders beyond decision variables reported an ATE of 0.456, demonstrating a significant bias. Specifically, the method in [19] failed to consider the absence of causal paths from D_1 to D_3; instead it produced an erroneous ATE based on correlation. Moreover, employing *CausalNex*[4], we provide the ATE of D_1 on D_2 in Table 2. Likewise, in this measurement the estimated ATE is significantly lower than the actual one, due to the ignorance of [19] to the confounding variable (V).

---

[2] https://doi.org/10.5281/zenodo.11399639
  https://github.com/aliash98/Causal-Discovery-for-Business-Processes.
[3] https://bayesian-ai.eecs.qmul.ac.uk/.
[4] https://github.com/mckinsey/causalnex.

Table 1. Generative probabilities and the final tiers only for the main process.

(a) Probabilities used to generate data

Expression	Probability
P_v	0.2
$P(d_1|v)$	$\begin{pmatrix} 0.2 \ 0.8 \end{pmatrix}$
$P(d_2|v, d_1)$	$\begin{pmatrix} 0.3 \ 0.8 \\ 0.8 \ 0.2 \end{pmatrix}$
$P(d_3|v)$	$\begin{pmatrix} 0.3 \ 0.9 \end{pmatrix}$

(b) Final tiers

tier(0)	tier(1)	tier(2)	tier(3)
V	D_1	D_2	D_3

5.2   Real-World Data

To demonstrate the applicability of our method to practical problems, we conducted a test on a real-world dataset. Specifically, we selected the 'BPI Challenge
2012' event log, published by the IEEE Task Force on Process Mining, which
pertains to a loan process [14]. Using the inductive miner, we obtained a process model depicted in Fig. 4, from which we extracted eight decision points.
The decision data corresponding to each case was then collected into a tabular dataset. Additionally, we included the static contextual variable, namely
'Amount Requested', and discretized it into four labels along with the dataset.

Next, following the method described in Sect. 4, we extracted three tiers,
with $\{D_3 \rightarrow D_2\}$ being a forbidden edge constraint. Again utilizing the Bayesys
package, we identified two causal arcs, namely $discrtized\_amount\_req \rightarrow D_1$
and $D_2 \rightarrow D_3$.

From [14] and the description of the BPI Challenge 2012 reports, we know
that D_1 plays a crucial role in determining which main branch the case will traverse. Returning to Fig. 4, branch C indicates approval of the loan request, while
branch D represents a rejection thereof. We observe that as the requested amount
increases, the likelihood of rejection decreases. This trend could be interpreted as
the bank's inclination toward approving larger loan applications. In Table 3, we
present a summary of the results of intervening on the discretized_amount_req
variable and its impact on the distribution of D_1. Additionally, D_2 indicates
whether another round of 'Assessing the application' is needed, while D_3 determines the same for 'Seeking additional information during the assessment phase'.
The causal influence of D_2 on D_3, with an Average Treatment Effect (ATE) of
0.513, signifies that the repetition of the assessing task increases the likelihood

Table 2. Causal effect table only for the main process.

	Process-Aware Causal Discovery	Approach from [19]
ATE of D_1 on D_3	0.0	0.456
ATE of D_1 on D_2	−0.283	−0.165

of seeking additional information. The above serves as a proof-of-concept and demonstrates that the approach can be easily applied to real-world datasets.

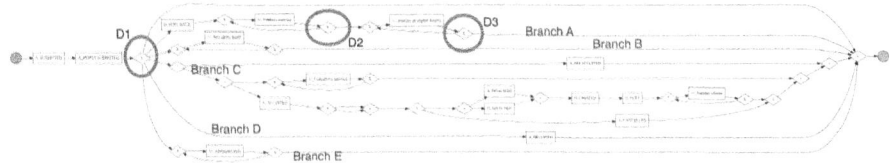

Fig. 4. BPMN model obtained by running inductive miner on BPIC 2012 dataset.

Table 3. Distribution of D_1 after intervening on discrtized_amount_req

DISC_AMNT_REQ	Branch of D_1				
	A	B	C	D	E
0 (from 0 to 5500)	0.1426	0.0072	0.2412	0.6088	0.0003
1 (from 5500 to 9500)	0.2538	0.0039	0.3170	0.4253	0.0
2 (from 9500 to 19000)	0.2966	0.0043	0.3357	0.3634	0.0
3 (above 19000)	0.3199	0.0058	0.3762	0.2981	0.0

6   Related Work

Our work relates to data-driven decision support and prescriptive process monitoring. Below, we provide an overview of the respective literature while emphasizing the research gaps that we aim to address.

Decision support for business processes received much attention recently, fueled by novel formalisms for decision modeling, such as Decision Modeling and Notation (DMN) [6], and advancements in process mining [28]. Early work in the area discovers decision trees from event data to predict the next activity and explain historical control-flow choices [21,25]. Yet, the proposed models do not include causal structures, and focus on the prediction task, using only features available in the data. Other work considers decisions related to resource assignments in processes [26]. Albeit probabilistic, the approach does not consider interventional queries.

Models of business process may be refined with models of their decision logic (via DMN) that are derived from historical data [3,5]. These ideas have later been refined to incorporate uncertainty related to decision outcomes through the discovery of fuzzy decision models [4,13]. However, since these models are not causal, they cannot capture interventions appropriately. To improve decision making, one may also adopt specialized Bayesian networks, known as influence diagrams, and treat them as causal models [2]. The approach assumes deterministic decision-making, and does not support conditioning on contextual variables to assess interventions.

Recently, first approaches that aim at learning causal dependencies between variables that may influence decisions have been proposed [19]. This framework ignores contextual variables, which we demonstrate to be crucial in analyzing decisions. Moreover, their solution only focuses on control-flow decisions, and overlook resource assignments, which are naturally embedded into our approach.

Beyond decision support, causal modeling and reasoning was employed in business processes to facilitate root-cause analysis for business processes [8,24]. Bozorgi et al. [8] showed how uplift trees can be mined to explain process outcomes, while Qafari and van der Aalst [24] presented a general technique for causal inference in processes. Moreover, prescriptive analytics that strives for interventions to improve process outcomes adopts causal models [9,10,27,30]. Here, causal dependencies are leveraged to answer prescriptive questions, e.g., on the next action to minimize cycle times [9,10]. However, their focus on processes that do not involve decisions stems from the fact that these methods assume continuous random variables, which makes them unsuitable for analyzing the impact of categorical variables on decisions. Additionally, a framework using Structural Causal Models codifies cause-effect assumptions, controls confounding, and performs counterfactual reasoning with observational data, addressing process improvement challenges without extensive trials but lacking multiple contextual variable measurements [22]. Another study enhances process discovery by using causal relationships among execution activity timestamps [15]. Both studies do not consider causal decision support, and do not aim to discover the underlying causal structure from the data.

7 Conclusion

In this paper, we argued that improving decision-making is essential for the success of organizations and proposed a framework for causal decision support that overcomes substantial drawbacks of existing approaches. Specifically, we showed how the framework enables causal reasoning beyond the state of the art, e.g., in flow analysis. We also introduced a new method for process-aware causal discovery to facilitate the identification of causal graphs from event data. Finally, we demonstrated the feasibility of our approach through synthetic experiments, and a real-world dataset application.

In future work, we intend to expand our framework by explicitly incorporating objective functions and utility variables that can then be optimized. Additionally, we aim at adding declarative temporal constraints into constraint-based discovery to improve the results of the causal discovery approach. Lastly, we wish to test the approach on a real-world dataset where the ground-truth causal graph and the true BPMN model are known. This would demonstrate the accuracy and usefulness both of the proposed framework and of the discovery algorithm.

References

1. Augusto, A., et al.: Automated discovery of process models from event logs: review and benchmark. IEEE Trans. Knowl. Data Eng. **31**(4), 686–705 (2019). https://doi.org/10.1109/TKDE.2018.2841877

2. Batoulis, K., Baumgraß, A., Herzberg, N., Weske, M.: Enabling dynamic decision making in business processes with DMN. In: Reichert, M., Reijers, H.A. (eds.) BPM 2015. LNBIP, vol. 256, pp. 418–431. Springer, Cham (2016). https://doi.org/10.1007/978-3-319-42887-1_34

3. Bazhenova, E., Buelow, S., Weske, M.: Discovering decision models from event logs. In: Abramowicz, W., Alt, R., Franczyk, B. (eds.) BIS 2016. LNBIP, vol. 255, p. Bazhenova, E., Buelow, S., Weske, M.: Discovering decision models from event logs. In: Business Information Systems: 19th International Conference, BIS 2016, Leipzig, Germany, July, 6-8, 2016, Proceedings 19. pp. 237-251. Springer (2016)-251. Springer, Cham (2016). https://doi.org/10.1007/978-3-319-39426-8_19

4. Bazhenova, E., Haarmann, S., Ihde, S., Solti, A., Weske, M.: Discovery of fuzzy DMN decision models from event logs. In: Dubois, E., Pohl, K. (eds.) CAiSE 2017. LNCS, vol. 10253, pp. 629–647. Springer, Cham (2017). https://doi.org/10.1007/978-3-319-59536-8_39

5. Bazhenova, E., Weske, M.: Deriving decision models from process models by enhanced decision mining. In: Reichert, M., Reijers, H.A. (eds.) BPM 2015. LNBIP, vol. 256, pp. 444–457. Springer, Cham (2016). https://doi.org/10.1007/978-3-319-42887-1_36

6. Bazhenova, E., Zerbato, F., Oliboni, B., Weske, M.: From BPMN process models to DMN decision models. Inf. Syst. **83**, 69–88 (2019)

7. Berti, A., van Zelst, S.J., Schuster, D.: PM4Py: a process mining library for Python. Softw. Impacts **17**, 100556 (2023). https://doi.org/10.1016/J.SIMPA.2023.100556

8. Bozorgi, Z.D., Teinemaa, I., Dumas, M., La Rosa, M., Polyvyanyy, A.: Process mining meets causal machine learning: discovering causal rules from event logs. In: 2020 2nd International Conference on Process Mining (ICPM), pp. 129–136. IEEE (2020)

9. Bozorgi, Z.D., Teinemaa, I., Dumas, M., La Rosa, M., Polyvyanyy, A.: Prescriptive process monitoring for cost-aware cycle time reduction. In: 2021 3rd International Conference on Process Mining (ICPM), pp. 96–103. IEEE (2021)

10. Bozorgi, Z.D., Teinemaa, I., Dumas, M., La Rosa, M., Polyvyanyy, A.: Prescriptive process monitoring based on causal effect estimation. Inf. Syst. **116**, 102198 (2023)

11. Burstein, F., W Holsapple, C., van der Aalst, W.M.: Decision support based on process mining. In: Handbook on Decision Support Systems 1: Basic Themes, pp. 637–657 (2008)

12. Constantinou, A.C., Guo, Z., Kitson, N.K.: The impact of prior knowledge on causal structure learning. Knowl. Inf. Syst. **65**(8), 3385–3434 (4 2023). https://doi.org/10.1007/s10115-023-01858-x

13. De Smedt, J., Hasić, F., vanden Broucke, S.K., Vanthienen, J.: Holistic discovery of decision models from process execution data. Knowl.-Based Syst. **183**, 104866 (2019)

14. van Dongen, B.: BPI challenge 2012 (2012). https://doi.org/10.4121/uuid:3926db30-f712-4394-aebc-75976070e91f

15. Fournier, F., Limonad, L., Skarbovsky, I., David, Y.: The WHY in business processes: discovery of causal execution dependencies (2024)

16. Guo, R., Cheng, L., Li, J., Hahn, P.R., Liu, H.: A survey of learning causality with data: problems and methods. ACM Comput. Surv. (CSUR) **53**(4), 1–37 (2020)
17. van Hulzen, G.A.W.M., Li, C., Martin, N., van Zelst, S.J., Depaire, B.: Mining context-aware resource profiles in the presence of multitasking. Artif. Intell. Med. **134**, 102434 (2022). https://doi.org/10.1016/J.ARTMED.2022.102434
18. Kourani, H., van Zelst, S.J.: POWL: partially ordered workflow language. In: Di Francescomarino, C., Burattin, A., Janiesch, C., Sadiq, S. (eds.) BPM 2023. LNCS, vol. 14159, pp. 92–108. Springer, Cham (2023). https://doi.org/10.1007/978-3-031-41620-0_6
19. Leemans, S.J., Tax, N.: Causal reasoning over control-flow decisions in process models. In: Franch, X., Poels, G., Gailly, F., Snoeck, M. (eds.) CAiSE 2022. LNCS, vol. 13295, pp. 183–200. Springer, Cham (2022). https://doi.org/10.1007/978-3-031-07472-1_11
20. Liu, S., Duffy, A.H., Whitfield, R.I., Boyle, I.M.: Integration of decision support systems to improve decision support performance. Knowl. Inf. Syst. **22**, 261–286 (2010)
21. Mannhardt, F., de Leoni, M., Reijers, H.A., van der Aalst, W.M.P.: Decision mining revisited - discovering overlapping rules. In: Nurcan, S., Soffer, P., Bajec, M., Eder, J. (eds.) CAiSE 2016. LNCS, vol. 9694, pp. 377–392. Springer, Cham (2016). https://doi.org/10.1007/978-3-319-39696-5_23
22. Narendra, T., Agarwal, P., Gupta, M., Dechu, S.: Counterfactual reasoning for process optimization using structural causal models. In: Hildebrandt, T., van Dongen, B.F., Röglinger, M., Mendling, J. (eds.) BPM 2019. LNBIP, vol. 360, pp. 91–106. Springer, Cham (2019). https://doi.org/10.1007/978-3-030-26643-1_6
23. Pearl, J.: Causality: Models, reasoning, and inference. Cambridge University Press (2009)
24. Qafari, M.S., van der Aalst, W.: Root cause analysis in process mining using structural equation models. In: Del Río Ortega, A., Leopold, H., Santoro, F.M. (eds.) BPM 2020. LNBIP, vol. 397, pp. 155–167. Springer, Cham (2020). https://doi.org/10.1007/978-3-030-66498-5_12
25. Rozinat, A., van der Aalst, W.M.: Decision mining in ProM. Bus. Process Manage. **4102**, 420–425 (2006)
26. Senderovich, A., Weidlich, M., Gal, A., Mandelbaum, A.: Mining resource scheduling protocols. In: Sadiq, S., Soffer, P., Völzer, H. (eds.) BPM 2014. LNCS, vol. 8659, pp. 200–216. Springer, Cham (2014). https://doi.org/10.1007/978-3-319-10172-9_13
27. Teinemaa, I., Tax, N., de Leoni, M., Dumas, M., Maggi, F.M.: Alarm-based prescriptive process monitoring. In: Weske, M., Montali, M., Weber, I., vom Brocke, J. (eds.) BPM 2018. LNBIP, vol. 329, pp. 91–107. Springer, Cham (2018). https://doi.org/10.1007/978-3-319-98651-7_6
28. Van Der Aalst, W.: Process Mining: Data Science in Action, vol. 2. Springer, Heidelberg (2016). https://doi.org/10.1007/978-3-662-49851-4
29. Verenich, I., Nguyen, H., La Rosa, M., Dumas, M.: White-box prediction of process performance indicators via flow analysis. In: Proceedings of the 2017 International Conference on Software and System Process, pp. 85–94 (2017)
30. Weinzierl, S., Dunzer, S., Zilker, S., Matzner, M.: Prescriptive business process monitoring for recommending next best actions. In: Fahland, D., Ghidini, C., Becker, J., Dumas, M. (eds.) BPM 2020. LNBIP, vol. 392, pp. 193–209. Springer, Cham (2020). https://doi.org/10.1007/978-3-030-58638-6_12

TOTeM: Temporal Object Type Model for Object-Centric Process Mining

Lukas Liss[(✉)][iD], Jan Niklas Adams[iD], and Wil M. P. van der Aalst[iD]

RWTH Aachen University, Aachen, Germany
{liss,niklas.adams,wvdaalst}@pads.rwth-aachen.de

Abstract. System behavior emerges from multiple subprocesses operating on interacting objects of different types. The relations between types of subprocesses are essential to understand the system's behavior. This includes temporal relations as well as cardinalities of relationships. Whether a product is produced before or after a customer order, which is essential with respect to lean management, would be an example of a temporal relationship. Conversely, the number of products for each customer order would be an example of a cardinality constraint. Current object-centric process modeling approaches focus on precedence constraints between activities, which is not sufficient to capture temporal and cardinality relationships between types. This paper introduces the temporal object type model (TOTeM) to model and discover process-specific type-level relations. We propose three type-level relations: a temporal relation, an overall log cardinality, and an event cardinality. The contributions of this paper include a definition of the temporal object type model, an algorithm to compute them, a publicly available implementation, and an evaluation.

Keywords: Process model · Object-centric process mining · Relationship model

1 Introduction

Real-world processes are complex orchestrations of interacting subprocesses. Each subprocesses instantiation is identified by an object of the subprocesses' type. The interaction between objects is defined by shared actions between subprocesses. In traditional process mining, a process is assumed to only consist of one single subprocess, i.e., the case identifier. As a result, one can not differentiate the different objects and their interactions that make up a process. To enable this distinction, process mining shifted towards an object-centric perspective on processes [14]. There, individual objects of different types can be tracked throughout the process. Each event has an activity, a timestamp, and a set of objects that are involved in the event via an event-to-object relation. Objects can also be connected without sharing events. These relations can be quantified and captured in explicit object-to-object relations [10]. Current models for

© The Author(s), under exclusive license to Springer Nature Switzerland AG 2024
A. Marrella et al. (Eds.): BPM 2024 Forum, LNBIP 526, pp. 107–123, 2024.
https://doi.org/10.1007/978-3-031-70418-5_7

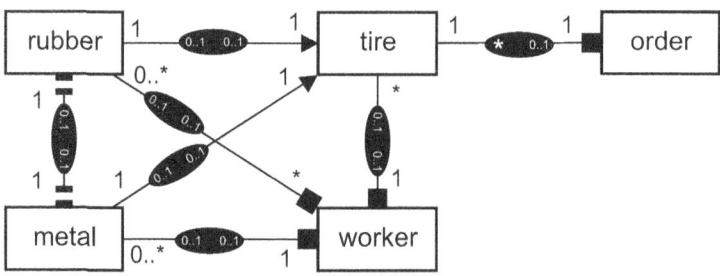

Fig. 1. Temporal object type model for the example of a tire production process.

object-centric processes have limitations in describing the temporal and cardinality relations between types. Thus, we proposed a new model called TOTeM.

Let us consider an object-centric example process that produces tires. It involves five object types: orders, tires, rubber, metal, and workers. The workers produce multiple tires per day. A tire is built by exactly one worker using one unit of metal and one unit of rubber. Both the metal and the rubber are prepared before. Orders can be connected to multiple tires, but a tire is always connected to exactly one order. The company follows the lean management pull principle [18], so they only produce tires if there is a demand for them. So, a new tire is only built after its order has been received. All the characteristics mentioned above are modeled in the TOTeM model in Fig. 1 for the running example. In Sect. 4, we introduce TOTeM models and explain how the characteristics are represented. Here, we first show that these characteristics are contained in event logs, but current models are not sufficient to describe all of them.

An example excerpt of the object-centric event log in the OCEL 2.0 format [10] from this process is presented in Table 1. This contains the event-to-object relations as well as explicit object-to-object relations. Some events in the process only involve objects from one type, like *receive order* that only involves an order object. Other involve objects of multiple types, like *build tire*, which involves a worker, metal, rubber, and a tire.

Table 1. Object-centric event log for the example tire production process. The event-to-object relations are on the left and the object-to-object relations are on the right.

eventId	activity	time	object types					source object	target object	qualifier
			order	tire	rubber	metal	worker	...		
e1	receive order	1.3.24 14:10	o1					r2	w1	handled by
...								r2	w2	handled by
e12	prepare rubber	1.3.24 15:15			r2		w1	m2	w1	handled by
e13	wash rubber	1.3.24 15:25			r2		w2	m2	t2	used for
e14	prepare metal	1.3.24 15:50				m2	w1	r2	t2	used for
e15	form metal	1.3.24 16:10				m2	w1	t2	w1	build by
e16	build tire	1.3.24 16:50		t2	r2	m2	w3	t2	o1	assigned to
e17	ship order	1.3.24 17:00	o1	t1,t2				w3	w1	supervises
e18	receive payment	5.3.24 13:00	o1					w3	w2	supervises
...								...		

An object-centric event log in the OCEL 2.0 format [10] contains four types of information about activities and object types. First, the event log contains information about the temporal ordering of activities. Second, the cardinality for how many objects per object type are involved in an activity is contained in the event log. We refer to this as the event cardinality. Third, the information on how object types are related temporally. For example, tires only come into existence after the order object exists. Fourth, the event log contains information about the cardinalities between object types. For example, metal is always handled by exactly one worker, but rubber can be connected to multiple workers. We refer to this as the log-level cardinality. All types of information are valuable in understanding and analyzing a process. Note that an object-centric event log is not limited to these four types of information since there can be, for example, additional attributes, but all object-centric event logs in the OCEL 2.0 format [10] contain these four.

Current approaches model object-centric processes with object-centric Petri nets [5] or class diagrams [8]. Class diagrams model only the fourth type of information, the cardinalities between object types. They use neither temporal information about activities nor object types. On the other hand, object-centric Petri nets focus on the first type of information, the precedence constraints of activities, and the second type. So the object-centric Petri net for the running example in Fig. 2 shows that *prepare rubber* has to precede *wash rubber* and that *ship order* can involve multiple tires but just one order. The arcs with a triple arrowhead in the object-centric Petri net are variable arcs that show that the connected activity can involve multiple objects. However, none of the models describe the third type of information, the temporal ordering of object types. It is also impossible to reliably derive one type of information from another, as shown by the example object-centric Petri net in Fig. 2.

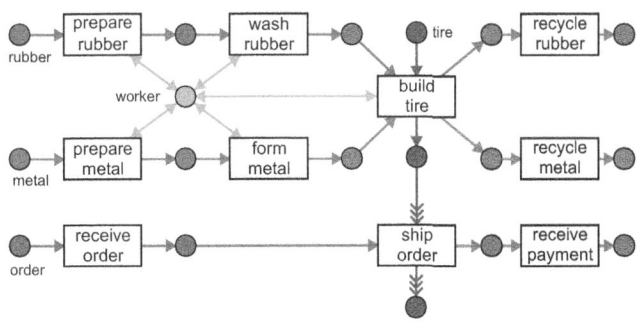

Fig. 2. Object-centric Petri net for the example tire production process.

For example, in the object-centric Petri net for the running example in Fig. 2, one can not differentiate whether the pull principle [18] is followed or not. In the context of the running example, this means that tires only come into existence when the order is placed, so the customer demand already exists. This information on the temporal relation between types is contained in the event log but not discovered or modeled by object-centric Petri nets. Since the tire and the order only connect when the order is shipped, there is no temporal relation between *build tire* and *receive order* modeled in the object-centric Petri net.

Another important characteristic of a process is the work distribution. This relates to information about the cardinalities between object types. For example, in the tire process, it makes a huge difference whether there is an assembly line where multiple workers are connected to (work on) a material or just one. In the example process, multiple workers are connected to one *rubber*, but for each *metal* object, there is only one *worker* connected. This difference is not visible in the object-centric Petri net. This example shows the limitations of current modeling approaches for object-centric processes. To close that gap, we propose the temporal object type model (TOTeM).

This paper presents four contributions to enable the data-based discovery of the described object-type relations that characterize object-centric processes.

- We propose a definition for temporal object type models (TOTeM) and the three types of object type relations.
- We describe an algorithm to discover TOTeM models from object-centric event logs [10].
- We implement the algorithm and publish the open-source code.
- We use the implementation to perform an evaluation of our approach on a publicly accessible event log.

The paper is structured in the following way. In Sect. 2, we present the related work. Afterward, in Sect. 3, we describe the preliminaries. Section 4 elaborate on the three types of relations and how they are combined to TOTeM models. A definition of the proposed algorithm to compute TOTeM models is given in Sect. 5. Section 6 evaluates the proposed model and algorithm with a qualitative and quantitative evaluation. In Sect. 7, we conclude the paper and give directions for future research.

2   Related Work

Process Mining aims to generate insights into business processes [1]. The main three tasks in process mining are discovery, conformance checking, and enhancement. In process discovery, one creates a process model given an event log that describes the process. To better represent real-world processes with multiple objects of different types, object-centric process mining was introduced [14]. This allows distinguishing multiple objects within a process and solves representation problems of traditional process mining like deficiency, convergence, and divergence [2]. Our proposed TOTeM model and the described discovery algorithm, therefore, belong to the field of object-centric process discovery. The goal of TOTeM models is to give insights into the process by understanding the temporal relations and cardinalities between object types. Current discovery approaches focus on mining dependency constraints between events [5].

There exist a variety of process models for traditional event logs, like directly follows graphs, Petri Nets [24], BPMN [12], Causal Nets [3], and Process Trees [21]. Since they are limited to a single case identifier, they do not model any object type relations.

There are also object-centric process models. For example, object-centric Petri nets [5] and object-centric directly follows graphs [9] can be discovered from object-centric event logs [10]. An object-centric directly follows graph shows the directly follows relations between activities per object type [9]. Object-centric Petri nets and a discovery approach for them are introduced by Berti and van der Aalst [5]. Other approaches like Petri nets with identifiers [26] and COA-nets [16] also extend Petri nets to track objects and their types in a process. These models describe process behavior by modeling precedence constraints of activities. As shown in the introduction, one can not always derive clear insights about the type-level behavior from the activity level constraints. Object-centric behavior constrains [4] also model temporal relations between activities and additionally model the cardinality of object types for the activities as well as the cardinality relation between object types. But similar to object-centric Petri nets [5], the temporal ordering of object types is only indirectly modeled via precedence constraints for activities, which is not sufficient, as shown in the introduction. Other approaches like PHILharmonicFlows [20] and the MERODE approach [25] combine multiple models to represent object-centric processes. They use state models to represent the lifecycle of object types and class diagrams to model the relations between them. The temporal relations between object types can be represented indirectly via state change dependencies but must not be explicitly modeled. But they can not be discovered from event data but must be modeled by an domain expert and they distribute the information over multiple models.

Besides the object-centric process models, one can also use class diagrams to describe how object types are related to each other [8]. This modeling is mainly done by hand and is not discovered, which hinders the ability to discover unknown behavior in the process. Many different class diagrams exist, e.g., UML-Class diagrams [8] or Entity-Relationship models [11]. Since they are not specifically designed for processes, they do not support the event-level cardinality and do not model temporal relation between types. There are multiple extensions that try to integrate time into class models [17]. However, those approaches focus on the time that modeled information is valid. Ferg for example, extended ER diagrams to represent the start and end date of the validity of attributes of classes [15]. Elmasri and Wuu extended ER models to model the lifespan of individual classes [13]. However, they do not model how the lifespans of two classes relate to each other, which is relevant for processes.

So far, a process-specific model notation describing the temporal relations and the cardinality relations between types that can be discovered from process data is missing. To close that gap, we propose TOTeM models.

3   Preliminaries

We assume object-centric event data stored in, for example, OCEL 2.0 format [10]. Events are activities that happen at a timestamp for a set of objects of different types. $\mathbb{U}_{event}$ is the universe of event identifiers. The universe $\mathbb{U}_{act}$ contains all visible activities. $\mathbb{U}_{type}$ is the universe of all object types. The universe of objects is $\mathbb{U}_{obj}$. Each object has exactly one type associated with it $\pi_{type} : \mathbb{U}_{obj} \rightarrow \mathbb{U}_{type}$. $\mathbb{U}_{time}$ is the universe of all timestamps.

Definition 1 (Object-Centric Event Log). $L = (E, O, OT, O2O, \pi_{act}, \pi_{obj}, \pi_{time})$ *is an event log with:*

- *$E \subseteq \mathbb{U}_{event}$ is a set of events, $O \subseteq \mathbb{U}_{obj}$ is a set of objects,*
- *$OT = \{\pi_{type}(o) | o \in O\}$ is a set of object types,*
- *$O2O \subseteq \{(o_1, o_2) | o_1, o_2 \in O \wedge o_1 \neq o_2\}$ is the set of object to object relations,*
- *$\pi_{act} : E \rightarrow \mathbb{U}_{act}$ maps each event to an activity,*
- *$\pi_{obj} : E \rightarrow \mathcal{P}(\mathbb{U}_{obj}) \setminus \{\emptyset\}$ maps each event to at least one object,*
- *$\pi_{time} : E \rightarrow \mathbb{U}_{time}$ maps each event to a timestamp.*

Note that we abstract from qualifiers and attribute values, i.e. we use a subset of OCEL 2.0 [10]. The tabular representation of an object-centric event log consists of one table showing the event-to-object relations and another table showing the object-to-object relations. Table 1 shows the event to object relations for the example event log. There, each event is connected to one activity, a timestamp, and some objects of the five types of the example tire production process. Table 1 shows the object-to-object relations for the exemplary tire production process. Often, two objects are connected via an object-to-object relation if they share an event. For example, the objects *r2* and *w1* of the first object-to-object relation in Table 1 are both involved in event *e2*. However, a shared event is not required if there is an explicit object-to-object relationship. For example, there is an object-to-object relation between *w3* and *w2*, although they never share an event because, for example, *w3* supervises *w2*. Older formats of object-centric event logs do not have explicit object-to-object relations. To be able to use these event logs as well and have one source of truth for which objects are connected, we add objects that share events to the potentially empty set of object-to-object relations.

Definition 2 (Object-to-Object Relations from Event-to-Object Relations). *Given event log $L = (E, O, OT, O2O, \pi_{act}, \pi_{obj}, \pi_{time})$ the event log with all connected objects represented in the object-to-object relations is given by: $L' = (E, O, OT, O2O', \pi_{act}, \pi_{obj}, \pi_{time})$ with $O2O' = O2O \cup \{(o_1, o_2) \in O \times O | o_1 \neq o_2 \wedge \exists_{e \in E} \{o_1, o_2\} \subseteq \pi_{obj}(e)\}$.*

In the following we assume, that all event logs are event logs that represent all connected objects in the object-to-object relation. We call two objects connected if they are part of an object-to-object relation and we call two types related if there is at least one object of each type that is connected to an object of the other type.

Definition 3 (Connected Objects and Types). *Given event log* $L = (E, O, OT, O2O, \pi_{act}, \pi_{obj}, \pi_{time})$, *the set of connected objects* $\overline{O2O} = \{(o_1, o_2) \in O \times O | (o_1, o_2) \in O2O \vee (o_2, o_1) \in O2O\}$ *contains all pairs of objects for which at least one direction is part of* $O2O$.

We reduce a set of objects $O' \subseteq O$ *to objects of type* $t \in OT$ *with the following notation:* $O'_{\downarrow t} = \{o \in O' | \pi_{type}(o) = t\}$.

We reduce the connected objects to pairs of type $t_1, t_2 \in OT$ *with* $\overline{O2O}_{\downarrow t_1, t_2} = \{(o_1, o_2) \in \overline{O2O} | o_1 \in O_{\downarrow t_1} \wedge o_2 \in O_{\downarrow t_2}\}$.

Two types $t_1, t_2 \in OT$ *are connected iff* $\exists_{o_1 \in O_{\downarrow t_1}, o_2 \in O_{\downarrow t_2}} (o_1, o_2) \in \overline{O2O}$.

For all the example objects in Table 1, the reduction to the type worker results in $\{o1, t2, r2, m2, w1, w2, w3\}_{\downarrow worker} = \{w1, w2, w3\}$. The types *rubber* and *worker* are connected because, for example, $r2$ and $w1$ are connected.

For the temporal type relation, we need to determine the lifespan of objects within a process. We assume that the lifespan of an object, in the context of the given process, is defined by the first and last appearance of that object in the event-to-object table.

Definition 4 (Object Lifespan Interval). *Given an event log* $L = (E, O, OT, O2O, \pi_{act}, \pi_{obj}, \pi_{time})$ *and object* $o \in O$. *The start of the lifespan of object* o *is* $ots_{L,start}(o) = min(\{t \in \mathbb{U}_{time} | \exists_{e \in E} \pi_{obj}(e) = o \wedge \pi_{time}(e) = t\})$. *The end of the lifespan of an object* o *is* $ots_{L,end}(o) = max(\{t \in \mathbb{U}_{time} | \exists_{e \in E} \pi_{obj}(e) = o \wedge \pi_{time}(e) = t\})$. *The lifespan of an object is the interval* $ols(o) = \{t \in \mathbb{U}_{time} | ots_{L,start}(o) \leq t \leq ots_{L,end}(o)\}$.

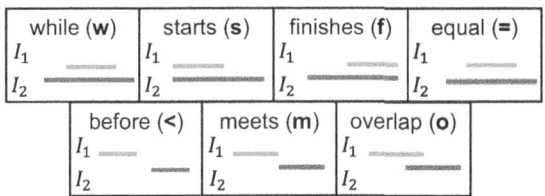

Fig. 3. Allen's interval relations [7].

For example the lifespan for object $r2$ starts at 1.3.24 15:15 and ends at 1.3.24 16:50 because event $e12$ is the first and event $e16$ is the last event that involves $r2$.

We use Allen's interval algebra [7] on the object lifespan intervals. Figure 3 shows all the relations that intervals can have. For example, the first one shows I_1 w I_2, which means that the interval for I_1 is *while*(**w**) interval I_2. Each of the relations (except *equal*(=)) has an inverse relation as well. So, for example, if I_1 is *while*(**w**) I_2, then I_2 is *while inverse*(**w**$^{-1}$) to I_1. In the following, we will focus on non-inverse relations. For example, the lifespan of $r2$ is *while*(**w**) the lifespan of $o1$ because the lifespan of $o1$ starts earlier and ends later. So the *while*(**w**) relation holds: $ols(r2)$ **w** $ols(o1)$. The concrete definitions for the relations are given in [7]. Note that Allen calls the *while*(**w**) relation *during*, but to avoid confusion with the *during*(■) relation proposed in this paper we use another name.

4   Temporal Object Type Model

In this section, we describe the proposed TOTeM model. First, we introduce each of the three relation types separately. Then, we explain how they are combined together in the TOTeM model.

4.1   Temporal Relation

The goal of the temporal relation is to describe how the lifespan of connected objects from two types relate to each other. The relation of one object's lifespan to another can be described using Allen's 13 interval algebra relations. Some of these relations are similar. We group these similar relations together and use these groups to define temporal relations over object types.

For example, the *while*(**w**), *starts*(**s**), *finishes*(**f**), and *equal*(=) interval relations have in common that the first interval only exists if the second one exists. This group of lifespan interval relations belongs to the temporal relation *during*(■). A *during*(■) relation from *type 1* to *type 2* describes that all pairs of connected objects from *type 1* to *type 2* have only lifespan interval relations that belong to the described group. So the lifespan of a *type 1* object either is *while*(**w**), *starts*(**s**), *finishes*(**f**), or is *equal*(=) to the lifespan of a connected *type 2* object. Figure 4 shows the temporal relations and the related groups of interval relations. In the running example, the *during*(■) from the tire to the order means that all tires only exist during their connected order object also exists and never earlier or later. Also, one can differentiate long-term assets like the workers and short-term assets like *rubber*, *metal*, and *tire*. In the TOTeM model, all the connected short-term assets have a *during*(■) relation to the *worker* type.

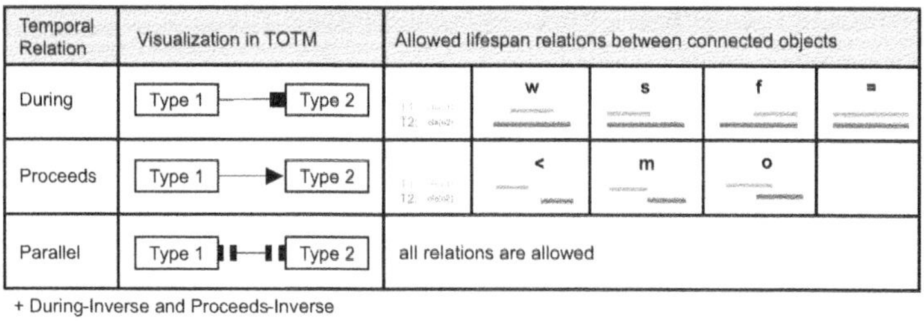

Temporal Relation	Visualization in TOTM	Allowed lifespan relations between connected objects			
			w	s	f	=
During	Type 1 ——■ Type 2					
			<	m	o	
Proceeds	Type 1 ——► Type 2					
Parallel	Type 1 ▮——▮ Type 2	all relations are allowed				

+ During-Inverse and Proceeds-Inverse

Fig. 4. The temporal relation *during*(■), *precedes*(►), and *parallel*(‖) (based on Allen's interval algebra [7]). For each temporal relation, the allowed lifespan relations for connected objects of *Type 1* (in orange) and *Type 2* (in green) are shown.

We also grouped together the *before*(ι), *meets*(**m**), and *overlaps*(**o**) interval relations because for all of them, the first interval starts before the second one starts and ends before the second one ends. This group relates to the *precedes*(▶) temporal relation. Both the *during*(■) and *precedes*(▶) temporal relation can be inversed by changing the target and source type. In the running example, there are *precedes*(▶) relations from *rubber* and *metal* to the tire. Since these materials are used to produce the tire, they exist before the *tire*.

The parallel(∥) temporal relation allows for all interval relations. This means that there is no clear order in the relationship between lifespans. The parallel relation between metal and rubber in the TOTeM model for the running example shows that the lifespans of objects of these types are not ordered.

Temporal relations are defined in Definition 5 by requiring that all lifespan interval relations for connected objects of the types belong to the described groups for the temporal relation. The definition introduces a scoring function that computes the fraction of object pairs from the two types that fulfill the requirements of the temporal relation. This scoring function is also used in the algorithm in Sect. 5 together with a noise parameter for filtering undesired infrequent behavior to determine possible temporal relations.

Definition 5 (Temporal Relations). *Given event log $\log L = (E, O, OT, O2O,$ $\pi_{act}, \pi_{obj}, \pi_{time})$ we define the score for $(t_1, t_2) \in OT \times OT$ for the temporal relations as:*

$$sc_{\blacksquare}(t_1, t_2) = \frac{|\{(o_1, o_2) \in \overline{O2O}_{\downarrow t1, t2} | ols(o_1) \; \boldsymbol{d} \; ols(o_2) \vee ols(o_1) \; \boldsymbol{s} \; ols(o_2) \vee ols(o_1) \; \boldsymbol{f} \; ols(o_2) \vee ols(o_1) = ols(o_2)\}|}{|\overline{O2O}_{\downarrow t1, t2}|}$$

$$sc_{\blacksquare_{inv}}(t_1, t_2) = sc_{\blacksquare}(t_2, t_1)$$

$$sc_{\blacktriangleright}(t_1, t_2) = \frac{|\{(o_1, o_2) \in \overline{O2O}_{\downarrow t1, t2} | ols(o_1) < ols(o_2) \vee ols(o_1) \; \boldsymbol{m} \; ols(o_2) \vee ols(o_1) \; \boldsymbol{o} \; ols(o_2)\}|}{|\overline{O2O}_{\downarrow t1, t2}|}$$

$$sc_{\blacktriangleright_{inv}}(t_1, t_2) = sc_{\blacktriangleright}(t_2, t_1)$$

$$sc_{\parallel}(t_1, t_2) = \frac{|\overline{O2O}_{\downarrow t1, t2}|}{|\overline{O2O}_{\downarrow t1, t2}|} = 1$$

A temporal relation holds for two types if the related score is equal to 1.

4.2   Log Cardinality

The log cardinality describes how many objects of a target type are associated with objects of the source type over the whole lifespan of an object. They are similar to cardinalities described by class diagrams [8]. Types are only connected when there is at least one pair of objects from the two types that are connected. Therefore, we do not need to consider the case that no objects of a type are connected to any object of the other type. The cardinality of connected objects can therefore be *1* (exactly 1), *0..1* (none or 1), *1..** (1 or many) *0..** (none, 1, or many). Figure 5 shows the possible cardinalities from *type 1* to *type 2* and what type of structures this relation allows in the connected object graph. The indicator for how many objects of *type 2* are connected to objects of *type 1* is located on the connection line between the types close to the box that represents *type 2*. Definition 6 defines log cardinalities using a scoring function that represents the fraction of objects of the source type that match a log cardinality by being connected to the right amount of objects from the source type. If all objects of the source type match the cardinality requirement, the log cardinality holds for the given types.

Definition 6 (Log Cardinality). *Given event log* $L = (E, O, OT, O2O, \pi_{act},$ $\pi_{obj}, \pi_{time})$ *and filter parameter* τ, *we detect temporal relation for a pair of connected types* $(t_1, t_2) \in OT \times OT$:

$$sc_{l-1}(t_1, t_2) = \frac{|\{o_1 \in O_{\downarrow t_1} | |\{o_2 \in O_{\downarrow t_2} | (o_1, o_2) \in \overline{O2O}\}| = 1\}|}{|O_{\downarrow t_1}|}$$

$$sc_{l-0..1}(t_1, t_2) = \frac{|\{o_1 \in O_{\downarrow t_1} | |\{o_2 \in O_{\downarrow t_2} | (o_1, o_2) \in \overline{O2O}\}| \in \{0,1\}\}|}{|O_{\downarrow t_1}|}$$

$$sc_{l-1..*}(t_1, t_2) = \frac{|\{o_1 \in O_{\downarrow t_1} | |\{o_2 \in O_{\downarrow t_2} | (o_1, o_2) \in \overline{O2O}\}| \geq 1\}|}{|O_{\downarrow t_1}|}$$

$$sc_{l-0..*}(t_1, t_2) = \frac{|\{o_1 \in O_{\downarrow t_1} | |\{o_2 \in O_{\downarrow t_2} | (o_1, o_2) \in \overline{O2O}\}| \geq 0\}|}{|O_{\downarrow t_1}|} = 1$$

A log cardinality relation holds for two types if the related score is equal to 1.

The log cardinality can give process owners and analysts insights into how objects of a given type relate to objects of other types. For example, Fig. 1 shows that tire objects are always connected to exactly one rubber and one metal object. This shows that these types of materials are required for the tire. If the rubber would be optional for the tire, one would find a *0..1* log cardinality there.

Some log cardinalities are stricter special cases of other cardinalities. For example, all tuples of types that fulfill the *1* log cardinality also fulfill all other log cardinalities because they are all less restrictive. Therefore, one should model the most restrictive log cardinality that applies.

One is interested in selecting the most restrictive cardinality because multiple ones can fit. For example, tires are always connected to only one order. This is allowed for the cardinalities *1*, *0..1*, *1..* *, and *0..* *. Since cardinality *1* is the most specific, one wants to select that one.

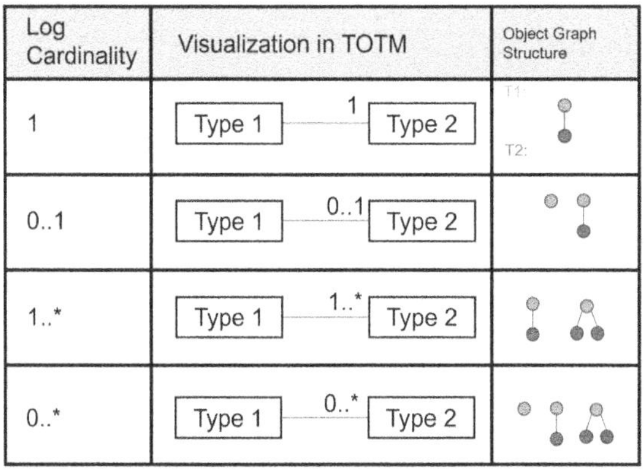

Log Cardinality	Visualization in TOTM	Object Graph Structure
1	Type 1 —— 1 —— Type 2	T1: / T2:
0..1	Type 1 —— 0..1 —— Type 2	
1..*	Type 1 —— 1..* —— Type 2	
0..*	Type 1 —— 0..* —— Type 2	

Fig. 5. The four log cardinality relations that describe how many objects of the types can be connected to how many objects of the other type in the log.

4.3 Event Cardinality

The event cardinality describes how many objects of the target type are involved in events that involve at least one object of the source type. So the event cardinality for *type 1* as the source and *type 2* as the target describes how many objects are involved in events that involve at least one object of *type 1*. Since objects that are related via object-to-object relations do not need to share events, there can be either always 0, exactly 1, 0 or 1, multiple (which includes 1), 0 or multiple objects of the target type involved. Figure 6 shows the event cardinalities, how they are visualized, and sketches events with their related objects from the source type (*type 1*) and target type (*type 2*). The cardinality is visualized in a black ellipsis that is on the connection line between two connected types. Since there are two directions between connected types, in a TOTeM model, there are always two event cardinalities in the black ellipses that are separated by a dash. The event cardinality is always placed closer to the target type. So in the first row in Fig. 6, the 0 in the black ellipsis is closer located to *type 2*. This shows that it is the cardinality that has *type 1* as the source and *type 2* as the target. Definition 7 defines event cardinalities by introducing scoring functions that compute the fractions of events that belong to a certain cardinality for the two types given. If all events of the given source type follow the cardinality in regards to the target type, the event cardinality holds for them.

Definition 7 (Event Cardinality). *Given event log $L = (E, O, OT, O2O, \pi_{act}, \pi_{obj}, \pi_{time})$ and filter parameter τ, we detect temporal relation for a pair of connected types $(t_1, t_2) \in OT \times OT$:*

$$sc_{e-0}(t_1, t_2) = \frac{|\{e \in E | \pi_{obj}(e)_{\downarrow t_1} \neq \emptyset \wedge |\pi_{obj}(e)_{\downarrow t_2}| = 0\}|}{|\{e \in E | \pi_{obj}(e)_{\downarrow t_1} \neq \emptyset\}|}$$

$$sc_{e-1}(t_1, t_2) = \frac{|\{e \in E | \pi_{obj}(e)_{\downarrow t_1} \neq \emptyset \wedge |\pi_{obj}(e)_{\downarrow t_2}| = 1\}|}{|\{e \in E | \pi_{obj}(e)_{\downarrow t_1} \neq \emptyset\}|}$$

$$sc_{e-0..1}(t_1, t_2) = \frac{|\{e \in E | \pi_{obj}(e)_{\downarrow t_1} \neq \emptyset \wedge |\pi_{obj}(e)_{\downarrow t_2}| \in \{0,1\}\}|}{|\{e \in E | \pi_{obj}(e)_{\downarrow t_1} \neq \emptyset\}|}$$

$$sc_{e-1..*}(t_1, t_2) = \frac{|\{e \in E | \pi_{obj}(e)_{\downarrow t_1} \neq \emptyset \wedge |\pi_{obj}(e)_{\downarrow t_2}| \geq 1\}|}{|\{e \in E | \pi_{obj}(e)_{\downarrow t_1} \neq \emptyset\}|}$$

$$sc_{e-0..*}(t_1, t_2) = \frac{|\{e \in E | \pi_{obj}(e)_{\downarrow t_1} \neq \emptyset \wedge |\pi_{obj}(e)_{\downarrow t_2}| \geq 0\}|}{|\{e \in E | \pi_{obj}(e)_{\downarrow t_1} \neq \emptyset\}|} = 1$$

Event cardinality relations hold for two types if the related score is equal to 1.

For example the TOTeM model for the tire process in Fig. 1 shows an event cardinality of *0..** from *order* to *tire*. This means that there are events involving orders that have no tire involved, but there are also events that involve multiple tires. The event cardinality is especially interesting in combination with the log cardinality. We can see, for example, that the worker is connected to multiple metal objects in total but handles at most one at a time.

4.4 TOTeM: Temporal Object Type Model Combined Model

The TOTeM model describes the overall behavior-based relations of types in a process. All three relations mentioned above, temporal relations, log cardinality,

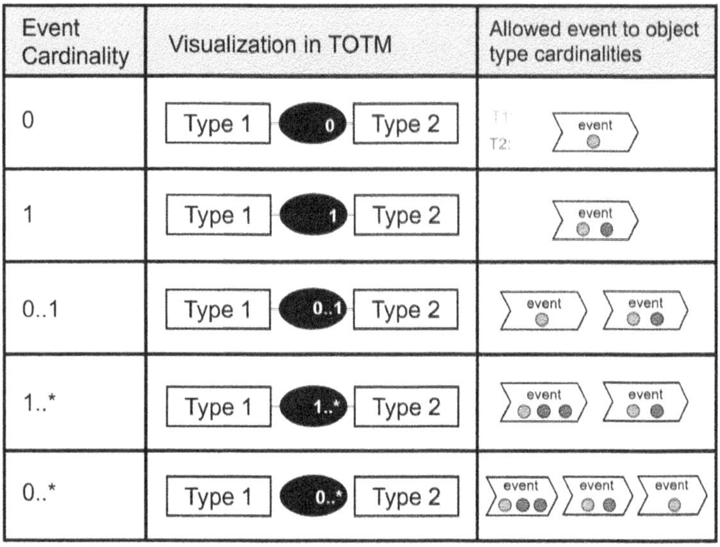

Event Cardinality	Visualization in TOTM	Allowed event to object type cardinalities
0	Type 1 — (0) — Type 2	T1: [event] / T2: [event ◉]
1	Type 1 — (1) — Type 2	[event ◉ ◉]
0..1	Type 1 — (0..1) — Type 2	[event ◉] [event ◉ ◉]
1..*	Type 1 — (1..*) — Type 2	[event ◉ ◉ ◉] [event ◉ ◉]
0..*	Type 1 — (0..*) — Type 2	[event ◉ ◉ ◉] [event ◉ ◉] [event ◉]

Fig. 6. The three event cardinality relations describe how many objects of the types participate in shared events. Note that we only visualized events with one object of *Type 1*, but there can also be more.

and event cardinality, are combined in the TOTeM model. In the TOTeM model, all object types of the object-centric event log are represented as boxes exactly once. Two types are related if there is at least one connection between objects of the two types. For the example tire process, that means that all the five types are represented as boxes in Fig. 1. Since there is no object-to-object relation between order objects and rubber objects, these two types are not connected. Whereas for example, metal and worker types are connected because there are workers that are connected with metal objects.

All three type relations are directed between two connected types, assuming a source and target type. For two connected types, we show six relations simultaneously on the connection line (all three relation types for both directions). For example, on the connection between rubber and tire, we show all three relation types, once assuming tire is the source type and once assuming rubber is the source type. So, the log cardinality from tire to rubber is shown, as well as the log cardinality from rubber to tire. The same holds for the event cardinality, which is shown in the black ellipsis on the connection line. Both types of cardinalities are visualized closer to the target type, as explained before, so one can differentiate the cardinalities for both directions. The *during*(■), and *precedes*(▶) temporal relations can be inversed and the *parallel*(∥) relation always holds for both directions. Therefore by visualizing either a *during*(■), *precedes*(▶), or *parallel*(∥) relation between two types, the temporal relations for both directions are defined. The definition of the TOTeM model is given in Definition 8.

Definition 8 (Temporal Object Type Model). *A temporal object type model is a directed graph* $(T, C, \pi_{tr}, \pi_{cc}, \pi_{oc})$ *where:*

- $T \subseteq \mathbb{U}_{type}$ *is a finite set of object types as nodes;*
- $C \subseteq T \times T$ *is a set of directed edges between connected types;*
- $\pi_{tr} : C \rightarrow \{\blacksquare, \blacksquare_{inv}, \blacktriangleright, \blacktriangleright_{inv}, \|\}$ *labels each edge with a temporal relation that holds;*
- $\pi_{lc} : C \rightarrow \{1, 0..1, 1..*, 0..*\}$ *labels each edge with a log cardinality that holds;*
- $\pi_{ec} : C \rightarrow \{0, 1, 0..1, 1..*, 0..*\}$ *labels each edge with an event cardinality that holds.*

5 Mining Algorithm

In this section, we define the algorithm to discover TOTeM models from object-centric event logs. The implementation code is publicly accessible on GitHub[1].

Real-world data, including event data, often contains noise. Process owners want to discover the process, so the influence of noise should be minimized. Also, process variants are often not equally frequent. Sometimes, one is interested in capturing infrequent behavior, but sometimes, one is only interested in mainstream behavior. This motivates the use of a parameter in the model discovery that allows users to adjust how strictly the discovered model should incorporate infrequent behavior. Therefore, we use the parameter $0 \leq \tau \leq 1$, which represents the fraction of conformity, to detect a relation for a tuple of types.

The algorithm starts with importing the object-centric event log. For that purpose, we use the open-source library ocpa [6]. First, all object types and the object-to-object relations are determined. Then, the object-to-object relations are supplemented with connections based on event-to-object relations. Definition 2 defines how the object-to-object relations are computed from the event-to-object relations. After that, we mine the temporal relations, the event cardinality, and the log cardinalities.

We use the scoring function defined in Definition 5, Definition 6, and Definition 7 to mine potential relations for each pair of types. Instead of requiring that the scoring function is equal to one, we require it to be higher or equal to the parameter τ. So for $\tau = 1$, only relations that hold for all objects and events are considered as potential relations. For a τ smaller than 1, we include relations that hold for a fraction of the behavior that is at least as big as τ. For example, if $\tau = 0.8$ then two types $(t1, t2)$ have a *during*($\blacksquare$) relation if at least 80 % of the connected objects of these types fulfill the during requirements.

Out of the potential relations, we select the most specific relation. As mentioned before, it is possible that a tuple is part of multiple relations of each of the three relation types. For example, a type tuple can belong to the *during*($\blacksquare$) relation and also to the *parallel*($\|$) relation. Also, all type tuples that have *1* as a potential log cardinality also have *1..** as a potential log cardinality. We construct the TOTeM model only with the most specific relation for each of the

---

[1] https://github.com/LukasLiss/TOTeM-temporal-object-type-model.

three types of relations for each tuple. Thus, the most precise model given the τ parameter is computed using the relation selection in Definition 9.

Definition 9 (Relations Selection). *Given event log* $L = (E, O, OT, O2O,$ $\pi_{act}, \pi_{obj}, \pi_{time})$ *and filter parameter* τ, *the functions* $\pi_{tr,\tau}$, $\pi_{lc,\tau}$, *and* $\pi_{ec,\tau}$ *select the relations for a pair of connected types* $(t_1, t_2) \in OT \times OT$:

$$\pi_{tr,\tau} = \begin{cases} \blacksquare, & \text{if } sc_{\blacksquare}(t_1, t_2) \geq \tau \\ \blacksquare_{inv}, & \text{else if } sc_{\blacksquare_{inv}}(t_1, t_2) \geq \tau \\ \blacktriangleright, & \text{else if } sc_{\blacktriangleright}(t_1, t_2) \geq \tau \\ \blacktriangleright_{inv}, & \text{else if } sc_{\blacktriangleright_{inv}}(t_1, t_2) \geq \tau \\ \|, & \text{else if } sc_{\|}(t_1, t_2) \geq \tau \end{cases}$$

$$\pi_{lc,\tau} = \begin{cases} 1, & \text{if } sc_{l-1}(t_1, t_2) \geq \tau \\ 0..1, & \text{else if } sc_{l-0..1}(t_1, t_2) \geq \tau \\ 1..*, & \text{else if } sc_{l-*}(t_1, t_2) \geq \tau \\ 0..*, & \text{else if } sc_{l-0..*}(t_1, t_2) \geq \tau \end{cases}$$

$$\pi_{ec,\tau} = \begin{cases} 0, & \text{if } sc_{e-0}(t_1, t_2) \geq \tau \\ 1, & \text{else if } sc_{e-1}(t_1, t_2) \geq \tau \\ 0..1, & \text{else if } sc_{e-0..1}(t_1, t_2) \geq \tau \\ 1..*, & \text{else if } sc_{e-*}(t_1, t_2) \geq \tau \\ 0..*, & \text{else if } sc_{e-0..*}(t_1, t_2) \geq \tau \end{cases}$$

6 Evaluation

In this section, we describe the evaluation results. We use a publicly accessible event log [19] to compare the TOTeM model and the object-centric Petri net. This object-centric event log describes the management of customer orders in a company. The log contains 21.008 events and 10.840 objects of 6 types. Figure 7 shows the object-centric Petri net for the order management.

We computed the TOTeM model with a publicly accessible implementation of the algorithm described in Sect. 5. The computation of the three relations took 3.14 s on average (95% confidence interval from 3.09 s to 3.19 with a sample size of 100) on an i7 2.8 GHz processor with 16 GB RAM. We selected a τ parameter of 0.9 to allow for some noise, infrequent behavior, and unfinished process executions. Figure 8 shows the resulting TOTeM model.

To evaluate whether the TOTeM model provides additional insights compared to existing approaches like object-centric Petri nets [5], we analyzed which insights present in the TOTeM model are missing in the object-centric Petri net.

For example, the TOTeM model shows that all products are ordered at least once because there is a log cardinality of *1..** from products to orders. The object-centric Petri net gives no information about the concrete objects used in the event log. This also leaves it unclear whether items are sold, although

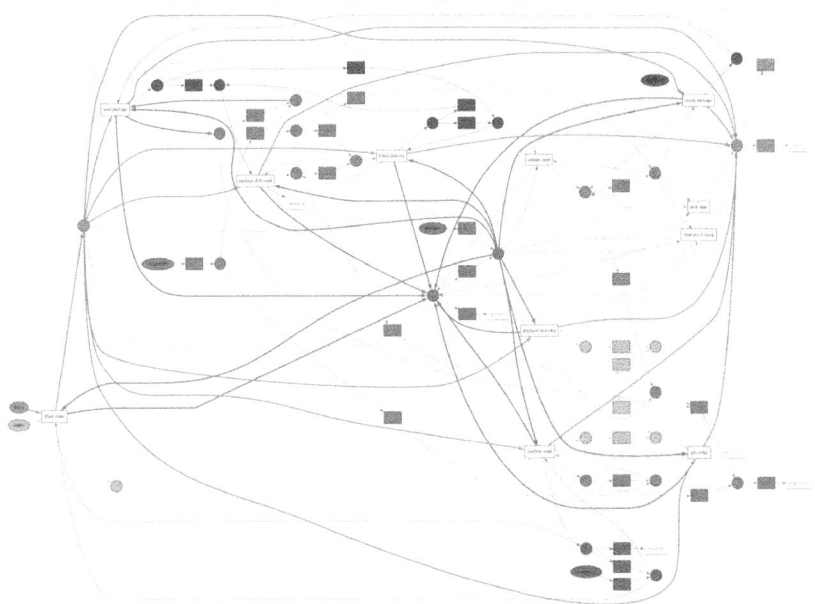

Fig. 7. Object-Centric Petri net for the order management example log [19]. Each color represents an object type. Ellipsis with text are initial places for a certain object type. The final places are represented by underlined text in the color of the object type.

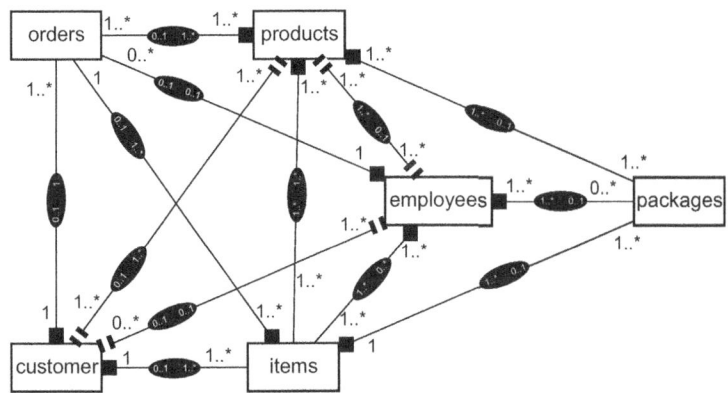

Fig. 8. TOTeM model for the order management process with $\tau = 0.9$.

the matching product no longer exists. We can not answer that question with the object-centric Petri net. However, the TOTeM model shows a *during*(■) relation between *items* and *products*, which shows that items only appear while their related product still exists. Moreover, the TOTeM model structures the types into a group with long lifespans (*products, customer*, and *employees*) and one with shorter lifespan (*orders, items*, and *packages*), such that all with shorter lifespans happen *during*(■) those with longer lifespans. Since most types' start and end activities are separate in the object-centric Petri net, it does not contain this information.

7   Conclusion

This paper presents four contributions to model and mine object-centric processes. First, we defined TOTeM models with temporal relations, log cardinalities, and event cardinalities. Second, we proposed an algorithm to compute TOTeM models from object-centric event logs. The algorithm's τ parameter allows it to handle noise and infrequent behavior. Third, we implemented the algorithm and published it on GitHub. Finally, in the evaluation, we showed that TOTeM models can give new insights into object-centric processes that are not captured in object-centric Petri nets.

This work also has limitations. We assume that the lifespan of objects is defined by their first and last events. Therefore, logging issues that affect these events can strongly affect the result. The robustness of the presented algorithm for given τ needs to be investigated in more detail. Currently, the TOTeM models algorithm does not consider object-to-object qualifiers, which is additional information that could be used to improve the discovered model. Also, the evaluation is limited to simulated data.

Therefore, one direction for future work is to perform real-world case studies to evaluate the insights analysts can get from TOTeM models. Another direction is to investigate the usage of TOTeM models for other process mining tasks like object-centric conformance alignments [23] and model enhancement [22].

Acknowledgments. The authors gratefully acknowledge the financial support by the Federal Ministry of Education and Research (BMBF) for the joint project Bridging AI (grant no. 16DHBKI023).

SPONSORED BY THE

Federal Ministry
of Education
and Research

References

1. van der Aalst, W.M.P.: Process mining. Commun. ACM **55**(8), 76–83 (2012)
2. van der Aalst, W.M.P.: Object-centric process mining: Dealing with divergence and convergence in event data. In: SEFM. pp. 3–25. Springer (2019)
3. van der Aalst, W.M.P., Adriansyah, A., van Dongen, B.F.: Causal nets: A modeling language tailored towards process discovery. In: CONCUR. pp. 28–42. Springer (2011)
4. van der Aalst, W.M.P., Artale, A., Montali, M., Tritini, S.: Object-centric behavioral constraints: Integrating data and declarative process modelling. In: Proceedings of the 30th International Workshop on Description Logics. vol. 1879. CEUR-WS.org (2017)

5. van der Aalst, W.M.P., Berti, A.: Discovering object-centric petri nets. Fundam. Informaticae **175**(1–4), 1–40 (2020)
6. Adams, J.N., Park, G., van der Aalst, W.M.P.: ocpa: A Python library for object-centric process analysis. Software Impacts **14**, 100438 (2022)
7. Allen, J.F.: Maintaining knowledge about temporal intervals. Commun. ACM **26**(11), 832–843 (1983)
8. Ashbacher, C.: The unified modeling language reference manual, Second Edition, by Rumbaugh J. Journal of Object Technol. **3**(10), 193–195 (2004)
9. Berti, A., van der Aalst, W.M.P.: Extracting multiple viewpoint models from relational databases. In: IFIP WG. pp. 24–51. Springer (2019)
10. Berti, A., Koren, I., Adams, J.N., Park, G., Knopp, B., Graves, N., Rafiei, M., Liß, L., Unterberg, L.T.G., Zhang, Y.: OCEL (Object-Centric Event Log) 2.0 Specification (2023)
11. Chen, P.P.: Entity-relationship modeling: Historical events, future trends, and lessons learned. In: Software Pioneers, pp. 296–310. Springer Berlin Heidelberg (2002)
12. Chinosi, M., Trombetta, A.: BPMN: an introduction to the standard. Comput. Stand. Interfaces **34**(1), 124–134 (2012)
13. Elmasri, R., Wuu, G.T.J.: A temporal model and query language for ER databases. In: ICDE. pp. 76–83. IEEE (1990)
14. Fahland, D.: Process mining over multiple behavioral dimensions with event knowledge graphs. In: Process Mining Handbook, Lecture Notes in Business Information Processing, vol. 448, pp. 274–319. Springer (2022)
15. Ferg, S.: Modelling the time dimension in an entity-relationship diagram. In: ER. pp. 280–286. IEEE (1985)
16. Ghilardi, S., Gianola, A., Montali, M., Rivkin, A.: Petri net-based object-centric processes with read-only data. Inf. Syst. **107**, 102011 (2022)
17. Gregersen, H., Jensen, C.S.: Temporal entity-relationship models - A survey. IEEE Trans. Knowl. Data Eng. **11**(3), 464–497 (1999)
18. Hopp, W.J., Spearman, M.L.: To pull or not to pull: What is the question? Manuf. Serv. Oper. Manag. **6**(2), 133–148 (2004)
19. Knopp, B., van der Aalst, W.M.P.: Order Management Object-centric Event Log in OCEL 2.0 Standard (Sep 2023)
20. Künzle, V., Reichert, M.: Philharmonicflows: towards a framework for object-aware process management. J. Softw. Maintenance Res. Pract. **23**(4), 205–244 (2011)
21. Leemans, S.J.J., Fahland, D., van der Aalst, W.M.P.: Discovering block-structured process models from event logs - A constructive approach. In: PETRI NETS. pp. 311–329. Springer (2013)
22. de Leoni, M.: Foundations of process enhancement. In: Process Mining Handbook, Lecture Notes in Business Information Processing, vol. 448, pp. 243–273. Springer (2022)
23. Liss, L., Adams, J.N., van der Aalst, W.M.P.: Object-centric alignments. In: ER 2023. pp. 201–219. Springer (2023)
24. Peterson, J.L.: Petri nets. ACM Comput. Surv. **9**(3), 223–252 (1977)
25. Snoeck, M.: Enterprise Information Systems Engineering - The MERODE Approach. Springer, The Enterprise Engineering Series (2014)
26. van der Werf, J.M.E.M., Rivkin, A., Polyvyanyy, A., Montali, M.: Data and Process Resonance - Identifier Soundness for Models of Information Systems. In: PETRI NETS 2022. vol. 13288, pp. 369–392. Springer (2022)

Engineering

NL2ProcessOps: Towards LLM-Guided Code Generation for Process Execution

Flavia Monti[1]([✉])[iD], Francesco Leotta[1][iD], Juergen Mangler[2][iD],
Massimo Mecella[1][iD], and Stefanie Rinderle-Ma[2][iD]

[1] Sapienza Università di Roma, Rome, Italy
{monti,leotta,mecella}@diag.uniroma1.it
[2] Technical University of Munich, Munich, Germany
{juergen.mangler,stefanie.rinderle-ma}@tum.de

Abstract. The optimization and automation of process deployment operations play a key role in enhancing efficiency and adaptability in Business Process Management. This paper presents NL2ProcessOps, a novel approach leveraging Large Language Models (LLMs) and concepts such as Retrieval Augmented Generation (RAG), agents, and tools for code generation to streamline process deployment operations. The proposed approach is designed to work with textual process descriptions and focuses on the various operations of process deployment, from extracting the control flow in terms of a process model, to retrieving required tools associated with each task, and generating executable code for manual refinement purposes and deployment in a process execution engine. The paper discusses the underlying principles of LLMs, the design and implementation of the approach, and its evaluation using a set of process descriptions. It demonstrates the effectiveness of NL2ProcessOps in generating high-quality code to support process deployment operations through both human and automated assessments. The paper concludes with a discussion of potential applications and future work.

Keywords: Process deployment operations · Large Language Models · Code generation · ProcessOps

1 Introduction

Process Execution Engines (PEEs) represent a vital part of Business Process Management (BPM) as they enable the enactment and monitoring of business processes. They are employed not only in traditional organizational scenarios but also in smart manufacturing [44] and coopetitive contexts.

PEEs encompass both open source (e.g., CPEE [31]) and commercial solutions (e.g., Bizagi, Camunda). They take process models as input, often in the form of BPMN diagrams or similar notations. As evidenced by the growing interest in process mining, the availability of well-defined process documentation, in the form of process models, is often challenging due to various reasons including

A. Marrella et al. (Eds.): BPM 2024 Forum, LNBIP 526, pp. 127–143, 2024.
https://doi.org/10.1007/978-3-031-70418-5_8

the evolution of processes over time. Process mining [1] addresses this issue by relying on the availability of event logs tracing the execution of the process.

However, organizations willing to adapt to changes rapidly and effectively may encounter sudden needs to support new processes, for which a clear definition is not available, and supporting event logs have not yet been produced. The challenge is how to combine fast deployment operations, necessary to cope with ever changing requirements and cloudified infrastructures, with the principled approaches of BPM, which are based on modeling processes and deriving enactable versions of them according to standard methods and techniques.

In this paper, we introduce the novel concept of ProcessOps, derived from *process* and *operations*, in analogy to DevOps [18]. ProcessOps can be defined as the agile approach in which new processes can be easily deployed and modified to accommodate changing requirements. In ProcessOps, a viable approach is to have process models in the form of scripts, similar to what happened in the recent past with orchestrations and choreographies expressed through scripting languages such as WS-BPEL and WS-CDL [33], and then to manually deploy them to PEEs, with minimal interventions aimed mostly at transforming abstract primitives in the script into runnable runtime calls to the PEE. This way, modifying the process upon changing requirements is quite easy and fast, as the script can be changed and then quickly deployed in the runtime environment.

In particular, we propose an approach based on *Large Language Models* (LLMs), called NL2ProcessOps, able to work with textual process descriptions and to support the generation of the different operations of process deployment, from extracting the control flow in terms of a process model, to retrieving required tools (e.g., services) associated with each task, and generating executable code for manual refinement purposes and deployment in PEEs. We propose a pipeline capable of generating an executable script representing the process (referred to as *process script*), which a designer can easily edit to deploy the process into the PEE. The proposed approach is quantitatively and qualitatively evaluated through human and automated assessments, demonstrating improvements over GitHub Copilot, a state-of-the-art LLM-based tool.

The paper is organized as follows: Sect. 2 introduces the basic notions we are building upon, specifically discussing LLMs principles and related concepts. Section 3 presents the proposed architecture and pipeline and describes how a prototype has been realized. Section 4 outlines and discusses the results of quantitative and qualitative evaluations. Finally, Sect. 5 discusses related works and Sect. 6 draws conclusions and outlines future work directions.

2   Background

The proposed approach relies on LLMs, advanced computational models capable of operating on and generating sentences in natural language [12]. These models are built upon the transformer (encoder-decoder) architecture and utilize the self-attention mechanism to capture dependencies and relevance of different parts of the input when generating the output [46]. LLMs are trained using massive datasets and contain hundreds of billions of parameters. Due to the resemblance between natural language and source code, LLMs find extensive utility in software engineering tasks like code understanding and generation [14].

State-of-the-art large pre-trained language models (PLMs) such as BERT [17], GPT [2,8] and LLaMA [45] are not domain-specific, and their output is often wrong or ambiguous (the term *hallucination* is often used) when applied to specialized applications like ours. Developing a domain-specific model requires a proper dataset and significant computing power for LLM training (or fine-tuning, transfer-learning) which may not always be available. In this context, the *in-context learning* ability of LLMs paired with the *prompt engineering* approach, becomes useful, enabling zero-shot or few-shot learning without additional training data [16,51]. Users design specific prompt text, sometimes with few examples (*few-shot prompting*), to guide an LLM in generating desired responses.

To enhance the accuracy of generated responses, *Retrieval-Augmented Generation* (RAG) has emerged [30]. This approach dynamically selects relevant external sources integrated into the LLM prompt. Typically, external knowledge is provided as a corpus of documents, aiming to retrieve a small subset to assist in generating a correct response [15]. The core idea of RAG is to convert both documents and LLM queries into embeddings (vector representations), usually via a transformer-based encoder, and then retrieve the documents most similar to the query. Embeddings similarity is computed via metrics such as dot product, cosine similarity, or Euclidean distance.

Another emerging approach involves developing LLMs as *agent* with the ability to use *tools* [23,24,37]. This allows LLMs to solve complex problems and mitigate hallucinations by generating, acting as an agent, action plans containing the invocation of external services (tools). A tool can access external information (e.g., documents, calendars) or affect the virtual or physical world (e.g., robotic arm) [34]. The general technique involves leveraging the in-context learning ability of LLMs by including in the prompt information about what an API can do (documentation) and, optionally, how it should be called (demonstration) [23].

To enhance LLMs performances, an important technique to consider is *prompt chaining*, which is beneficial for accomplishing complex tasks [49,50]. Detailed prompts may cause issues for an LLM when processing responses. The 'divide et impera' strategy can be applied to split a complex task into sub-tasks and create a chain of prompt operations, with each prompt responsible for a sub-task, thereby achieving better performance.

We develop an LLM-based approach for generating a process script from a textual process description. In practice, the script is a Python code containing invocations to external tools for executing the tasks of the given process. Each tool is characterized by a description that provides a textual representation of operations offered to execute the tasks. Tool documentations are prompted to the LLM to generate the script. To overcome the limitation of the input context length of LLMs, which cannot incorporate too much information, we consider the most appropriate tool descriptions and fit them within the prompt length.

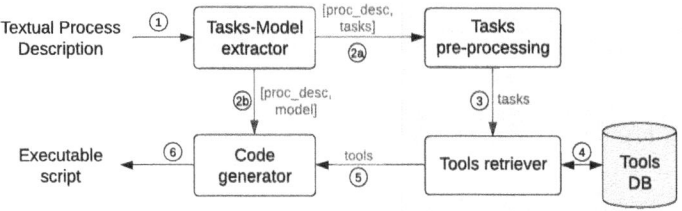

Fig. 1. Architecture of the NL2ProcessOps.

3   The NL2ProcessOps Architecture and Pipeline

The goal of NL2ProcessOps is to provide a solution for process operations (ProcessOps), aiming to simplify the development and deployment of processes, similar to how DevOps simplifies the development and deployment of general software. Specifically, once a new process definition is available, the involved operations include *(a)* extracting a process model from the description, defining the legal execution traces similarly to a program, *(b)* binding of each task to a software module implementing that task, and *(c)* defining the data flow, i.e., how data artifacts are manipulated by the tasks.

The generation of the *process script* from a textual process description is broken down into multiple *stages* supported by LLMs and chained together as follows: *(i)* extraction of tasks and control flow from the textual process description (operation *(a)* of ProcessOps), *(ii)* retrieval of relevant tools corresponding to the extracted tasks (operation *(b)* of ProcessOps), and *(iii)* generation of the process script implementing the process (operation *(c)* of ProcessOps). Figure 1 illustrates the components of NL2ProcessOps and their interactions. The numbers in the circles represent the order of the performed operations.

Stage *(i)* consists of a textual process description given as input ① to the *Tasks-Model extractor*. This component is an LLM prompted to extract the tasks and the control flow of the process and generates the model representation as a Mermaid.js [27,28]. The list of tasks paired with the textual process description (i.e., [proc_desc, tasks]) is input ②a to the *Tasks pre-processing* component. Concurrently, the process model and the textual process description (i.e., [proc_desc, model]) are input ②b to the *Code generator* component.

Stage *(ii)* (depicted in gray in Fig. 1) is inspired by the RAG concept to retrieve the relevant tools for the particular textual process description. The *Tasks pre-processing* component employs an LLM to refine the descriptions of the extracted tasks based on the textual process description. The refined list of tasks is then processed ③ by the *Tools retriever* component. This component interacts ④ with the vector database *Tools DB* and retrieves the most similar embedded tools offering the most suitable operation for each embedded task. *Tools DB* stores vectors consisting of the embeddings of the descriptions of the tools.

The list of retrieved tools is fed ⑤ into the *Code generator* component to initiate stage *(iii)*. The *Code generator* LLM. The *Code generator*, given the textual process description, process model and the list of tools (and their operations) implementing the process tasks, generates ⑥ a Python code – *process script* – embedding the control and data flows and implementing the process.

The integration of LLMs within the proposed approach is essential for several reasons. LLMs excel at processing natural language, making them ideal for extracting tasks and control flow from textual process descriptions. This first stage is critical as it forms the basis for the subsequent stages of the pipeline. Without accurate extraction of tasks and control flow, subsequent stages would lack the necessary information for generating meaningful outputs. The retrieval of relevant tools highly relies on the quality of the extracted tasks. An incorrect set of tools would affect the generation of the process script, leading to incorrect data and control flows. Finally, LLMs excel in generating high-quality code from a description. In our case, the textual process description guides the code generation, supported by the control flow and tools information derived from the previous stages.

3.1   A Running Example

Let's consider a real-world example in the Smart Manufacturing domain. Smart Manufacturing is a modern trend where cutting-edge technologies such as Industrial Internet of Things (IIoT) and Artificial Intelligence (AI) play pivotal roles in enabling quality enhancement, optimization and automation of production processes [38]. In this domain, the integration of the proposed solution, paired with PEE and enterprise systems like Manufacturing Execution System (MES) and Enterprise Resource Planning (ERP) enables the orchestration and execution of specific processes in a quick and efficient way [44].

> **Example** The automatic calibration process of cardboard production consists of continuously capturing a photo of the cardboard being produced. Each photo is analyzed to check if all the markers identified are ok. If markers are not ok, the calibration process continues. If the markers are ok, the speed of the die-cutting machine is set to 10000 RPM and the process ends.

The *Example* describes the automatic calibration process in cardboard production. Cardboard production is a manufacturing process that involves a die-cutting machine for the transformation of raw cardboard into printed cut-out cardboard sheets for the packaging industry. When starting a new order, calibration is needed to guarantee quality before proceeding with the production.

Figure 2 depicts the input-output of each of the NL2ProcessOps components over the *Example*. The numbers in circles are tightly connected to those in Fig. 1. Figure 2 reports the process model represented as Mermaid.js (green colored) and the list of (refined) tasks (blue and red colored) extracted from *Example*, the list of tools operations (purple colored) retrieved, and the process script – Python code (black colored). All the artifacts are available at the provided link[1].

---

[1] Cf. https://github.com/iaiamomo/NL2ProcessOps.

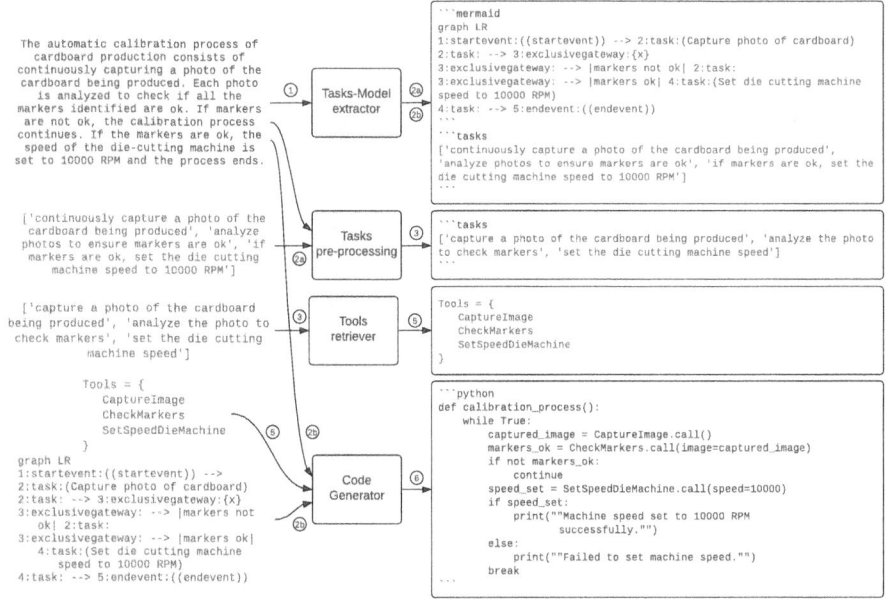

Fig. 2. Input-output of each component of NL2ProcessOps on *Example* (Color figure online)

3.2   Realization

We have realized a prototype (see footnote 1) of NL2ProcessOps [35]. The prototype is developed in Python and built on top of Langchain[2], a framework for constructing LLM-powered applications facilitating the creation, management and chaining of prompts. The base LLM model utilized for the three LLM-based components is GPT-4 (`gpt-4-0125-preview`) from OpenAI [2] with temperature set to 0, implying a more deterministic LLM mode.

Tools Retriever. Tools are central to our approach as they support task execution in a service-oriented way, i.e., offering operations for specific tasks. As proposed in [34], the LLM relies solely on the documentation of these tools which outlines their capabilities, usage instructions and outcomes. Figure 3 provides an excerpt from the documentation for the `SetSpeedDieMachine` tool of the *Example*. It consists of a general description of its unique operation, parameters details, and outcome description.

The *Tools DB* is a vector database containing the vector representations of the descriptions of the available tools. Each vector constitutes the embedding of the following information: `tool_name` and `description`, where the `description` is extracted from the tool documentation. An embedding is a sequence of numbers that represent information and enable better comprehension of relationships

---

[2] Cf. https://www.langchain.com/.

```
SetSpeedDieMachine = {
    "description": "Set the speed of the die-cutting machine.",
    "more_details": "It takes as input the speed. It returns a boolean value,
         True if the speed has been set, False otherwise.",
    "input_parameters": ["speed:int"],
    "output_parameters": ["speed_set:bool"],
    "tool": "die_machine" }
```

Fig. 3. Example of tool documentation

between similar information. We utilize the `text-embedding-ada-002` model from OpenAI[3] to compute these vectors. The embeddings are usually managed in vector DBs which enable a set of capabilities, including indexing, distance metrics and similarity search. As vector DB we rely on ChromaDB[4] which is open-source and well integrated with Langchain. ChromaDB enables the implementation of RAG-based approach basing similarity search on cosine distance (where a lower score indicates better similarity).

In the *Example*, for the refined extracted task description *"set the speed of the die cutting machine"*, the similarity search identifies `SetSpeedDieMachine` tool operation (cf. Fig. 3) as the most similar. Consequently, the *Die machine* tool is the most suitable for task execution.

The similarity search outputs a set of most similar tools with corresponding score results. The module guarantees that the useful tools are among those retrieved. In practice, given a task, the *Tool retriever* selects one or more tools based on their score values. The set of tools is then processed by the *Code generator* LLM that considers only those necessary for the specific case.

Prompt Engineering. The prompt is a guide for the model, instructing it on relevant information and desired output formatting. The quality of the LLM output directly correlates with the quality of the provided prompt [48].

The proposed approach consists of three different LLM-based components, each specialized in a particular task, i.e., extraction of the process tasks and model, pre-processing of the extracted tasks and generation of Python code. Each component is characterized by a specific prompt. Each prompt is characterized by all (or some) of the following parts: *(a)* the role the LLM plays that helps in controlling the output style [41], *(b)* a clear description of the task to be performed, *(c)* additional information (context) to aid the LLM in generating better responses, *(d)* few examples to teach the LLM, *(e)* type and format of the desired output and *(f)* the input data used by the LLM to compute the response. Table 1 presents detailed information regarding each part of the prompt for the three LLM-based components. All the prompts are available at the provided link (see footnote 1).

---

[3] Cf. https://platform.openai.com/docs/models/embeddings.
[4] Cf. https://www.trychroma.com/.

Table 1. Prompt information of the LLM-based components.

	Tasks-Model extractor	Tasks pre-processing	Code generator
(a)	BPM expert	BPM expert	–
(b)	Extract the control flow in terms of a process model and the list of tasks	Rephrase the tasks descriptions	Generate a Python code
(c)	Description of BPMN elements	–	Tools descriptions and guidelines
(d)	Yes	Yes	No
(e)	Set of custom rules for the process model representation	–	Python program structure
(f)	Textual process description	Textual process description and extracted tasks	Textual process description and process model

4   Evaluation

To quantitatively and qualitatively evaluate the proposed approach, we conducted separate assessments for the three stages, i.e., process model (and list of tasks) extraction, retrieval of appropriate tools, and executable script generation. Specifically, we assess (see footnote 1) the correctness and completeness of the retrieved tools and the generated Python code. We excluded the evaluation of the process model generation stage as it is based on [27, 28], where the solution has already been evaluated based on quantitative and qualitative assessments.

The retrieval stage is quantitatively evaluated using a synthetic dataset generated with an LLM. The dataset contains pairs of textual process descriptions (the semantics of which we do not care about) and their associated set of tools. The evaluation aims to determine whether the tools retrieved for a specific process are appropriate.

The code generation stage is both quantitatively and qualitatively evaluated. We utilized a dataset consisting of pairs of textual process descriptions extracted from existing datasets and their corresponding reference executable code. Code-BLEU metric was used for quantitative evaluation, while human assessments were conducted to determine the quality of the generated output.

4.1   Tools Retrieval Evaluation

The retrieval stage is evaluated using a synthetic dataset comprising 30 textual process descriptions. Each description is generated as follows: a certain number of operations is randomly selected from the *Tools DB* (which contains different tools, each including one or more operations, for a total of 62 operations) and their descriptions are prompted to an LLM to produce a random textual process description. The control flow of the generated process is not of interest here, as we only want to evaluate the ability of our system to correctly identify the right tools from the textual process description. Please note that the generated process description does not trivially include the operation documentation, which is instead re-elaborated by the LLM. The resulting dataset includes 30 process descriptions along with the respective ground truth consisting of the originally selected operations. Each process description is generated from a random number of operations ranging from 5 to 15.

We evaluate the effectiveness of the *Tools retriever* component in extracting the list of operations and corresponding tools. In particular, we evaluate three different strategies for extraction, denoted as *(R1)*, *(R2)* and *(R3)*, employing both GPT-3.5 and GPT-4 models. Strategy *(R1)* involves the LLM extracting the list of tasks from the textual process description using a few-shot prompting approach. In *(R2)* strategy, the LLM extracts both the control flow and the list of tasks from the textual process description using the prompt of the *Tasks-Model extractor*). In *(R3)* strategy, two separate LLMs extract the control flow and the list of tasks, followed by refinement of the extracted tasks descriptions.

Metrics commonly employed in evaluating an information retrieval system are precision and recall [40]. A high precision value indicates a low rate of false positives, while a high recall value corresponds to a low rate of false negatives. In our case, recall is of the utmost importance because we aim to extract all relevant tools. As explained in Sect. 3.2, non relevant tools are filtered out by the *Code generation* LLM.

The evaluation results of the three strategies show similar scores for GPT-3.5 and GPT-4 models, with precision ≈ 0.75 and recall ≈ 0.76 (results are averaged over the 30 processes). Results are available, for fairness and repeatability, at the provided link (see footnote 1). This similarity in scores makes it difficult to choose the final strategy based solely on quantitative analysis. However, from a qualitative perspective (see Sect. 4.2), strategy *(R3)* appears to be the optimal choice. The discrepancy between the qualitative analysis and the quantitative results may stem from how the LLM generates the dataset. The employed prompts for *(R3)* are described in Table 1 alongside the *Tasks-Model extractor* and *Tasks pre-processing* components.

4.2   Process Code Generation Evaluation

The code generation phase has been evaluated using two distinct approaches: an automated evaluation based on the CodeBLEU metric [39] and a human evaluation conducted over a group of academic users from the BPM area and proficient in the Python programming language.

For the evaluation, we used a dataset comprising of 10 textual process descriptions[5]. The descriptions illustrate processes with at least one start event and end event, exclusive and/or parallel gateways and at minimum of 5 tasks. Each description is accompanied by manually provided Python code implementing the respective process.

---

[5] Textual process descriptions are taken from https://zenodo.org/records/7783492 and https://huggingface.co/datasets/patriziobellan/PET.

Automated Evaluation. Various metrics have been proposed to evaluate generated code. These metrics include those used in machine translation (e.g., BLEU [36]), code comparison (e.g., CodeBLEU [39]), and unit testing. In this study, we utilize CodeBLEU, a metric designed to compare two pieces of code that demonstrates high coherence with human evaluation scores [39]. CodeBLEU is defined as a weighted combination of four components:

$$CodeBLEU = \alpha \cdot BLEU + \beta \cdot BLEU_{weight} + \gamma \cdot Match_{ast} + \delta \cdot Match_{df}$$

Here, *(i)* $BLEU$ represents the text match calculated from the n-grams overlapping between the hypothesis and reference codes, *(ii)* $BLEU_{weight}$ denotes the weighted n-gram match obtained by comparing the tokens of the hypothesis and reference codes with different weights, *(iii)* $Match_{ast}$ indicates the syntactic AST (abstract syntax tree) match exploring the syntactic information of code and *(iv)* $Match_{df}$ measures the semantic data flow match considering the semantic similarity between the hypothesis and the reference. $BLEU_{weight}$ and $Match_{ast}$ are used to measure grammatical correctness and $Match_{df}$ is used to calculate logic correctness.

CodeBLEU[6] returns a value between 0 (indicating no match) and 1 (representing a perfect match). In our evaluation, we compare the code generated by our approach with the reference code in the dataset. Following the suggestion of [39] we set $(\alpha, \beta, \gamma, \delta) = (0.1, 0.1, 0.4, 0.4)$, giving more weight to the AST and data-flow matches.

We assess three different strategies: in *(G1)*, the prompt of *Code generator* component does not include the process model generated by the *Tasks-Model extractor*; in *(G2)*, the retrieval part does not involve the *Task pre-processing* component; and in *(G3)*, the retrieval part includes the *Task pre-processing* component. Strategies *(G2)* and *(G3)* are closely related to the *(R2)* and *(R3)* strategies described in Sect. 4.1. While *(G1)* was evaluated using only GPT-4, *(G2)* and *(G3)* were evaluated using both GPT-3.5 and GPT-4 for the *Tasks-Model extractor* and *Task pre-processing* components and GPT-4 for the *Code generator* component. GPT-3.5 was also tested as a model for the *Code generator* component, but results are not reported as it generated malformed code, thus preventing the complete automation of the solution.

As a baseline, we compute the CodeBLUE score between the reference code for each process in the dataset and the code generated by GitHub Copilot[7], a code generation tool developed by GitHub and OpenAI to assist developers. Specifically, Copilot is utilized through the chat mode provided by Visual Studio Code[8]. Given a workspace containing the set of implemented tools, we invoke the @workspace agent, which leverages a meta-prompt to determine what information to collect from the workspace to help answer a question. The prompt input to GitHub Copilot is as follows: "Based on the available

---

[6] Cf. https://github.com/k4black/codebleu.
[7] Cf. https://github.com/features/copilot.
[8] Cf. https://code.visualstudio.com/docs/copilot/overview.

Table 2. Average CodeBLEU scores of the three strategies and GitHub Copilot outputs for the code generation evaluation.

Strategy	Model	$CodeBLEU$	$BLEU$	$BLEU_{weight}$	$Match_{AST}$	$Match_{df}$
(G1)	GPT-4	0.54	0.09	0.25	0.63	0.65
(G2)	GPT-4	0.60	0.07	0.25	0.69	0.73
	GPT-3.5-4	0.60	0.07	0.25	0.68	0.73
(G3)	GPT-4	0.61	0.08	0.27	0.70	0.75
	GPT-3.5-4	0.60	0.08	0.28	0.70	0.71
Baseline	Copilot	0.37	0.04	0.09	0.47	0.43

Table 3. NL2ProcessOps, GitHub Copilot evaluation results.

	(G3) – NL2ProcessOps	Copilot
Criterion 1	4.36 ± 0.61	3.80 ± 1.35
Criterion 2	4.53 ± 0.86	3.94 ± 1.33
Criterion 3	4.09 ± 0.98	4.17 ± 0.95

```
tools, generate a Python code that implements the following process
"""<textual_process_description>""".
```

Table 2 presents the final and component scores obtained by measuring Code-BLEU using the different strategies. The results from the *(G1)* strategy suggest that the inclusion of the process model in the prompt positively impacts the quality of the generated code. *(G2)* and *(G3)*, on the other hand, yield similar CodeBLEU scores. However, $Match_{AST}$ and $Match_{df}$ metrics, which we prioritize based on the chosen weights, indicate better performance when using GPT-4 alone. Notably, the results show how our approach, *(G3)* based on GPT-4, significantly outperforms the Copilot baseline.

Human Evaluation. We conduct a human evaluation following the guidelines proposed in [29]. We collect human feedback on the quality of the code generated by the *(G3)* strategy (referred to as NL2ProcessOps), which achieved the highest CodeBLEU score, and the baseline Copilot. The selection of questions is designed to emphasize the ability of the proposed approach to produce high-quality code. We identify various criteria and provide clear definitions to assist users in understanding and mitigating potential confusion. The criteria focus on: *(1)* the correctness of control and data flows, *(2)* the correctness of the code (i.e., absence of syntax errors) and *(3)* the presence of unnecessary function definitions. We utilize a 5-point Likert scale and provide its interpretation alongside the questions. Our sample comprises 14 users, including academic users from the BPM area and proficient in the Python programming language. The full questionnaire is available in the repository at (see footnote 1).

Table 3 displays the results of the assessment in terms of the mean and variance of the answers to the questionnaire. Overall, the results indicate that NL2ProcessOps outperforms Copilot. Additionally, results obtained with Copilot show higher variance. Upon qualitative analysis of individual responses, NL2ProcessOps was found to better extract control and data flows, while also producing error-free scripts.

5 Related Works

This work stands at the crossroads of BPM and service composition. As explained in [6], service composition addresses the situation when a complex task, also called *target service*, cannot be realized by calling a single service, but a composite service, obtained by combining "parts of" available *component services*, might be used. Despite the apparent similarity between a complex task and business processes, differences emerge in their formalization and temporal aspects. Business processes typically have longer time spans, treating tasks as black boxes, whereas service composition emphasizes the behavior of component services.

In service composition, target and component services are often modeled as state machines, with automated synthesis techniques employed to compute an *exact* solution to the problem. The main limitation of classical service composition methods is that the formal definition of the target and component services is a demanding task that must be done manually. Also, symbolic formalisms employed usually make it hard to model complex behavior, especially when the data flow is considered. Additionally, the computational complexity of the solving methods is exponential, quickly making solutions hard to compute. As pointed out in [3], LLMs can help address many of these limitations, even though the possible fuzziness of provided solutions must be taken into account.

On the BPM side, the employment of LLMs to overcome the burden of manually defining business processes, especially when the user is not expert in the process modeling formalism, has been explored, for example, in [27,28]. In particular, authors in [28] show the suitability of LLM-based chatbots to extract process models from textual descriptions. In the present work, even though – as shown by the experimental evaluation – process model extraction from text remains fundamental to make results reliable, the goal is to produce an executable script that can be deployed in a process engine. This makes it fundamental to also take into account the description of the composing tasks, in a way similar to what service composition does.

With the continuous growth of LLM capabilities each year, there is a growing interest in exploring the potential advantages that could be obtained for the entire BPM domain [4,9,25,26,47]. In a more narrow context than the one addressed by this paper, LLM-based chatbots raised interest in process mining and process modeling, i.e., the generation of new process models with event logs or unstructured text as input, respectively. In [7], for example, LLMs are applied for achieving an information abstraction of standard process mining artifacts such as traditional event logs, directly-follows graphs, or Petri nets. Authors in [5] propose the extraction of process elements and relations using prompts with different levels of pre-knowledge. In [21], the generation of an entire model using a specific level of abstraction is presented. Authors in [20] utilize the JSON format to improve LLMs capabilities to generate BPMN models, Entity-Relationship (ER), and UML Class Diagrams. The author in [19] uses a fine-tuned LLM for generating a recommendation for the next process element, i.e., its type and label in an unfinished BPMN process model.

Our approach can be considered an example of LLMs applied to software engineering [14,22], and in particular it is very much related, as already discussed, to automated code generation from text. As demonstrated in Sect. 4, though, our approach shows better results than plain code generation (considering Copilot as a baseline) by including the process control flow into the prompt.

Even though the goals are different, our work is also very much related to scientific workflow extraction using LLMs [43], especially for what concerns the focus on the data flow. Here authors start from an already defined scientific workflow and use LLMs to answer questions related to workflow comprehension, workflow modification, and workflow extension, without considering the possibility of extracting the full workflow from the textual description.

As one of the goals of the approach proposed in this paper is to extract an executable script for experiments and simulation, Business Process Simulation (BPS) models represent a related topic. As described in [13], a BPS model consists of a stochastic control-flow model, an activity performance model, and an arrival and congestion model. Authors in [32] explain how process mining can be used for BPS. Authors in [10], for example, extract a BPS model from data by applying deep learning techniques. These works are complementary to our approach as we do not use of event logs, which makes it impossible to extract the full BPS model. This consideration applies, more in general, to the whole research area that applies data mining, machine learning or deep learning to perform process discovery from event log data (see for example [42]).

6   Discussion and Future Works

In this paper, we introduce an approach based on the utilization of LLMs. Given natural language process descriptions and the available tools to be mapped to process tasks, our approach extracts an executable script that realizes the process for evaluation purposes and that can be easily deployed in an execution engine. We evaluate the approach both qualitatively, through questionnaires, and quantitatively by applying standard measures for assessing the quality of automatically generated code, including the data flow analysis. The evaluation demonstrates that our proposed multi-step approach overcomes the limitations of a direct application of LLMs by integrating external knowledge.

As discussed in the introduction, such an approach may be useful when new processes must be implemented suddenly without precise documentation or supporting data. This is exemplified in smart manufacturing where multiple actors (i.e., machines, humans, entire organizations) must be orchestrated on the fly to achieve agile manufacturing goals [11].

Due to space constraints, this paper does not delve deeply into the deployment phase within the PEE. However, our proposed pipeline does incorporate a crucial final step: converting the generated information (process script) into a deployable asset readable by a PEE, such as CPEE [31]. The generated process script may lack uniformity, as it can potentially utilize various existing libraries, resulting in different implementations achieving the same semantic. In contrast, PEEs rely on simpler syntactic programming languages, which limit semantic expressiveness, making them easy to analyze. Consequently, the challenge arises of transforming from a generic programming language into a highly specialized and limited programming language like BPMN. Currently, this transformation is facilitated by an additional LLM, but future works aim for native integration.

Despite the positive results, our approach should not be intended for fully automate use. Instead, the generated script can be inspected by a human expert and either manually modified or discarded, necessitating improvements to the description of the process or of the available tools.

Evaluation was performed by using GPT, with GPT-4 providing the best results. Although this model cannot be run locally, requiring the paid API by OpenAI, we can argue that *(i)* the number of calls required to run the technique on a single process description is limited, *(ii)* the approach is general enough to allow GPT-4 to be replaced with any current (e.g., LLAMA2) or future LLMs, and *(iii)* with the current trend of improvement in available hardware and open source networks, good performance will be achieved in the near future with offline LLMs.

Regarding evaluation, quantitative evaluation of the retrieval part is performed on a dataset synthetically generated by the LLM itself. While this is a common evaluation practice in the recent years, it may pose a threat to validity. For this reason, it is used in this paper only as confirmation of a corresponding qualitative evaluation performed by humans.

A potential limitation of the proposed pipeline is the tool repository. Although populating such a database does not inherently require a huge effort, the obtained results are sensitive to the quality of tool descriptions. Future work includes generating such descriptions directly from tool source code when available.

Finally, at the current stage, the approach is based solely on natural language description of the process and the available services (tools) to be employed for the tasks. As discussed in Sect. 5, approaches exist that apply deep learning to obtain all the components needed for a BPS model from an event log. Hybrid approaches could be employed where the execution and deployment scripts extracted through natural language descriptions are complemented with statistical information needed to drive the simulation.

Acknowledgments. The work of Flavia Monti was supported by the agreement on "Agile&Secure Digital Twins (A&S-DT)". The work of Francesco Leotta was partially supported by project SERICS (PE00000014) Extended Partnership under the PNRR

MUR program funded by the EU - NextGenerationEU. The work of Massimo Mecella is partially funded by MICS (Made in Italy—Circular and Sus-tainable) (PE00000004) Extended Partnership (CUP B53C22004130001) funded by the EU – NextGenerationEU PNRR MUR.

References

1. van der Aalst, W.M.P.: Process Mining - Data Science in Action, Second Edition. Springer, Berlin, Heidelberg (2016). https://doi.org/10.1007/978-3-662-49851-4_1
2. Achiam, J., et al.: GPT-4 technical report. arXiv preprint arXiv:2303.08774 (2023)
3. Aiello, M., Georgievski, I.: Service composition in the chatgpt era. SOCA (2023)
4. Beheshti, A., et al.: Processgpt: transforming business process management with generative artificial intelligence. arXiv preprint arXiv:2306.01771 (2023)
5. Bellan, P., Dragoni, M., Ghidini, C.: Extracting business process entities and relations from text using pre-trained language models and in-context learning. In: EDOC 2022. Springer (2022)
6. Berardi, D., Calvanese, D., De Giacomo, G., Lenzerini, M., Mecella, M.: Automatic service composition based on behavioral descriptions. Int. J. Coop. Inf. Syst. (2005)
7. Berti, A., Qafari, M.S.: Leveraging large language models (LLMs) for process mining (technical report). arXiv preprint arXiv:2307.12701 (2023)
8. Brown, T., et al.: Language models are few-shot learners. NeurIPS 2020 (2020)
9. Busch, K., Rochlitzer, A., Sola, D., Leopold, H.: Just tell me: prompt engineering in business process management. In: van der Aa, H., Bork, D., Proper, H.A., Schmidt, R. (eds.) Enterprise, Business-Process and Information Systems Modeling. BPMDS EMMSAD 2023 2023. LNBIP, vol. 479, pp. 3–11. Springer, Cham (2023). https://doi.org/10.1007/978-3-031-34241-7_1
10. Camargo, M., Báron, D., Dumas, M., González-Rojas, O.: Learning business process simulation models: a hybrid process mining and deep learning approach. Inf. Syst. (2023)
11. Catarci, T., Firmani, D., Leotta, F., Mandreoli, F., Mecella, M., Sapio, F.: A conceptual architecture and model for smart manufacturing relying on service-based digital twins. In: ICWS 2019. IEEE (2019)
12. Chang, Y., et al.: A survey on evaluation of large language models. ACM TIST (2023)
13. Chapela-Campa, D., Benchekroun, I., Baron, O., Dumas, M., Krass, D., Senderovich, A.: Can i trust my simulation model? Measuring the quality of business process simulation models. In: Di Francescomarino, C., Burattin, A., Janiesch, C., Sadiq, S. (eds.) Business Process Management. BPM 2023. LNCS, vol. 14159, pp. 20–37. Springer, Cham (2023). https://doi.org/10.1007/978-3-031-41620-0_2
14. Chirkova, N., Troshin, S.: Empirical study of transformers for source code. In: Proceedings of ESEC/FSE 2021 (2021)
15. Cuconasu, F., et al.: The power of noise: redefining retrieval for rag systems. arXiv preprint arXiv:2401.14887 (2024)
16. Dai, D., Sun, Y., Dong, L., Hao, Y., Sui, Z., Wei, F.: Why can GPT learn in-context? Language models secretly perform gradient descent as meta optimizers. arXiv preprint arXiv:2212.10559 (2022)
17. Devlin, J., Chang, M.W., Lee, K., Toutanova, K.: Bert: pre-training of deep bidirectional transformers for language understanding. arXiv preprint arXiv:1810.04805 (2018)

18. Ebert, C., Gallardo, G., Hernantes, J., Serrano, N.: Devops. IEEE software (2016)
19. Farkas, V.: Towards a machine learning-based approach for recommending next elements in BPMN models. In: NLP4BPM 2023 (2023)
20. Fill, H.G., Fettke, P., Köpke, J.: Conceptual modeling and large language models: impressions from first experiments with chatgpt. EMISAJ 2023 (2023)
21. Grohs, M., Abb, L., Elsayed, N., Rehse, J.R.: Large language models can accomplish business process management tasks. In: De Weerdt, J., Pufahl, L. (eds.) Business Process Management Workshops. BPM 2023. LNBIP, vol. 492, pp. 453–465. Springer, Cham (2024). https://doi.org/10.1007/978-3-031-50974-2_34
22. Hou, X., et al.: Large language models for software engineering: a systematic literature review. arXiv preprint arXiv:2308.10620 (2024)
23. Hsieh, C.Y., et al.: Tool documentation enables zero-shot tool-usage with large language models. arXiv preprint arXiv:2308.00675 (2023)
24. Huang, W., Abbeel, P., Pathak, D., Mordatch, I.: Language models as zero-shot planners: extracting actionable knowledge for embodied agents. In: ICML (2022)
25. Jessen, U., Sroka, M., Fahland, D.: Chit-chat or deep talk: prompt engineering for process mining. arXiv preprint arXiv:2307.09909 (2023)
26. Kampik, T., et al.: Large process models: business process management in the age of generative AI. arXiv preprint arXiv:2309.00900 (2023)
27. Klievtsova, N., Benzin, J.V., Kampik, T., Mangler, J., Rinderle-Ma, S.: Conversational process modelling: state of the art, applications, and implications in practice. In: BPM 2023 Forum (2023)
28. Klievtsova, N., Benzin, J.V., Kampik, T., Mangler, J., Rinderle-Ma, S.: Conversational process modeling: can generative AI empower domain experts in creating and redesigning process models? arXiv preprint arXiv:2304.11065 (2024)
29. van der Lee, C., Gatt, A., van Miltenburg, E., Krahmer, E.: Human evaluation of automatically generated text: current trends and best practice guidelines. Comput. Speech Lang. (2021)
30. Lewis, P., et al.: Retrieval-augmented generation for knowledge-intensive NLP tasks. NeurIPS 2020 (2020)
31. Mangler, J., Rinderle-Ma, S.: Cloud process execution engine: architecture and interfaces. arXiv preprint arXiv:2208.12214 (2022)
32. Martin, N., Depaire, B., Caris, A.: The use of process mining in business process simulation model construction: structuring the field. BISE (2016)
33. Mendling, J., Hafner, M.: From WS-CDL choreography to BPEL process orchestration. J. Enterp. Inf. Manag. (2008)
34. Mialon, G., et al.: Augmented language models: a survey. arXiv preprint arXiv:2302.07842 (2023)
35. Monti, F., Leotta, F., Mangler, J., Mecella, M., Rinderle-Ma, S.: NL2ProcessOps (2024). https://doi.org/10.5281/zenodo.11219809
36. Papineni, K., Roukos, S., Ward, T., Zhu, W.J.: Bleu: a method for automatic evaluation of machine translation. In: ACL 2002 (2002)
37. Patil, S.G., Zhang, T., Wang, X., Gonzalez, J.E.: Gorilla: large language model connected with massive apis. arXiv preprint arXiv:2305.15334 (2023)
38. Popkova, E.G., Ragulina, Y.V., Bogoviz, A.V. (eds.): Industry 4.0: Industrial Revolution of the 21st Century. SSDC, vol. 169. Springer, Cham (2019). https://doi.org/10.1007/978-3-319-94310-7
39. Ren, S., et al.: Codebleu: a method for automatic evaluation of code synthesis. arXiv preprint arXiv:2009.10297 (2020)
40. Sanderson, M., Zobel, J.: Information retrieval system evaluation: effort, sensitivity, and reliability. In: ACM SIGIR 2025 (2005)

41. Shanahan, M., McDonell, K., Reynolds, L.: Role play with large language models. Nature (2023)
42. Sommers, D., Menkovski, V., Fahland, D.: Supervised learning of process discovery techniques using graph neural networks. Inf. Syst. (2023)
43. Sänger, M., et al.: Large language models to the rescue: reducing the complexity in scientific workflow development using ChatGPT (2023)
44. Thalmann, S., et al.: Data analytics for industrial process improvement a vision paper. In: CBI 2018. IEEE (2018)
45. Touvron, H., et al.: Llama: open and efficient foundation language models. arXiv preprint arXiv:2302.13971 (2023)
46. Vaswani, A., et al.: Attention is all you need. NeurIPS 2017 (2017)
47. Vidgof, M., Bachhofner, S., Mendling, J.: Large language models for business process management: opportunities and challenges. In: Di Francescomarino, C., Burattin, A., Janiesch, C., Sadiq, S. (eds.) Business Process Management Forum. BPM 2023. LNBIP, vol. 490, pp. 107–123. Springer, Cham (2023). https://doi.org/10.1007/978-3-031-41623-1_7
48. White, J., et al.: A prompt pattern catalog to enhance prompt engineering with chatgpt. arXiv preprint arXiv:2302.11382 (2023)
49. Wu, T., et al.: Promptchainer: chaining large language model prompts through visual programming. In: CHI EA 2022 (2022)
50. Wu, T., Terry, M., Cai, C.J.: AI chains: transparent and controllable Human-AI interaction by chaining large language model prompts. In: CHI 2022 (2022)
51. Zhao, H., et al.: Explainability for large language models: a survey. ACM TIST (2023)

CMMN-Based Modeling and Customization of Declarative Business Process Families

Felipe Castellanos, Nicolás Navascués, Daniel Calegari[(✉)][iD],
and Andrea Delgado[iD]

Instituto de Computación, Facultad de Ingeniería, Universidad de la República,
Montevideo 11300, Uruguay
{felipe.castellanos,nicolas.navascues,dcalegar,adelgado}@fing.edu.uy

Abstract. A process family represents a set of processes with common aspects and variable parts based on an organization's specific business requirements. It can be expressed using concrete modeling languages and then customized at design time into a specific process variant to enact according to particular configuration requirements. Most process family modeling proposals focus on imperative processes, such as extensions of the Business Process Model and Notation (BPMN) standard. Some proposals also focus on declarative languages that express more flexible process families. However, no proposal focuses on the Case Management Model and Notation (CMMN) standard, a declarative proposal that complements BPMN for expressing hybrid processes. In this article, we study declarative process family languages, and we introduce CMM-Next, an approach for its modeling and automatic customization based on CMMN. We also present a supporting tool and compare CMMNext's potential concerning other declarative proposals.

Keywords: process family · variability · CMMN · declarative processes

1 Introduction

Process-Aware Information Systems (PAIS) [9] cover a broad spectrum of business processes, from entirely predictable and highly repetitive processes, such as product sales, to unpredictable and non-repetitive processes, such as innovation.

When handling predictable processes, it is possible to pre-specify the control flow of activities in business process models. Indeed, Business Process Management Systems (BPMS) [3] traditionally support this kind of process, for which imperative languages such as BPMN [12] are used. On the contrary, unpredictable processes introduce looseness as a flexibility requirement [16] since the

Partially supported by project "Ingeniería dirigida por modelos para la especificación y configuración de familias de procesos" funded by Comisión Sectorial de Investigación Científica (CSIC), Proyecto I+D 2022 "22520220100018UD", Uruguay.

course of actions is unknown primarily a priori. In this case, it is usually chosen to declare the execution restrictions between the activities (declarative process) instead of defining all possible execution paths (imperative process). To specify this type of unstructured process, declarative languages are used, such as Declare [14] and the Case Management Model and Notation (CMMN) [13].

Process design has a second flexibility requirement of interest in both cases: *variability* [16]. It arises when several processes share common elements while others differ depending on the context, e.g., sales for different products or markets. It leads to specific process variants, which are members of a so-called *business process family* (BP family) [17]. Although there is a common *base process*, the specific course of action differs depending on which specific *variants* are selected for the many *variation points* defined for the BP family. A customization process (manual or automatic) must be carried out to derive *process variants* from the family process model, selecting concrete variants for each variant point according to configuration parameters or context information.

The proposals for modeling process families have focused mainly on imperative processes [8,17], such as BPMNext [7]. As expressed in [20], variability settings for declarative languages differ from those for imperative languages. A variation point in an imperative language potentially extends/diversifies the behavior allowed by the base process. However, since a declarative language is based on an open-world assumption in which everything not explicitly forbidden is automatically allowed, a variation point in a declarative language potentially restricts the behavior allowed by the base process.

There are also proposals concerning declarative process families, such as BPFM [6], DECO [19], and Configurable Declare [20]. They all propose different process variability approaches and support of the process families' life cycle. These approaches can be systematically compared using the VIVACE framework [2], as done for BPMN-based approaches in [8]. As far as we know, no process family approach focuses on the CMMN standard. Moreover, although there are works about hybrid business process representations [1], no process family approaches connect the imperative and the declarative paradigms for the specification of hybrid business process families. In this sense, since CMMN is a standard declarative proposal defined to be consistent with and complementary to BPMN, it could be possible to connect existent BPMN-based business process families approaches [8] with a CMMN-based business process family approach.

This article introduces CMMNext, a CMMN-based approach for modeling and customizing declarative process families. This approach follows the same strategy for the BPMNext approach used in [7] for BPMN, which makes it compatible with specifying hybrid process families. We also present a supporting tool for defining and customizing CMMNext process variants [5]. Finally, we compare the different declarative process family approaches using the VIVACE framework to analyze their strengths and weaknesses.

The rest of the paper is structured as follows. In Sect. 2, we review declarative process family languages. Then, in Sect. 3, we briefly introduce CMMN, and in Sect. 4, we define CMMNext. In Sect. 5, we provide an example of use as an initial validation of the approach. In Sect. 6, we compare the existing approaches. In

Sect. 7, we discuss several aspects. Finally, in Sect. 8, we provide conclusions and an outline of future work.

2   Declarative Process Families

We conducted a literature review, based on [11], to answer the following research question: **which approaches exist to deal with declarative process families?** We identified candidate primary studies by searching within electronic databases using the following search string:

```
declar* AND ("business process family" OR "business process line"
             OR "configurable process model")
```

We first queried within titles, abstracts, and keywords. As summarized in Table 1, the search retrieved 80 papers from the following databases: Springer Link, ACM Digital Library, IEEE Xplore, ScienceDirect, and Wiley Online Library.

Table 1. Summary of papers

Source	Found	Relevant	Candidate	Primary
Springer Link	60	18	6	3
ACM Digital Library	4	2	0	0
IEEE Xplore	1	1	1	0
ScienceDirect	13	1	1	0
Wiley Online Library	3	1	0	0

We filtered the papers by considering inclusion and exclusion criteria, selecting 23 relevant papers. As inclusion criteria, we considered papers mentioning related terms about declarative process families, such as declarative, variability, configurable, family, and flexible. As exclusion criteria, we considered (a) papers not electronically available on the web, (b) not written in English, and (c) non-peer-reviewed publications, such as technical reports, books, book chapters, proceedings' prefaces, and journal editorials.

We refined the selection by reading their title, abstract, keywords, and introduction, if necessary, to decide. We selected eight candidate papers. Finally, we fully read the papers, finding three primary studies that describe three different approaches. The remaining five articles either referred to these approaches or made contributions of a theoretical nature that did not significantly contribute to the objective of this review of identifying works that introduce an approach.

We also reinforced the results by snowballing, i.e., tracking related works using the selected papers' bibliography and searching for authors' and conferences' web pages and the DBLP database. Finally, we referred to existent literature reviews about business process variability modeling, e.g., [17], and Hybrid

Business Process Representation (HBPR) [1] to strengthen the results by targeting hybrid approaches. However, no new process family approach was identified.

The papers considered are listed in a supplemental spreadsheet at [5].

2.1   Existing Approaches

In [6], the authors introduce the case-based process fragment modeling language called **Business Process Feature Model (BPFM)** that supports manual generation and refinement of generalized cases. The modeling language extends feature models whose features are BPMN elements. Each internal element represents a subprocess that can be refined, and leaves represent atomic tasks. As with feature models, constraints express variability (e.g., mandatory, optional, and alternative tasks to be selected in the process variant). As evaluated in [8], BPFM is supported by a tool that allows modeling, configuring, and automatically customizing process fragments (not a whole process variant).

In [20], the authors present **Configurable Declare**, an extension of the well-known declarative process language Declare, which is based on LTL formulas evaluated on traces of events. The extension allows the Declare model to express when an event (i.e., a process task) could be hidden or a constraint omitted. A configuration requires expressing whether hiding happens in a concrete process variant. Since events are associated with constraints, their hiding usually involves the generation of new constraints to keep consistency, which must be generated during customization. Moreover, certain constraints cannot be removed, so there are meta-rules that prevent this. The authors describe a (unavailable) supporting tool for modeling, configuring, and automatically customizing a process variant.

Finally, in [19], the authors present **Declarative Configurable specifications (DeCo)**, a BPMN-extension providing a mechanism for descriptive process modeling, formal analysis, and stepwise refinement from a configurable design to concrete process execution. Although based on BPMN, it keeps a declarative nature based on the Configurable Integrated EPC [18] notation and a semantical definition based on first-order logic. The language defines the notion of state concerning values of data objects related to a case at a given moment and pre/post conditions that define when a state is enabled and what the resulting data state is when performed, respectively. Tasks of a process can be associated with a set of states and enabled at runtime if pre-conditions are satisfied. As evaluated in [8], DeCo defines new elements to the BPMN notation, allowing tasks, roles, and data to be configured as variation points and connect with their variants in the same model (they must be elements of the same kind). The authors do not provide a tool or information on extending BPMN.

3   Case Management Model and Notation (CMMN)

CMMN [13] allows declarative modeling of case-based unstructured processes without a pre-defined mandatory control flow, as in BPMN. Figure 1 depicts elements of the CMMN notation (Fig. 1a) and an excerpt of their corresponding concepts in the metamodel (Fig. 1b).

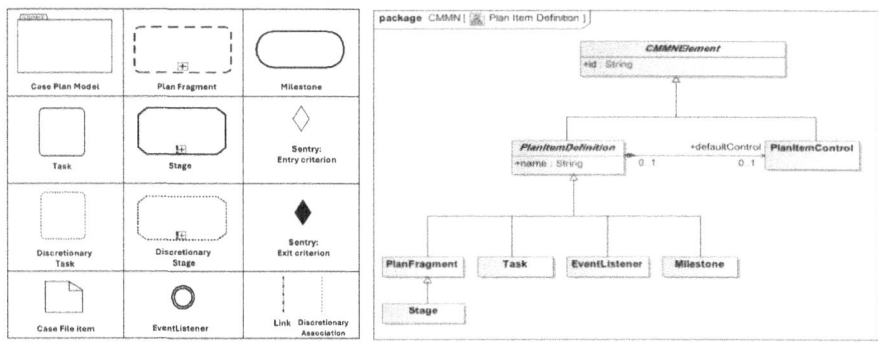

(a) CMMN notation elements   (b) CMMN metamodel

Fig. 1. CMMN notation elements and corresponding metamodel from [13]

A Case Plan Model defines the process, comprising different types of plan items: i) Task that can be of type human, process or case, defining the work to be done in the process; ii) Event Listener that can be of type user or timer and represent events that can occur during the case; iii) Milestone that is a target to be achieved within the execution of the case; and iv) Plan Fragment, specialized as Stage, representing phases in the case, the first with no execution behavior and the second similar to sub-processes in BPMN. Tasks and stages can be Discretionary, allowing the workers to execute them optionally. A condition can be added to elements as a Sentry: as an entry criterion of tasks, milestones, and stages to enable their execution and as an exit criterion to tasks, stages, and case plans to stop the work. Each case instance contains a single case file where all data is managed and stored as Case file items. Connectors between elements can be Link or Discretionary Association.

Figure 2 depicts a simple claim management example, where three milestones must be achieved based on three corresponding stages for identifying responsibilities, attaching base information, and processing the claim. The first stage depends on completing the "Identify Responsibilities" process, which is required to complete or terminate (! sign) before the stage can be completed. The following milestone can optionally perform a human task to change the responsibilities, which can be repeated (# sign). Once the milestone is completed, the second stage can be started, particularly with the user task "Create Claim Notification" enabled by the entry sentry connected from the milestone. It is also mandatory at such a stage. After the "Base Information Attached" milestone is completed, a final stage allows the creation of a claim. Isolated tasks can be performed to review documents and create letters. Finally, human event listeners allow the process to be ended by exiting it or through the "Claim Processed" milestone.

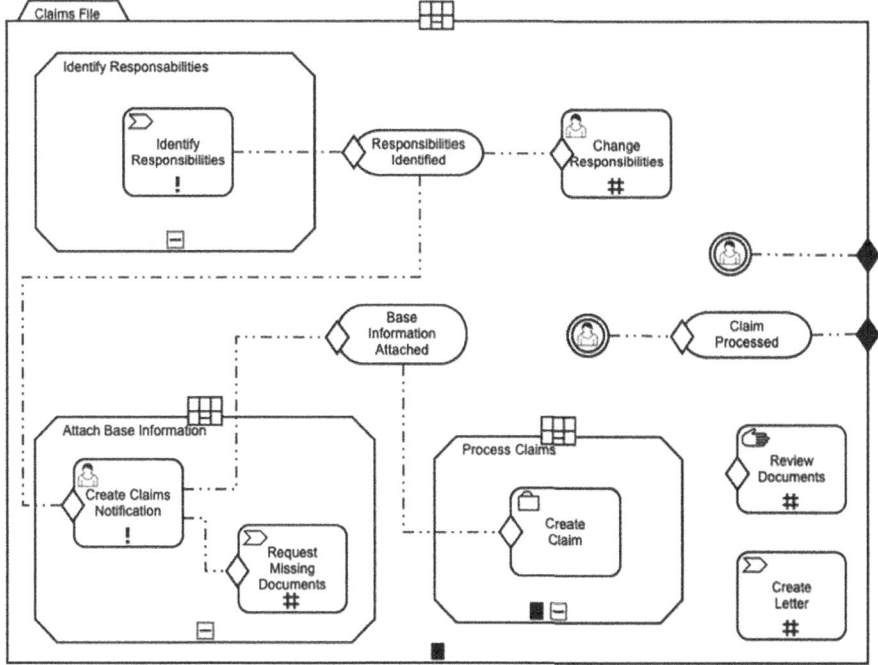

Fig. 2. Claims management example (from [13])

4   CMMNext Variability Approach

We defined a conservative extension of the CMMN language to allow compatibility with the base language, as with BPMN, and potentially with their process family approaches to express hybrid process families. CMMNext follows a model-driven approach [10] based on metamodels, models, and transformations between them to automate the support for managing declarative process families. The elements that comprise our approach are described below (available at [5]) and presented in the next sub-sections:

1. The CMMNext language is defined as an extension of the CMMN metamodel and its corresponding graphical notation.
2. The cmmn-js web tool[1] is extended to support the specification of CMMNext process families.
3. A configuration metamodel is integrated to support defining configuration models for specific variants.
4. A model-to-model (M2M) transformation is defined for taking a CMMNext process family and a configuration model and customizing a concrete CMMN process variant.

---

[1] CMMN web modeler, https://bpmn.io/toolkit/cmmn-js/.

4.1 The CMMNext Language

We followed the conceptual approach for modeling structured business process families with the BPMN language in [7]. We selected this approach over others as presented in [8], due to its focus on the BPMN metamodel, available at [12], which is extended with few elements for variability (variants and variation points), providing the basis for defining a model-driven approach to support the complete variability process. In this sense, the CMMNext extension becomes compatible with the former BPMNext proposal.

Figure 3 depicts the extension of the CMMN metamodel we defined, where elements in the center (blue) represent the original *CMMN metamodel* concepts. On the left side (pink), the VPoint hierarchy represents variation points in the base case (i.e., the base process). Moreover, on the right side (violet), the Variant hierarchy represents specific variant elements that can replace variation points when the customization process is performed for a concrete process variant. We defined specialized concepts for considering tasks, milestones, and stages as variation points (i.e., elements VPx) and their corresponding variants (i.e., elements Vx) that connect with the former elements.

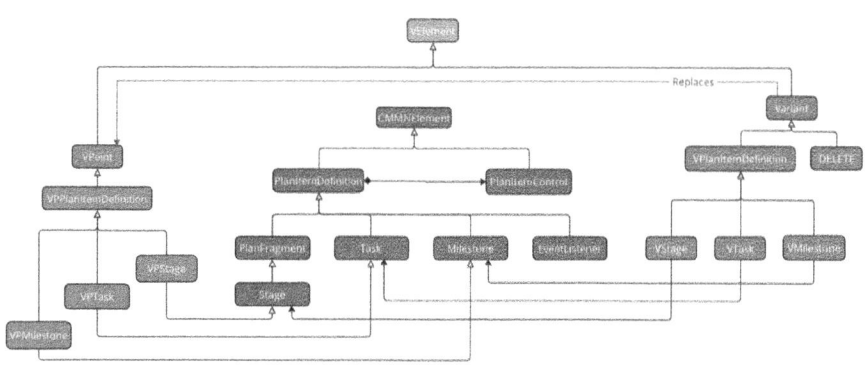

Fig. 3. Excerpt of the CMMNext metamodel extending CMMN metamodel (Color figure online)

We allow the replacement of a variation point with a variant of the same type (e.g., replacing a VPTask with a VTask within the many options that can be defined) and substitute tasks and stages with each other (e.g., a VPTask with a VStage). The latter provides more flexibility, tailored to the required complexity for representing that part of the process. Replacing VPMilestone with elements other than VMilestone is impossible. This decision stems from the understanding that, given the significance milestones hold in CMMN, it would not be meaningful to model such a variation point that a task or a stage could replace. We also allow the removal of a variation point using the variant DELETE, which also implies removing its associated items, e.g., entry/exit sentries.

We defined a graphical notation for each element based on adding decorators to the CMMN original shapes: a diamond on the left upper corner for variation points and a circle on the left upper corner for variants, as shown in Fig. 4a. An example of a variant definition for a task variation point with an entry sentry is shown in Fig. 4b. In this case, we defined sentries as not being part of variants but only of variation points and as being inherited when the variant occupies the variation point to maintain consistency. As shown, the first variant with the corresponding circle on the left corner will be valid (marked with a green tick), and the other two (marked with a red cross) will not be valid since they each also present an entry and an exit sentry, respectively.

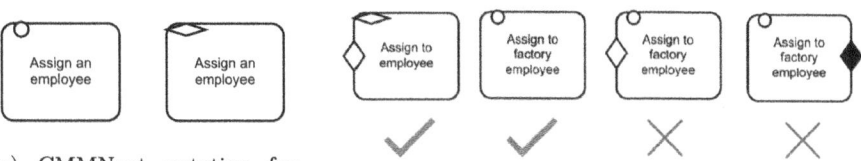

(a) CMMNext notation for variants (left) and variation points (right)

(b) Sentries are allowed in variations points, but not in variants

Fig. 4. CMMNext notation elements and variant definition examples

We have customized the `cmmn-js` web tool for specifying CMMNext models. An output model is conformed by a base process with their variation points and a set of variants that can replace the variation points plus the `DELETE` variant. We have also developed conversion algorithms back and forth from `.cmmn` files, used by the `cmmn-js` tool, to `.xmi` files, compliant with the CMMNext model and used for the customization model transformation.

4.2 Configuration Metamodel and Models

We defined a configuration metamodel, depicted in Fig. 5a, using Ecore[2] It allows specifying configurations for process variants. In such a model, `VarPoints` are occupied by specific valid `Variant` elements. Both elements refer to the corresponding variation points and variants in a CMMNext process family definition. Figure 5b depicts an example of a configuration model using Ecore tools[3]

4.3 CMMNext to CMMN Model Transformation

We define an M2M transformation from a CMMNext family model to a CMMN model corresponding to the configured process variant. We used the ATL language[4] The core aspect of the M2M transformation involves carrying out the necessary substitutions of the variation points with their corresponding variants

---

[2] Eclipse Modeling Framework (EMF), https://eclipse.dev/modeling/emf/.

[3] Ecore tools, https://eclipse.dev/ecoretools/.

[4] ATL, https://eclipse.dev/atl/.

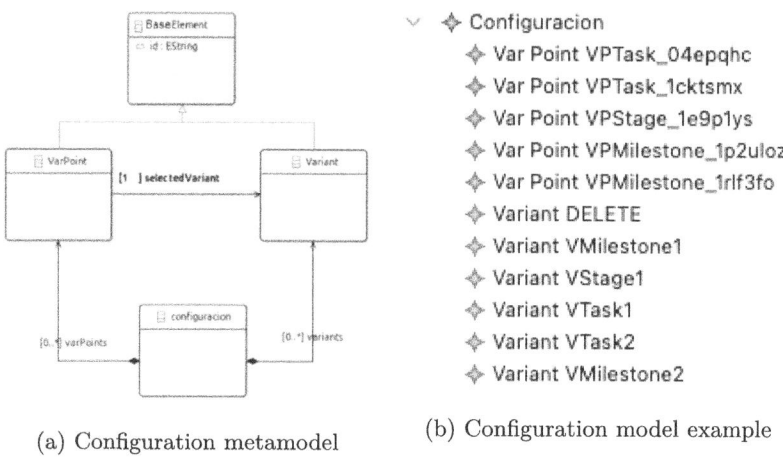

(a) Configuration metamodel   (b) Configuration model example

Fig. 5. Configuration metamodel and an example of a configuration model

(or deletion), considering the configuration model, and respecting the replacement rules already described in Sect. 4.1. The transformation excludes all other variants not used during customization, resulting in a CMMN model.

5   Example Application of CMMNext

To validate the approach, we analyzed the three existing approaches for declarative process families (BPFM, Configurable Declare, and DeCo) and represented in CMMNext the examples in the papers presenting such approaches. Due to space restrictions, we provide the example based on Configurable Declare.

Figure 6 depicts the application of Configurable Declare for specifying the process for requesting an excerpt from the civil registration in a Dutch municipality. The model shows events as rectangles, e.g., filling out a form, producing the excerpt, and signing it. The model also indicates behavioral constraints, e.g., the constraint init shows that the first task in the process must be "Fill in e-form", and the connecting arrow between "Produce extract and sign it" and "Archive" indicates that whenever the excerpt is produced and signed, it must eventually be archived later. Crosses represent variation points that must be kept or deleted during customization. For example, in the case of "Municipality B", it is unnecessary to indicate that the excerpt has already been paid or to archive it. The final Declare model, after customization, is depicted in Fig. 7.

We have analyzed how Declare could be represented using CMMN to represent this example. We defined the CMMNext process family using the modeling tool [5]. The base model is depicted in Fig. 8. We also describe the corresponding variants, shown in Fig. 9a (possibly multiple variants apply for a given variation

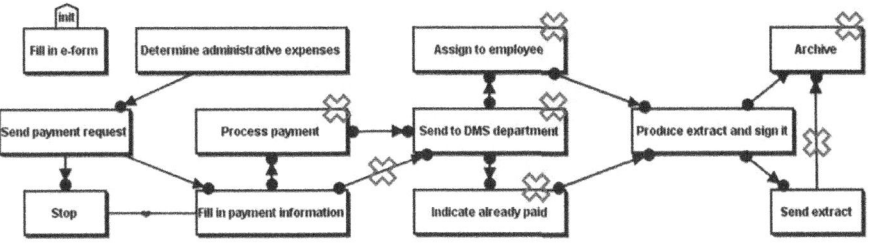

Fig. 6. Configurable Declare model for Dutch municipalities (from [20])

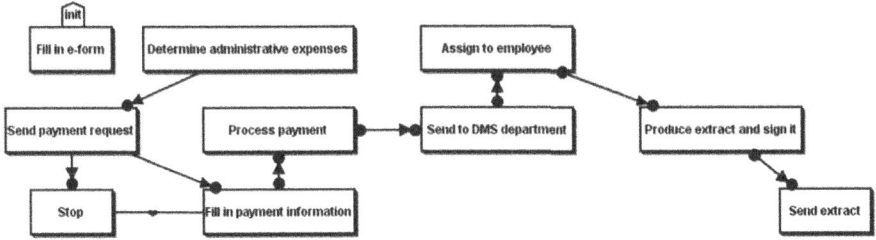

Fig. 7. Declare model for Municipality B (from [20])

point). Due to the nature of CMMN, it was not possible to represent the variability of the relationships between tasks, so only VPTasks were modeled, which can either be maintained (being replaced by a VTask), or be eliminated (by choosing DELETE as the variant).

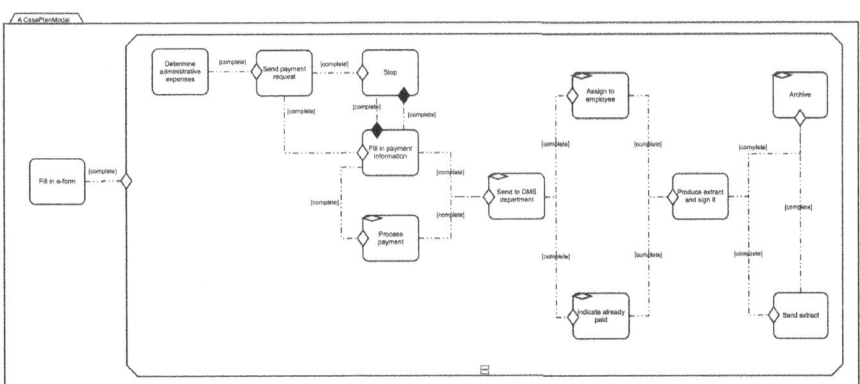

Fig. 8. Representation of the process family in the Configurable Declare article with CMMNext

A configuration model, depicted in Fig. 9a, represents the specific instance for Municipality B. The following configuration decisions were made:

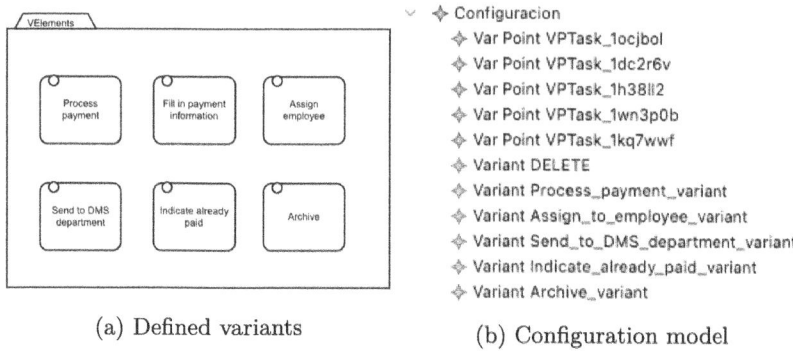

(a) Defined variants

(b) Configuration model

Fig. 9. Variants and configuration model for Municipality B

- VPTask **Process payment** is replaced by the variant
  Process_payment_variant.
- VPTask **Send to DMS department** is replaced by the variant
  Send_to_DMS_department_variant
- VPTask **Assign_to_employee** is replaced by the variant
  Assign_to_employee_variant
- VPTask **Indicate already paid** is replaced by the variant **DELETE**
- VPTask **Archive** is replaced by the variant **DELETE**

Finally, the M2M transformation was executed, resulting in a CMMN model shown in Fig. 10, which conforms to the CMMN metamodel.

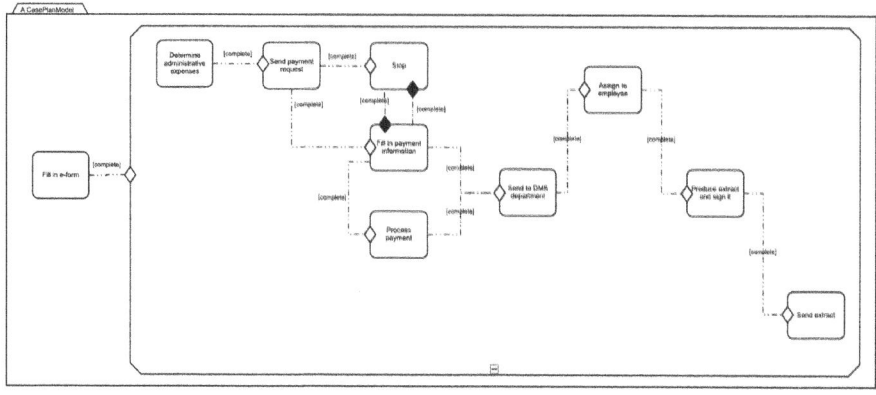

Fig. 10. CMMN process variant for Municipality B

6 Comparisson of Approaches

The VIVACE framework [2] allows assessing and comparing process variability approaches. It was refined in [8] for a more specific evaluation of BPMN-based proposals. It defines several evaluation features, depicted in Table 2.

Table 2. Refined VIVACE framework (from [8])

VIVACE		
(a) Modeling language used to represent process variability: extension of BPMN, BPMN-inspired, or language-independent		
(b) Technique used for building the process family		
(c) Method for modeling the process family: Node Configuration, Element Annotation, Activity Specialization, or Fragment Customization		
(d) Process perspectives covered		
Variability-specific language constructs	LC1 Configurable Region	
	LC2 Configuration Alternative	
	LC3 Configuration Context Condition	
	LC4 Configuration Constraint	
	LC5 Configurable Region Resolution Time	
Variability support features	**Analysis & Design phase**	
	F1.1 Modeling a configurable process model	
	F1.2 Verifying a configurable process model and its related process family	
	F1.3 Validating a configurable process model	
	F1.4 Evaluating the similarity of different process variants	
	F1.5 Merging process variants	
	Configuration phase	
	F2 Configuring specific regions of a process variant out of a configurable process model	
	Enactment phase	
	F3.1 Configuring specific regions of a process variant at enactment time	
	F3.2 Dynamically re-configuring an instance of a process variant at enactment time	
	Diagnosis	
	F4 Analyzing a collection of process variants	
	Evolution	
	F5.1 Versioning of configurable process model	
	F5.2 Propagating changes of a configurable process model to already configured process variants	
(e) Tool implementation: support not available (N/A); support for modeling (M), configuring (C), deriving variants (D)		
(f) Empirical evaluation		
(g) Application domain		

It considers general features of interest: (a) the modeling language, (b) the technique (if single or multiple artifacts are involved), (c) the method (variability mechanism) for expressing the BP family, (d) the process perspectives covered (functional, behavioral, organizational, informational, temporal, and operational), (e) its tool support (for modeling, configuring, and deriving), (f) the existent empirical evaluation, and (g) the application domain used. Concerning the variability mechanism, there are four options, i.e., *Node Configuration* where elements can be modified for any of their multiple possible variants, *Element Annotation* where an element is connected to a predicate over a domain model, *Activity Specialization* where activities in the base model, and not of other types of elements, can be replaced by others, and *Fragment Customization* when complete process fragments can be modified. Moreover, it considers if languages support specific features about language constructs, i.e., (LC1) configurable regions and (LC2) alternatives, i.e., variation points and a variant, (LC3) context conditions, (LC4) constraints, and (LC5) if customization can be resolved during process execution. Finally, it considers variability support features throughout the process life cycle (F1 to F5). For both language constructs and support features, VIVACE defines a scale for the support provided: no support [–], partial support [+/–], and full support [+].

BPFM and DeCo were evaluated in [8]. For completeness, we added the evaluation of Configurable Declare and CMMNext, as summarized in Table 3.

Table 3. VIVACE-based evaluation of declarative approaches

VIVACE		BPFM	CDeclare	CMMNext	DeCo
Language		BPMN-insp	Declare ext	CMMN ext	BPMN ext
Technique		single-artifact	single-artifact	single-artifact	single-artifact
Mechanism		Act. Special.	Node Conf.	Node Conf.	Node Conf.
Perspectives		F, B, I, Op	F, B, T	F, B, T	F, I, O
Variability-specific language constructs	LC1	+	+	+	+
	LC2	+	+	+	+
	LC3	-	+	-	-
	LC4	+	+	+	+
	LC5	-	-	-	-
Variability support features	**Analysis & Design phase**				
	F1.1	+	+	+	-
	F1.2	+	+	-	-
	F1.3	-	-	-	-
	F1.4	-	-	-	-
	F1.5	-	-	-	-
	Configuration phase				
	F2	+	+	+	-
	Enactment phase				
	F3.1	-	-	-	-
	F3.2	-	-	-	-
	Diagnosis				
	F4	-	-	-	-
	Evolution				
	F5.1	-	-	-	-
	F5.2	-	-	-	-
Tool		+ (M, C, D)	-	+ (M, C, D)	-
Evaluation		-	-	-	-
Domain		government	government	government	banking

6.1   Applying VIVACE to Configurable Declare

It uses a single-artifact technique with a node configuration mechanism. It considers the functional (F), behavioral (B), and temporal (T) perspectives since it allows varying tasks and temporal relations between them. Its configuration consists of selecting or deleting the variation points. Such configuration can be verified according to well-formed rules before deriving a process variant. The customization mechanism assures that the resulting Declare model is well-formed and consistent. The approach is supposed to be supported by an unavailable tool. The process shown in Sect. 5 illustrated the approach.

Concerning the variability-specific language constructs, it supports representing variation points (LC1 [+]) and variants (LC2 [+]) (existence or deletion of a variation point). Moreover, based on Declare, it allows defining constraints between tasks (LC4 [+]). The language also supports the definition of context conditions (LC3 [+]) as meta-constraints, i.e., logical expressions over configuration choices such as "if event A is hidden, then event B is not hidden".

Finally, concerning variability support features, although the tool is unavailable, the authors describe how a BP family can be graphically defined (F1.1[+]) and verified (F1.2 [+]) and a process variant automatically derived (F2 [+]).

6.2   Applying VIVACE to CMMNext

It uses a single-artifact technique with a node configuration mechanism. It considers the functional (F), behavioral (B), and temporal (T) perspectives since it allows for varying tasks, milestones, and stages and their precedences. Its configuration consists of replacing or deleting the variation points. The configuration needs to be consistent before customization. The approach is supported by a tool that allows modeling the BP family, defining a concrete configuration, and automatically producing a process variant (M, C, D). The process shown in Sect. 5, and other examples, illustrated the approach.

Concerning the variability-specific language constructs, it supports representing variation points (LC1 [+]), variants (LC2 [+]), and constraints between elements (LC4 [+]). However, the language does not support the definition of context conditions (LC3 [−]) or configurable region resolution time (LC5 [−]).

Finally, concerning variability support features, the toolchain allows graphically specifying a BP family (F1.1[+]) and customizing a process variant (F2 [+]). Other aspects, such as specification verification, still need to be developed.

7   Discussion

The approaches present a similar basis for representing declarative process families, focused on how modeling elements (primarily tasks) can be replaced by specific variants or suppressed. However, they present differences in their concrete technique, which strongly depends on their base modeling languages. The four cases are based on different settings, i.e., BPFM is based on feature models,

Configurable Declare in Declare, CMMNext in CMMN, and DeCo in BPMN. The **cognitive complexity** of modeling and representing a process family is a critical aspect when defining a new modeling language. Previous declarative approaches lack an evaluation of the cognitive complexity involved in the process family's modeling and configuration tasks. We left the comparative analysis with other approaches concerning their cognitive complexity for future studies, as existing experiments on imperative and declarative base languages (with no variability) have done, such as [15].

Nevertheless, as an initial evaluation, we conducted a brief empirical study in the context of the "Latin American School on Process Variability and Process Mining"[5] within 13 postgraduate students and researchers from Uruguay(9), Chile(3) and Costa Rica(1). We introduced the BPMNext and CMMNext languages and conducted a two-hour hands-on laboratory. We provided students with a text description of a business process family and asked them to model it using both languages. After completing the task, we asked them two questions (Likert scale 1-little to 5-much) to evaluate the variability approach's understandability and each family modeling language. We also asked them about their previous knowledge. Table 4 summarizes the answers.

Table 4. Answers collected from the empirical study

Criteria	1	2	3	4	5	min	max	mean	med	mod	std
BPMN knowledge	2	3	1	5	2	1	5	3,153	4	4	1,41
BPMNext variability approach	1	0	2	4	6	1	5	4,076	4	5	1,19
BPMNext modeling	1	0	3	5	4	1	5	3,846	4	4	1,14
CMMN knowledge	10	2	0	1	0	1	4	1,385	1	1	0,87
CMMNext variability approach	1	0	1	11	0	1	4	3,692	4	4	0,85
CMMNext modeling	1	0	4	8	0	1	4	3,462	4	4	0,88

Results showed that although attendants had more knowledge on BPMN (five marked 4 and two marked 5 as answer) than on CMMN (ten marked 1 as answer), the variability approach and the family modeling language are both evaluated primarily within 4 and 5 values in both cases with median 4. It could be because the same conceptual variability approach is used in both languages. Also, the tool support we provide for CMMNext is web-based and has an easier-to-understand interface than the one we provide for BPMNext, an extension of the Eclipse BPMN 2.0 modeler. As this is an initial evaluation, explicit elements of the family modeling tasks carried out were not assessed in detail (i.e., the variability and resulting models), and the low number of participants constitute limitations on the results.

VIVACE provides a high-level evaluation framework without deepening a critical and in-depth comparative analysis of the approaches. In particular, it leaves aside the **expressiveness** of the languages, i.e., how much they could

---

[5] https://www.fing.edu.uy/inco/eventos/bpmuy/.

express. The languages differ in the perspective from which a base process is defined. For example, Configurable Declare focuses primarily on the relationships between tasks, while CMMNext, BPFM, and DeCo focus more on the tasks themselves, not directly representing complex time constraints between them. As done with Declare, it is possible to find correspondences between the four primary languages. However, our first studies have shown us that carrying out a complete mapping, e.g., between Declare and CMMN, may not be possible, but that does not mean that a process would no longer be able to be represented by one or another notation. Further research is needed to study this aspect.

The proposal is a valuable step toward providing a tool integrating variability management for hybrid BPMNext and CMMNext models. Nevertheless, many **limitations** exist up to this point. First, we must study the co-variability of both parts of a hybrid model from semantic and pragmatic perspectives. Second, since CMMN is only partially considered, other CMMN elements could be extended with their variable counterparts, e.g., events and discretionary tasks. It also requires studying the pragmatics of having such variability language constructs.

Finally, a primary limitation is about **consistency checking**. Except for DeCo, the other approaches describe how a process variant can be automatically derived from a specific configuration. However, only BPFM and Configurable Declare pay special attention to consistency: BPFM allows for the verification of a configuration concerning well-formed rules before deriving a process variant. Configurable Declare checks whether constraints can be omitted by observing the meta-constraints. In the case of CMMNext, the consistency must be ensured by definition. We observe two challenges in this sense. First, an extension of the CMMNext language requires defining the correspondences between variation points and variants, which could lead to inconsistent specifications, e.g., is it reasonable to select a task in the place of an event variation point? Second, although a specification could be consistent, we must address the correctness of the variants generated through the model transformation process.

8   Conclusions

In this article, we explored declarative process family languages by performing a literature review and defining CMMNext, an approach for modeling and automatic customization. We exemplified its use and compared it with existing approaches as a first validation. However, further work is needed to evaluate its potential more deeply. Also, concerning the CMMN language used as a basis, we could extend CMMNext to consider the variability of other elements, such as a `Case File`, as well as considering constraints that must be checked before customization to ensure consistency of the resulting process variant.

We provided a toolchain that uses model-driven engineering to provide a conservative extension of the CMMN language and define model transformations to customize a process variant. Future work is needed to put the toolchain into an integrated environment, such as BPFM [4], allowing the complete flow to be carried out, from the specification of the family and its variants in a graphical environment, determining a desired configuration, automatically applying the successive transformations, and displaying the final result in a user-friendly way.

Since CMMN is compatible with the BPMN standard, CMMNext could be compatible with many BPMN-based proposals for specifying imperative process families. In this sense, further work is needed to study its compatibility with the specification of hybrid process families.

References

1. Andaloussi, A., Burattin, A., Slaats, T., Kindler, E., Weber, B.: On the declarative paradigm in hybrid business process representations: a conceptual framework and a systematic literature study. Inf. Syst. **91**, 101505 (2020)
2. Ayora, C., Torres, V., Weber, B., Reichert, M., Pelechano, V.: VIVACE: a framework for the systematic evaluation of variability support in process-aware information systems. Inf. Softw. Technol. **57**, 248–276 (2015)
3. Chang, J.: Business Process Management Systems: Strategy and Implementation. CRC Press, Boca Raton (2016)
4. COAL Group: Business Process Family Manager (BPFM). https://gitlab.fing.edu.uy/open-coal/bpfm
5. COAL Group: CMMNext. https://gitlab.fing.edu.uy/open-coal/cmmnext
6. Cognini, R., Hinkelmann, K., Martin, A.: A case modelling language for process variant management in case-based reasoning. In: Reichert, M., Reijers, H.A. (eds.) BPM 2015. LNBIP, vol. 256, pp. 30–42. Springer, Cham (2016). https://doi.org/10.1007/978-3-319-42887-1_3
7. Delgado, A., Calegari, D.: BPMN 2.0 based modeling and customization of variants in business process families. In: XLIII Latin American Computer Conference, CLEI, pp. 1–9. IEEE (2017)
8. Delgado, A., Calegari, D., García, F., Weber, B.: Model-driven management of BPMN-based business process families. Softw. Syst. Model. **21**(6), 2517–2553 (2022)
9. Dumas, M., van der Aalst, W.M.P., ter Hofstede, A.H.M. (eds.): Process-Aware Information Systems: Bridging People and Software Through Process Technology. Wiley, Hoboken (2005)
10. Kent, S.: Model driven engineering. In: Butler, M., Petre, L., Sere, K. (eds.) IFM 2002. LNCS, vol. 2335, pp. 286–298. Springer, Heidelberg (2002). https://doi.org/10.1007/3-540-47884-1_16
11. Kitchenham, B.: Procedures for performing systematic reviews. Technical report tr/se-0401, Keele University (2004)
12. OMG: Business Process Model and notation (BPMN) v2.0. Technical report, Object Management Group (OMG) (2013)
13. OMG: Case Management Model and Notation (CMMN) Version 1.1. Technical report, Object Management Group (OMG) (2016)
14. Pesic, M., Schonenberg, H., van der Aalst, W.M.P.: DECLARE: full support for loosely-structured processes. In: 11th IEEE International Enterprise Distributed Object Computing Conference, pp. 287–300. IEEE Computer Society (2007)
15. Pichler, P., Weber, B., Zugal, S., Pinggera, J., Mendling, J., Reijers, H.A.: Imperative versus declarative process modeling languages: an empirical investigation. In: Daniel, F., Barkaoui, K., Dustdar, S. (eds.) BPM 2011. LNBIP, vol. 99, pp. 383–394. Springer, Heidelberg (2012). https://doi.org/10.1007/978-3-642-28108-2_37

16. Reichert, M., Weber, B.: Enabling Flexibility in Process-Aware Information Systems - Challenges, Methods, Technologies. Springer, Berlin, Heidelberg (2012). https://doi.org/10.1007/978-3-642-30409-5
17. Rosa, M.L., van der Aalst, W.M.P., Dumas, M., Milani, F.: Business process variability modeling: a survey. ACM Comput. Surv. **50**(1), 2:1–2:45 (2017)
18. Rosemann, M., van der Aalst, W.M.P.: A configurable reference modelling language. Inf. Syst. **32**(1), 1–23 (2007)
19. Rychkova, I., Nurcan, S.: Towards adaptability and control for knowledge-intensive business processes: declarative configurable process specifications. In: 2011 44th Hawaii International Conference on System Sciences, pp. 1–10 (2011)
20. Schunselaar, D.M.M., Maggi, F.M., Sidorova, N., van der Aalst, W.M.P.: Configurable declare: designing customisable flexible process models. In: Meersman, R., et al. (eds.) OTM 2012. LNCS, vol. 7565, pp. 20–37. Springer, Heidelberg (2012). https://doi.org/10.1007/978-3-642-33606-5_3

Interactive Drift Visualization in Sensor Data Streams for Explainable Process Outcome Prediction

Matthias Ehrendorfer$^{(\boxtimes)}$ ⓘ, Jennifer Hebstreit, Juergen Mangler ⓘ, and Stefanie Rinderle-Ma ⓘ

Technical University of Munich, TUM School of Computation, Information, and Technology, Garching, Germany
{matthias.ehrendorfer,jennifer.hebstreit,juergen.mangler, stefanie.rinderle-ma}@tum.de

Abstract. In real-world process scenarios such as manufacturing and logistics, the process outcome is frequently predicted by IoT sensor data streams and their drifts, e.g., the quality of a product can be affected by the temperature during the production process. In particular, drifts can explain variations in the process outcome, and hence, their early detection can support the definition of mitigation actions, e.g., canceling the production of probably low quality products. As in most cases, humans have to define such actions, it is crucial to support them in making decisions in the context of outcome prediction. Hence, this paper aims to support the interactive visualization of drifts in specific points of sensor data streams to show the development of critical sensor measurement points over different traces. Furthermore, these critical points can be identified based on drifts between distinct groups of traces. Being able to visualize drifts and identify critical points enables (early) outcome prediction, ranging from the analysis of drifts at single points or at specific timestamps in a trace to the investigation of average time series of traces representing different outcomes, e.g., OK/NOK. Three different methods to derive and visualize drift points are presented and evaluated based on a prototypical implementation and a survey with users.

Keywords: BPM and IoT · Sensor Data Streams · Process Outcome Prediction · Drift Visualization · Drift Analysis

1 Introduction

Process models provide a blueprint on how and in which order the tasks of a process should be performed. However, when process models are enacted repeatedly, over time, there might be changes in the order of the tasks or how they are carried out. Such changes are called drifts which might ultimately lead to an adaptation of the process model [1]. Naturally, such drifts can happen not only in the control-flow perspective of the process (as described above) but also in

A. Marrella et al. (Eds.): BPM 2024 Forum, LNBIP 526, pp. 162–178, 2024.
https://doi.org/10.1007/978-3-031-70418-5_10

other perspectives, such as the organizational perspective, if roles carrying out the tasks change or the data perspective, when data elements obtained during the process (e.g., amount of a loan or age of an applicant) start to shift over the course of multiple enacted processes [11].

Another source for drifts to occur lies in the sensor data streams that are recorded during the execution of different traces of a process [12]. Figure 1 presents an example, how a sensor data stream of a temperature recording could look like for two traces. The drift between two traces here signifies a potential thermal runoff affecting the process outcome. Its detection would allow users to rectify the situation by increasing the machine downtime, to achieve a stable production procedure. *Visual drift detection* seems easy for two sensor data streams, but becomes challenging for multiple sensor data streams with multiple drifts between the traces. Hence, this work asks **how users can be supported in visually detecting drifts in sensor data streams and assessing their effects on the process outcome.**

In general, the detection of drifts in sensor data streams is important because it provides information on how a specific process instance progresses and allows to react if undesired behavior occurs. This can be done by either correcting the situation for the currently analyzed instance or by letting a domain expert decide on how and if such deviations/drifts could be avoided or handled in the future, which leads to an improved process model.

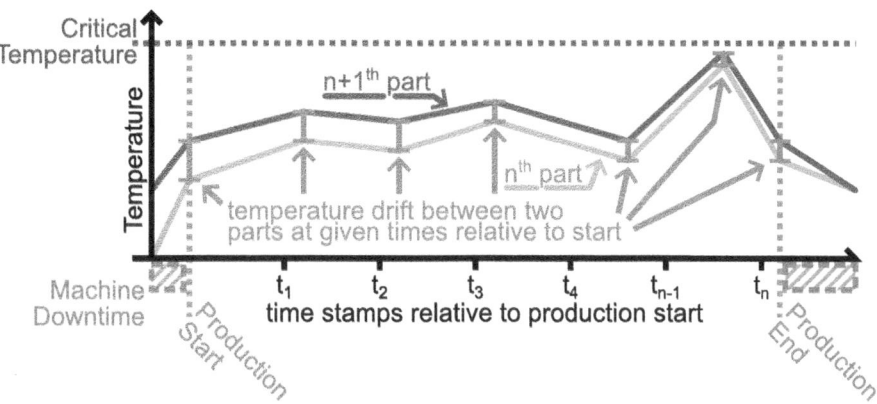

Fig. 1. Temperature Drift in a Production Process

The contribution of this paper is three-fold: **Contribution C1:** We propose an approach to visualize drifts for multiple traces in an easy-to-perceive way. **Contribution C2:** By using information about drifts between traces, we show how outcome prediction by considering drifts in different points of the sensor data streams, becomes possible. This method of outcome prediction has the advantage that it perfectly explains which points led to the decision and how the sensor data drifts in these points behave compared to other traces. **Contribution C3:** We evaluate the approach by implementation and conducting a survey with experts in the field.

The remainder of the paper is structured as follows: Sect. 2 discusses related work, Sect. 3 describes the presented approach, Sect. 4 provides a survey and the results for outcome prediction performed with the proposed methods, Sect. 5 discusses limitations, and Sect. 6 concludes the paper.

2   Related Work

The visualization of processes and their execution plays a crucial role in understanding and managing complex data flows and process behaviors. Business process management often employs BPMN to model processes graphically, enabling stakeholders to comprehend workflows comprehensively and identify inefficiencies. Utilizing process models can significantly enhance decision-making, potentially leading to improved operational efficiency and increased revenue [2]. On the execution side sensor event streams can play a crucial part in explaining the root cause of deviations from expected process behaviors [12]. To visualize these deviations this work aims to provide an interactive visualization framework for drifts in sensor event streams, enhancing the understanding of process outcomes and enabling more informed decision-making. [5] highlights the value of visualizing data from manufacturing execution systems, drawing special attention to how interactive visualizations empower users to selectively focus on data points relevant to their decision-making needs. One approach in ubiquitous computing visualizes time-series sensor data to aid in just-in-time adaptive intervention design [10]. However, the mentioned works fall short of addressing concept drifts and the challenges posed by concept drifts in process execution.

Concept drift and data drift affect the accuracy and reliability of numerous real-world applications exposed to dynamic environments [3,14]. Concept drifts are changes in the control-flow of a process, which are detected through analyzing process execution logs [11]. In existing work in machine learning, data drift is defined as the discrepancies between the data set initially used to train a machine learning model and the data subsequently collected by the model during real-world application [3]. This paper follows the definition of data drift as defined by [11]: it is the change in process data, e.g., when machine errors change machining parameters during production in manufacturing processes. Due to the different types of input data and detection methods employed, many different visualizations of concept drifts were developed in existing work across different domains. Extensive research exists on the visualization of drifts between process models [9], event logs [15], and multidimensional problem spaces [7]. Concept drifts have also been visualized using feature importance on streaming data [8]. Other fields, such as cloud computing, produced works on cluster-based data drift visualizations of CPU and memory resource usage [4]. However, the mentioned works do not examine external data such as IoT sensor data or visualize drifts over distinct groups of traces.

3 Approach

Three methods to visualize the drift between a specified point in a sensor data stream and another sensor data stream are presented in Sect. 3.1. Furthermore, an approach to automatically determine interesting points in sensor data streams which should be observed with the aforementioned methods is presented in Sect. 3.2.

3.1 Visualization Approaches

Data Preparation. Log files of process instances contain sensor event streams that capture data from external sensors monitoring the physical conditions under which the process is executed. Each data point corresponds to a single time-stamped measurement from the sensor, which, compiled, constitute time sequence data. This data needs to be resampled into discrete and equidistant time series data like in [12] to tackle the problem of unevenly distributed time sequences and obtain an even time scale over all available time sequences. Afterwards, the most recent time series are considered, and outliers (i.e., whole time series that are too long/short) within this predetermined window are excluded based on the criteria *"sequences with a duration shorter than the first quartile minus 1.5 times the IQR (Interquartile Range) or with a duration greater than the third quartile plus 1.5 times the IQR, similar to boxplots. The IQR is calculated here between third and first quartile."* [12]. This outlier detection method is adapted from [12] and modified to simultaneously detect outlier traces from all available time sequence data instead of performing the method trace by trace. Moreover, the method works only on already completed time series (because start and end events must be known) and when outliers are detected, the whole time series is excluded.

After outlier removal, traces are grouped to allow more efficient drift visualization with a large number of time series. For each group an average time series (ATS) can be calculated as described in [12], by utilizing the DTW Barycenter Averaging (DBA) [6] algorithm as distance measure. Averaging sequences using DBA involves an iterative refinement process that aims to minimize the squared distance (DTW) to averaged sequences, even if the initial average sequence is arbitrary [6]. As a result, the computation time for this technique is quadratic, as a DTW matrix must be generated for each iteration [12]. In this paper, two approaches for grouping the traces are used. In the first one, traces are split up based on their time of occurrence into fix-sized groups to detect drifts that happen over time, e.g., because of tool wear. In the second one, the traces are grouped based on their process outcome, i.e., whether the part produced in the process is OK or not OK.

Data Visualization. The three novel drift visualization methods presented comprise 90 degrees angle (*90Deg*), shortest distance (*SD*), and same timestamp (*ST*) drift visualization. They allow to compare a specified point in one time series (in the following called **analyzed time series**) to another time series (in the following called **target time series**). Both time series can either be a time series from an individual trace or an average time series obtained from multiple traces. Figure 2 shows an illustrative example with a time series of sensor measurements from an individual trace (i.e., black line) and average time series (i.e., red, blue, and green lines), each being derived from sensor measurement of multiple consecutive traces. Drifts between time series are shown with orange, pink, and purple arrows at different points in time (green dots). Techniques that scale the lengths of the lines with a user-selected factor and insert offsets to add space between overlaps (see pale purple line between pink and purple lines) are used to prevent overlapping lines and barely perceivable differences.

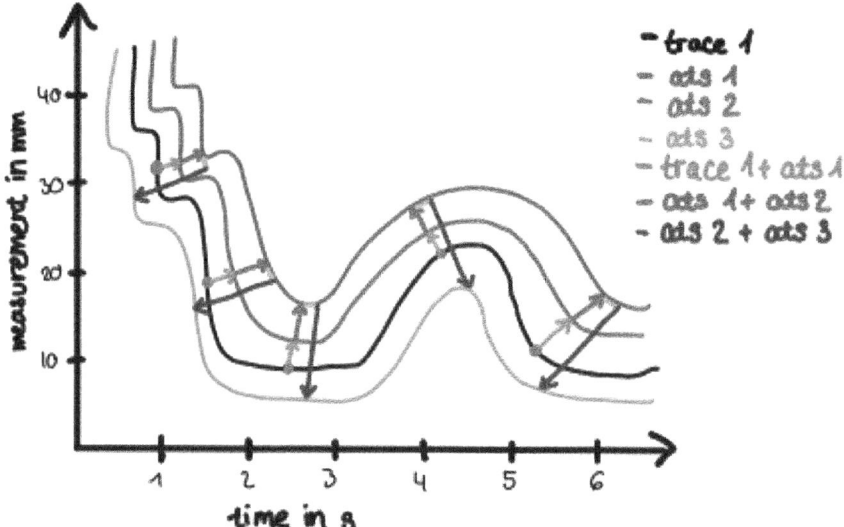

Fig. 2. Underlying Idea Behind the Drift Visualization Techniques

Figure 2 shows that calculating drifts between time series depends on which point on the target time series is used to measure the distance to the analyzed time series (i.e., distance between the defined point on the analyzed time series and the one chosen on the target time series). Different methods for finding the point on the target time series are possible:

– **90Deg Drift Visualization Method:** This method involves calculating a line perpendicular to the point on the analyzed time series by considering the slope of its neighboring points. Then, the nearest intersection point on the target time series is found, and the distance between the starting point on the analyzed time series and this point represents the drift. Algorithm 1 outlines the pseudo code for this method. It requires a time series, a list of average time series, a specific timestamp to be analyzed (`point_x`), a scaling factor, values for segment spacing, and neighbor point distance. First, lists for storing line segments and process execution are initialized and then the sensor measurement corresponding to `point_x` is determined (`get_point` function) and used as the starting point of the drift calculations. The starting point, alongside `distance` and the first average time series, is used to calculate intersection points with the next time series by finding neighboring points to construct the slope of a perpendicular line that crosses the next average time series. The `identify_intersection` function selects the closest intersection point to `p1`, or the closest point on the average time series if no intersection exists. The x-value of `p1` is logged in `proc_exec`. The method iterates through subsequent average time series to calculate new intersection points, which are then used as bases for the next calculations. Finally, the line segments are scaled and offset for improved visualization, returning the (modified) line segments and process execution identifiers. Figure 3a depicts the expected behavior of this visualization.

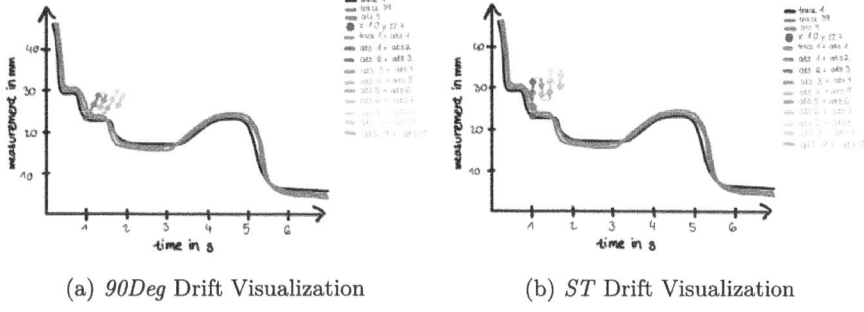

(a) *90Deg* Drift Visualization   (b) *ST* Drift Visualization

Fig. 3. Behavior of 90Deg, ST Drift Visualizations

– **SD Drift Visualization Method**: This method portrays drifts as the minimum distance between a point on the analyzed time series and the target time series. At the specified point, the smallest distance to a point on the target time series must be found. Algorithm 2 shows pseudo-code for this method. It calculates the shortest distance from the starting point to the next average time series, using the project method to find the minimum distance and then linear interpolation to identify the data point on the average time series. This point and starting point are added to `line_segments`, and its x-value to `proc_exec`. This is done for each average time series, using the last point found as the starting point for the next calculation. Line segments are scaled, and offsets are added to prevent overlap if set by the user. Afterwards, the (updated) line segments and `proc_exec` list are returned.

Input: ts, ats_list, point_x, scale_factor, offset, distance
Output: scaled_and_spaced_lengths, line_segments, proc_exec
line_segments, proc_exec = []
p1 = get_point(ts, point_x)
original = ts
proc_exec.append(p1.x)
for *ats in ats_list* **do**
    intersections = get_perp_line(original, p1, dist, ats)
    closest_point = identify_intersection(intersections, ats)
    line_segments.append([p1, closest_point])
    proc_exec.append(closest_point.x)
    p1 = closest_point, original = ats
end
scaled_lengths = scale_lengths(line_segments, scale_factor)
scaled_and_spaced_lengths = add_offset(scaled_lengths, offset)
return scaled_and_spaced_lengths, line_segments, proc_exec

Algorithm 1: Calculate *90Deg* Drift

Input: ts, ats_list, point_x, scale_factor, offset
Output: scaled_and_spaced_lengths, line_segments, proc_exec
line_segments, proc_exec = []
p1 = get_point(ts, point_x)
proc_exec.append(p1.x)
for *ats in ats_list* **do**
    closest_point = ats.interpolate(ats.project(p1))
    line_segments.append([p1, closest_point])
    proc_exec.append(closest_point.x)
    p1 = closest_point
end
scaled_lengths = scale_lengths(line_segments, scale_factor)
scaled_and_spaced_lengths = add_offset(scaled_lengths, offset)
return scaled_and_spaced_lengths, line_segments, proc_exec

Algorithm 2: Calculate *SD* Drift

– **ST Drift Visualization Method:** This method visualizes the drift between a point in the analyzed time series and the target time series by calculating the distances between measurements on these two time series at the same timestamp. Algorithm 3 details the pseudo code: The method uses linear interpolation to find a y-value corresponding to an x-value from the next average time series, forming line segments for drift visualization with the start point. The x-coordinate of this point is recorded for process information retrieval. This interpolation process repeats for each subsequent average time series, with the last intersection point as the new start. The resulting line segments are adjusted for scale and spacing according to the input, and the method returns this information. Figure 3b displays the expected behavior of this method.

Input: ts, ats_list, point_x, scale_factor, offset
Output: scaled_and_spaced_lengths, line_segments, proc_exec
line_segments, proc_exec = []
p1 = get_point(ts, point_x)
proc_exec.append(p1.x)
for *ats in ats_list* **do**
    ipvy = interpolate(p1.x, ats)
    closest_point = (p1.x, ipvy)
    proc_exec.append(closest_point.x)
    line_segments.append([p1, closest_point])
    p1 = closest_point
end
scaled_lengths = scale_lengths(line_segments, scale_factor)
scaled_and_spaced_lengths = add_offset(scaled_lengths, offset)
return *scaled_and_spaced_lengths, line_segments, proc_exec*

Algorithm 3: Calculate ST Drift

3.2 Point of Interest Detection

The methods described in Sect. 3.1 allow visualization and calculation of drifts in one point of an analyzed time series. However, which **point(s) of interest (POI(s))** are to be observed is not specified and in the following two approaches to find such POIs are proposed:

Point of Interest Detection Based on Individual Traces. Detecting POIs for individual traces uses only information from the analyzed trace itself by fitting a curve to the available points after the Data Preparation step (cf. Sect. 3.1). Afterwards, maximal and minimal points (i.e., extreme points), as well as points with a strong change (i.e., inflection points), are used as POIs of the series. Therefore, each individual trace has its own POIs. However, the time series need to already be completely recorded to allow curve fitting.

Point of Interest Detection Based on Analyzing Multiple Traces. The detection of POIs based on multiple traces is described in Algorithm 4. First, traces have to be split up into two classes based on some characteristic that identifies them (e.g., traces leading to OK vs. not OK parts). Then, for each class, the average time series can be calculated as described in Sect. 3.1. With these two average time series as input for every point in one of the average time series (**ats_A** in the algorithm and the remainder of this section) the distance to the other average time series (**ats_B** in the algorithm and the remainder of this section) is calculated based on one of the three methods described in Sect. 3.1. The results are then separated by points where the distance between ats_A to ats_B is zero or close to zero (i.e., where they intersect) to split the time series into segments where they deviate. Finally, the maximal distance isMaxInSegment in the algorithm, as well as extreme points (calculated by looking at upwards trends of minimal length min_trend_length and then choosing the highest point

out of the next points, denoted as isMaxInUpwardsTrend in the algorithm), are
used as POIs because in these timestamps ats_A deviates the most from ats_B
(under the used distance metric).

Input: (average time series) ats_A, ats_B, min_trend_length
Output: pois
segments, pois = []
for *point in ats_A* **do**
    if *calculate_distance(point,ats_B) ≤ 0.01* **then**
       | segments.append([])
    end
    segments.last.append([point.x, calculate_distance(point,ats_B)])
end
for *segment in segments* **do**
    for *point in segment* **do**
       if *isMaxInSegment(point,segment) OR*
       isMaxInUpwardsTrend(point,segment,min_trend_length) **then**
          | pois.append(point)
       end
    end
end
return *pois*

Algorithm 4: Determine POIs Based on Multiple Traces

4 Evaluation

The first part of the evaluation described in Sect. 4.1 assesses the proposed drift
visualization methods by conducting a survey comparing them (including their
additional features like getting rid of overlapping drifts) to the visualization of
raw sensor data streams. The second part described in Sect. 4.2 evaluates the
approaches for finding POIs by (1) using the automatically determined POIs to
predict outcomes for traces and (2) conducting a survey based on which criteria
experts would choose POIs and to which conclusion regarding the outcome of
traces they would come based on the chosen POIs.

    The online questionnaire created for the evaluation contains closed-ended
questions about the visualization methods using a six-point Likert scale (series
of statements with which respondents can state their level of agreement [13]) as
answer options range from "Strongly Disagree" to "Strongly Agree". According
to [13], using a six to seven-point rating scale is recommended. We opted for
a six-point scale to prevent neutral answers. Furthermore, open questions were
included to get additional insights. The second part of the questionnaire entailed
closed-ended questions inducing binary responses (yes/no) with some questions
also allowing "undecidable". The questions deal with which points in a trace
should be analyzed in-depth and what that means for predicting the outcome

of a trace. Additionally, open questions are asked to find out how respondents came up with their answers. The survey was completed by ten experts in the field of business process management (between 2 and 24 years experience) with different areas of expertise reaching from process modelling and process mining to resource allocation and IoT data integration.

The data set used for the questionnaire and evaluation contains logs of producing 37 parts in a manufacturing process. Each part is produced by a machine tool and measured directly afterwards by a fast, but imprecise measuring machine and finally by a slower, but more precise measuring machine. For the evaluation, the sensor data stream of the first measurement, which measures the diameter of the part while moved through the measuring machine, is used to detect drifts between different traces. The second measurement is used to define the outcome (18 parts are faulty while 19 are OK) of a trace.

The survey was performed using Google Forms and the questions are made available on Zenodo[1] together with the results. The code used to create the screenshots for the first part of the questionnaire focusing on the evaluation of drift visualization methods, including a description of how to execute it, is available on GitHub[2]. On two other GitHub repositories[3][4], the code utilized in the second part of the questionnaire that analyzed traces based on different drift visualization methods and used for the outcome prediction evaluation, can be found. The repositories also contain the used data sets.

4.1 Evaluation of the Proposed Visualization Methods

The conducted survey evaluates if the drift visualization methods presented in Sect. 3.1 are useful in displaying drifts happening in one sensor data stream over several traces and investigates details of the approaches. Therefore, for three different points in a trace, all three proposed methods for drift visualization are applied. The participants are presented with a picture of a specific method applied on a specific point together with a text explaining the parts of the picture. Five questions (described below in more detail) assessing how well the drift visualization works are asked with six answer options (from "Strongly Agree" to "Strongly Disagree") for each point/method combination.

For subfigures (a)–(e) in Fig. 4, we report on the number of positive responses (i.e., "Strongly Agree", "Agree", "Slightly Agree") to the questions. For the first question "Can you easily identify in which direction measurements have drifted?" (see Fig. 4a) 8 out of 10 responses are positive for all point/method combinations apart from the ST method at timestamp 4.75. The question "Can you easily identify the overall amount of drift that occurred?" (see Fig. 4b) also achieves at least

---

[1] https://doi.org/10.5281/zenodo.11654993, accessed on 14th June 2024.

[2] https://github.com/jennvheb/ba_drift_visualization, accessed on 14th June 2024.

[3] https://github.com/jennvheb/paper_drift_visualization, accessed on 14th June 2024.

[4] https://github.com/me33551/drift_visualization_based_outcome_prediction, accessed on 14th June 2024.

8 positive responses for all combinations except the ST method at timestamp 4.75. For question "Can you easily identify individual drift amounts between certain grouped traces (e.g., between traces 6–10 and 11–15)?" (see Fig. 4c) positive responses are lower. Again, ST at timestamp 4.75 got the lowest number of positive responses (4), and also for timestamp 1.55, only 5 participants respond positively to the question. Interestingly, the ST method at timestamp 1 receives the most positive responses. For the question "Are the details displayed for a certain drift (arrow) useful?" (see Fig. 4d) the ST method at timestamp 1 receives 9 positive responses while the same method at timestamp 4.75 receives the least number of positive responses (3). The question "Is the offset helpful in identifying individual drift amounts?" (see Fig. 4e) receives between 7 and 10 positive responses for all combinations apart from the ST method at timestamp 4.75. Overall, all participants, apart from one, disagree (i.e., "Strongly Disagree", "Disagree", or "Slightly Disagree") with the statement that "A concept drift is easier to spot with the raw data visualization than with ..." for every method shown (see Fig. 4f). However, disagreement is stronger for the ST and SD visualization methods compared to the $90Deg$ method.

Participants' answers in the open questions where they were asked (after seeing each method for a specific timestamp) to sum up the strengths and weaknesses of each method reveal that:

- For timestamp 1 participants agreed that they liked the ST method (because it is easy to understand) best. However, one participant also mentioned the $90Deg$ method as intuitive.
- For timestamp 1.55 participants describe that using approaches 1 and 2 (i.e., $90Deg$ and SD visualization method) makes more sense for them when looking at this timestamp compared to the first one because it also captures some kind of "temporal" shift.
- For timestamp 4.75 participants again describe that using approaches 1 and 2 (i.e., $90Deg$ and SD visualization method) makes more sense for them when looking at this timestamp. Additionally, they describe that the ST visualization method does not provide any real insight into the data.
- The ST visualization method is perceived as the most native/easiest to understand while the other ones are described as slightly confusing by some participants.
- Some participants state in their answers among different timestamps that offsetting the drifts in the visualizations is useful for them.

Overall, participants' answers to the open questions described above and the questions summarized in Fig. 4 show that (1) drifts can be spotted easier with the presented approaches than by just looking at the raw data, (2) even inside the same scenario depending on the timestamp that should be analyzed different visualization methods might be useful (3) preventing overlaps of the visualized drifts makes it easier to perceive drifts.

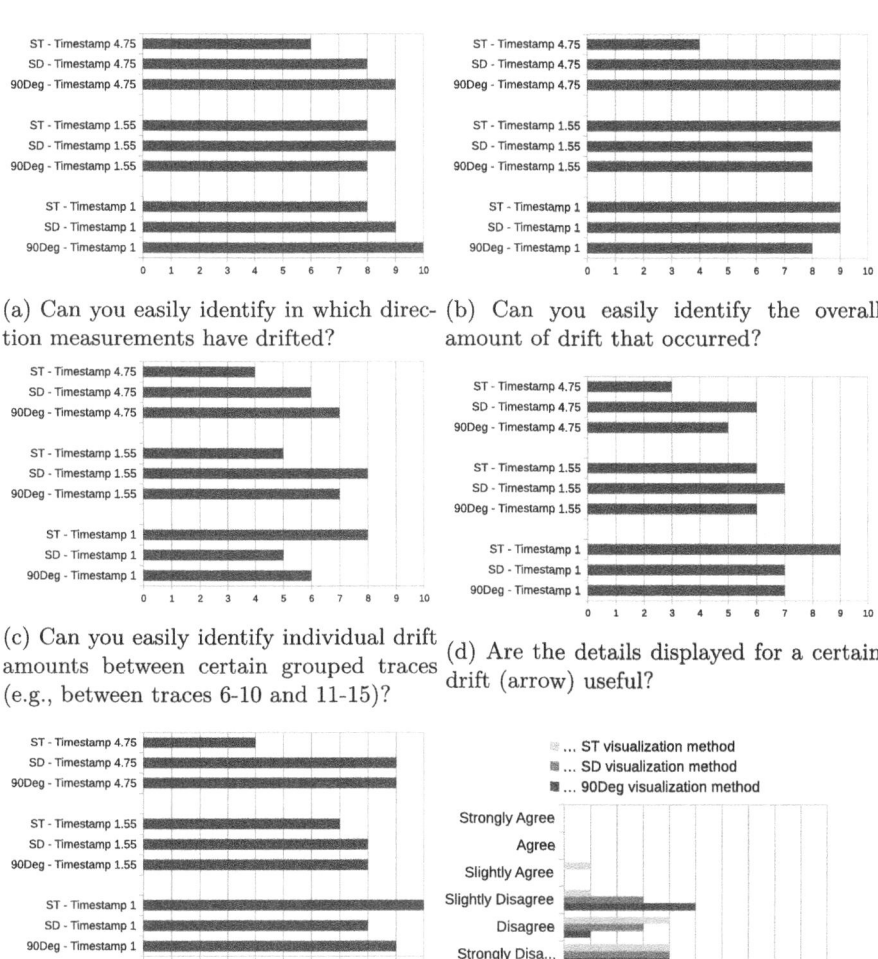

(a) Can you easily identify in which direction measurements have drifted?

(b) Can you easily identify the overall amount of drift that occurred?

(c) Can you easily identify individual drift amounts between certain grouped traces (e.g., between traces 6-10 and 11-15)?

(d) Are the details displayed for a certain drift (arrow) useful?

(e) Is the offset helpful in identifying individual drift amounts?

(f) A concept drift is easier to spot with the raw data visualization than with ...

Fig. 4. Summarized Survey Results for the Proposed Visualization Methods

4.2   Point of Interest Detection for Outcome Prediction

Evaluating the detection of POIs and subsequent outcome prediction is done by applying the techniques described in Sect. 3.2 and using the POIs found to predict the outcome of traces. The trace is predicted to belong to the class to which average time series more POIs of a trace are closer.

The results are presented in Table 1 and show in the leftmost column which method was used to obtain POIs (either "individual trace" for the approach described in Sect. 3.2 using individual traces for POI detection or *90Deg*, *SD*, or *ST* for the approach described in Sect. 3.2 based on multiple traces). Additionally, it is important if the distance is measured from a point in the average OK time series to the average not OK time series or the other way around (i.e., which one is ats_A and which one is ats_B in Algorithm 4), i.e.,"OK/NOK" or "NOK/OK". One row in the table uses POIs derived with the given method (with min_trend_length = 3) and calculates the distance to the average OK and not OK time series for each POI (i.e., each timestamp) in each trace. The three methods described in Sect. 3.1 are utilized (see the first line of the table) for this distance calculation. As explained above, the number of POIs closer to the OK or not OK average time series is used to predict the produced part quality of each trace. This is reported as the number of correctly classified traces (column labelled "correct"/"c"), number of incorrectly classified traces (column labelled "incorrect"/"i") and number of "undecidable"/"u" traces (e.g., the same number of POIs is closer to the average OK and average not OK time series). It can be seen in Table 1 that the approaches based on finding points using multiple traces (lines 4–9) are slightly better than the ones performed on each trace individually (line 3). The approaches to find POIs shown in lines 4–9 perform similarly even when using different distance metrics for measuring the distance from POIs to the average OK or not OK traces. Overall, in these approaches 18–21 traces out of 37 are correctly predicted ($\approx$ 48.65%–54.05%) while 4–8 traces remain undecidable ($\approx$ 10.81%–21.62%) and 9–15 are incorrectly classified ($\approx$ 24.32%–40.54%).

Table 1. Predicted Outcome of Traces Based on Automatic and Manual POIs

	90Deg			SD			ST		
	correct	undecidable	incorrect	c	u	i	c	u	i
individual trace	17	8	12	17	8	12	16	8	13
90Deg OK/NOK	19	4	14	20	4	13	21	4	12
90Deg NOK/OK	18	4	15	18	4	15	20	4	13
SD OK/NOK	18	6	13	19	5	13	20	8	9
SD NOK/OK	18	5	14	19	5	13	19	4	14
ST OK/NOK	18	4	15	19	4	14	20	4	13
ST NOK/OK	18	4	15	19	4	14	20	4	13
90Deg OK/NOK - Tr1	18	4	15	19	4	14	20	4	13
90Deg OK/NOK - Tr2	20	4	13	21	4	12	23	4	10
SD OK/NOK - Tr1	20	4	13	19	4	14	19	4	14
SD OK/NOK - Tr2	20	4	13	20	4	13	23	4	10
ST OK/NOK - Tr1	19	5	13	20	7	10	20	8	9
ST OK/NOK - Tr2	19	5	13	20	7	10	21	7	9

The survey carried out for this paper includes questions regarding which points experts would examine in more detail to determine if an individual trace is expected to produce an OK or not OK part. Figures 5b, 5d and 5f show the POIs found when using the three methods described in Sect. 3.1 as distance measure between the average OK and average not OK time series for Algorithm 4. Figures 5a, 5c and 5e show which points experts wanted to examine more closely - based on looking at two different traces for which the marked timestamps as well as their drifts to the average OK and not OK time series were shown.

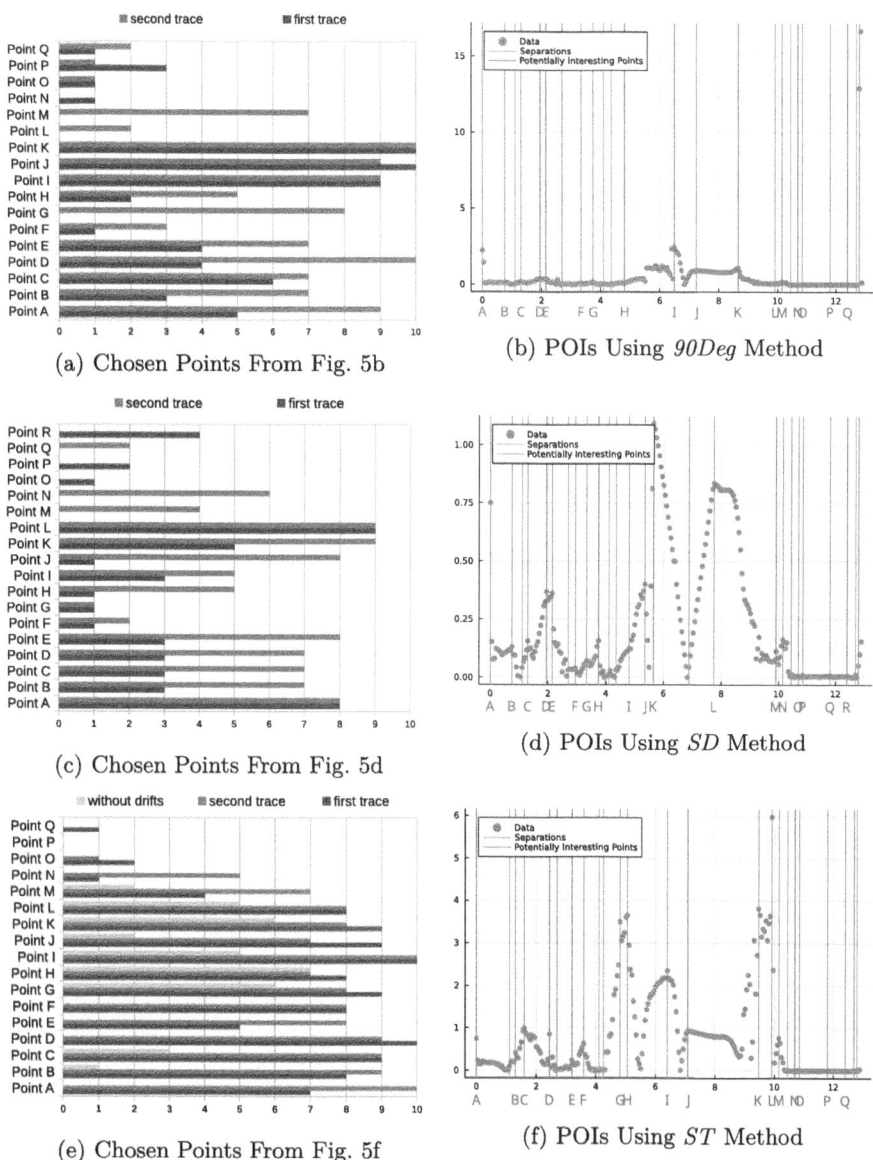

(a) Chosen Points From Fig. 5b

(b) POIs Using *90Deg* Method

(c) Chosen Points From Fig. 5d

(d) POIs Using *SD* Method

(e) Chosen Points From Fig. 5f

(f) POIs Using *ST* Method

Fig. 5. Point of Interest Detection Results for Different Distance Measures

The experts then had to choose for these two traces if they thought the trace represented a process run where the outcome would be OK or not OK. The results are shown in Fig. 6. As the first trace produces a correct part and the second trace produces a not OK part nearly all participants chose the correct option. Using only the points that have been selected by 5 or more participants in the survey (see Figs. 5a, 5c and 5e) and conducting the same analysis as for Table 1 rows 3–9 leads to the results shown in rows 10–15 - for each method the points selected for each trace are used ("Tr1"/"Tr2"). Overall, there is not much difference between the results achieved with the points selected by the experts in the survey and the automatically generated POIs when applied to all traces. Looking at the questions on how points that need to be analyzed in more depth are chosen reveals that experts choose the points based on (1) the length of the arrows/drifts, (2) where the difference between the arrows was sufficiently big, and/or (3) where arrows (drifts) go in different directions. Afterwards, the experts' prediction of the trace's class is based on which arrows are shorter (i.e., to which of the average time series chosen points are closer).

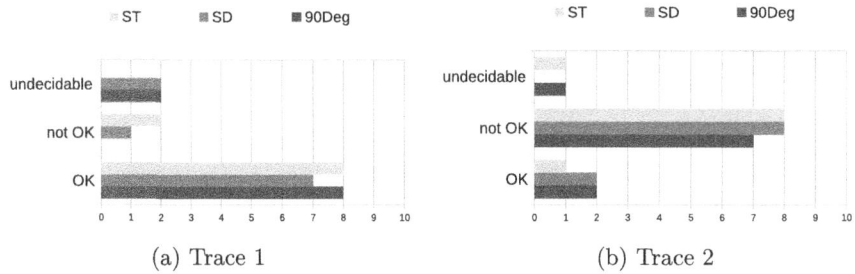

(a) Trace 1   (b) Trace 2

Fig. 6. Participants Prediction Results per Trace and Drift Visualization Method

5 Discussion

A limitation of the presented approach is that it can only be performed ex-post because detection of outlier traces (see Sect. 3.1), as well as calculation of average time series, depends on traces (or at least analyzed sensor data streams) being already completed. Additionally, the *90Deg* approach depends on data from timestamps to its left and right being available, which means it cannot be applied on the latest recorded point of the sensor data stream. Another limitation is that drifts are only visualized and analyzed for single sensor data streams. However, potentially multiple sensor data streams are recorded per trace which introduces the problem of determining which sensor data streams are important for detecting drifts for the whole process. An additional challenge in visualization is to consider the effects of large volumes of data (i.e., many traces being recorded). With the presented visualization approaches this can be tackled by utilizing grouping sequences of traces and adjusting how many traces are in a group together (therefore abstracting from information of the individual traces in favor of presenting properties of the groups) to allow for a comprehensible visualization.

The visualization approach presented in this paper shows drifts between different traces. This information is used to perform explainable outcome prediction based on the drifts at certain timestamps (the methods used to find these timestamps and determine the outcome of individual traces based on them is described in Sect. 4.2). However, while the results for predicting the outcome are not satisfactory, as the quality of parts could only be predicted with $\approx 50\%$ accuracy, the approach nonetheless shows that it is feasible to work towards finding points and thus better prediction results in future work. Finally, we opted for a qualitative study to gather insights directly from end users which helped us in identifying strengths and weaknesses. But one potential problem of the questionnaire is, that it tests the understanding of the experts, and thus may lead to biased results. We will thus in future work improve the assessment of how the visualisation furthers data understanding, by conducting a quantitative study. In addition, we plan to improve the visualization by including multiple sensor data streams in the detection and visualization of drifts. Furthermore, we plan to extend the methods to make them applicable for runtime/live scenarios.

6   Conclusion

The paper proposes three approaches for the visualization of drifts in sensor data streams of different traces: "90 Degrees" (*90Deg*), "Shortest Distance" (*SD*), and "Same Timestamp" (*ST*). Furthermore, explainable outcome prediction is performed by utilizing how far traces drifted from OK / not OK traces.

The applicability of the visualization approaches was tested through implementation (contribution C1) and through a survey with 10 experts (partially contribution C3). 10 out of 10 experts considered *90Deg* and *SD* as better suited for identifying drifts than a raw trace visualization. Only 1 expert considered the raw visualization slightly useful in comparison to *ST*. However, the best method for visualizing drifts also depends strongly on the observed timestamp.

Contribution C2 of this paper, a way to use the visualization techniques at automatically derived points in the time series to predict the process outcome, was evaluated in the same survey (thus completing contribution C3). Here, the results of the survey show that using this approach allows for correctly predicting some of the trace outcomes ($\approx 48.65\%$–54.05%) and label some as undecidable ($\approx 10.81\%$–21.62%). But also a significant number of traces ($\approx 24.32\%$–40.54%) was assigned the incorrect outcome. However, even if the outcome prediction results are not satisfactory, the results show that when the right points are chosen, explainable predictions become feasible. In future work, we will thus focus on proposing methods for automatically finding points that yield better results for the outcome prediction.

References

1. Bose, R.P.J.C., van der Aalst, W.M.P., Žliobaitė, I., Pechenizkiy, M.: Handling concept drift in process mining. In: Mouratidis, H., Rolland, C. (eds.) CAiSE 2011. LNCS, vol. 6741, pp. 391–405. Springer, Heidelberg (2011). https://doi.org/10.1007/978-3-642-21640-4_30
2. Figl, K.: Comprehension of procedural visual business process models. Bus. Inf. Syst. Eng. **59**, 41–67 (2017)
3. Mallick, A., Hsieh, K., Arzani, B., Joshi, G.: Matchmaker: Data drift mitigation in machine learning for large-scale systems. In: Machine Learning and Systems, vol. 4, pp. 77–94 (2022)
4. Mehmood, T., Latif, S.: Dynamic big data drift visualization of CPU and memory resource usage in cloud computing. In: Artificial Intelligence Applications and Innovations, pp. 27–36 (2022)
5. O'Neill, M., Morgan, J., Burke, K.: Process visualization of manufacturing execution system (MES) data. In: SmartWorld, Ubiquitous Intelligence & Computing, Advanced & Trusted Computing, Scalable Computing & Communications, Internet of People and Smart City Innovation, pp. 659–664 (2021)
6. Petitjean, F., Ketterlin, A., Gançarski, P.: A global averaging method for dynamic time warping, with applications to clustering. Pattern Recogn. **44**(3), 678–693 (2011)
7. Pratt, K.B., Tschapek, G.: Visualizing concept drift. In: Knowledge Discovery and Data Mining, pp. 735–740. KDD '03 (2003)
8. Sarnovský, M.: Concept drift visualization using feature importance on the streaming data. In: 2022 IEEE 20th Jubilee World Symposium on Applied Machine Intelligence and Informatics (SAMI), pp. 000449–000454 (2022)
9. Sato, D.M.V., Barddal, J.P., Scalabrin, E.E.: Interactive process drift detection framework. In: Rutkowski, L., Scherer, R., Korytkowski, M., Pedrycz, W., Tadeusiewicz, R., Zurada, J.M. (eds.) ICAISC 2021. LNCS (LNAI), vol. 12855, pp. 192–204. Springer, Cham (2021). https://doi.org/10.1007/978-3-030-87897-9_18
10. Sharmin, M., et al.: Visualization of time-series sensor data to inform the design of just-in-time adaptive stress interventions. In: Pervasive and Ubiquitous Computing, pp. 505–516. UbiComp '15 (2015)
11. Stertz, F., Rinderle-Ma, S.: Detecting and identifying data drifts in process event streams based on process histories. In: Information Systems Engineering in Responsible Information Systems - CAiSE Forum, pp. 240–252 (2019)
12. Stertz, F., Rinderle-Ma, S., Mangler, J.: Analyzing process concept drifts based on sensor event streams during runtime. In: Fahland, D., Ghidini, C., Becker, J., Dumas, M. (eds.) BPM 2020. LNCS, vol. 12168, pp. 202–219. Springer, Cham (2020). https://doi.org/10.1007/978-3-030-58666-9_12
13. Taherdoost, H.: What is the best response scale for survey and questionnaire design; review of different lengths of rating scale/attitude scale/Likert scale (2019)
14. Webb, G.I., Lee, L.K., Petitjean, F., Goethals, B.: Understanding concept drift (2017)
15. Yeshchenko, A., Ciccio, C.D., Mendling, J., Polyvyanyy, A.: Comprehensive process drift detection with visual analytics. CoRR **abs/1907.06386** (2019)

Conformance Checking and Performance Analysis Using Object-Centric Directly-Follows Graphs

Gyunam Park$^{(\boxtimes)}$, Jan Niklas Adams, and Wil M. P. van der Aalst

Process and Data Science, RWTH Aachen University, Aachen, Germany
{gnpark,niklas.adams,wvdaalst}@pads.rwth-aachen.de

Abstract. Traditional process mining often simplifies the multi-object reality of business processes, linking each event to a single object called "the case". This oversimplification may lead to inaccurate analyses, caused by missing interactions between object types. In contrast, object-centric process mining permits events to be tied to multiple objects, capturing complex interactions and providing a more accurate representation of business processes. This paper introduces an approach for supporting object-centric process mining utilizing Object-Centric Directly-Follows Graphs (OC-DFGs). Despite their advantages, e.g., simplicity, OC-DFGs have been relatively untapped for essential process mining tasks, such as conformance checking and performance analysis. In order to address this, our research presents a comprehensive approach for OC-DFG-based conformance checking and performance analysis. We fully implement the proposed approach as a web application and demonstrate the use of OC-DFGs for these tasks within a case study of a real-life loan application process.

Keywords: Object-Centric Process Mining · Object-Centric Directly-Follows Graphs · Conformance Checking · Performance Analysis

1 Introduction

Conventional process mining necessitates every event to associate with a singular object type (i.e., a single case notion), and each change of case perspective requires new data extraction and configuration [12]. This requirement can inadvertently duplicate events (convergence) and displace causal information between events (divergence) [1]. Furthermore, the single-object focus can overlook interactions with other object types in the process, leading to potentially inaccurate and misleading analysis. These limitations are due to the oversimplified representation of processes that involve multiple objects, essentially flattening a 3-dimensional reality into 2-dimensional event logs [2].

Similarly to traditional process mining, process models play a pivotal role in object-centric process mining. The primary phase involves the discovery of process models from object-centric event logs. Additionally, conformance checking,

A. Marrella et al. (Eds.): BPM 2024 Forum, LNBIP 526, pp. 179–196, 2024.
https://doi.org/10.1007/978-3-031-70418-5_11

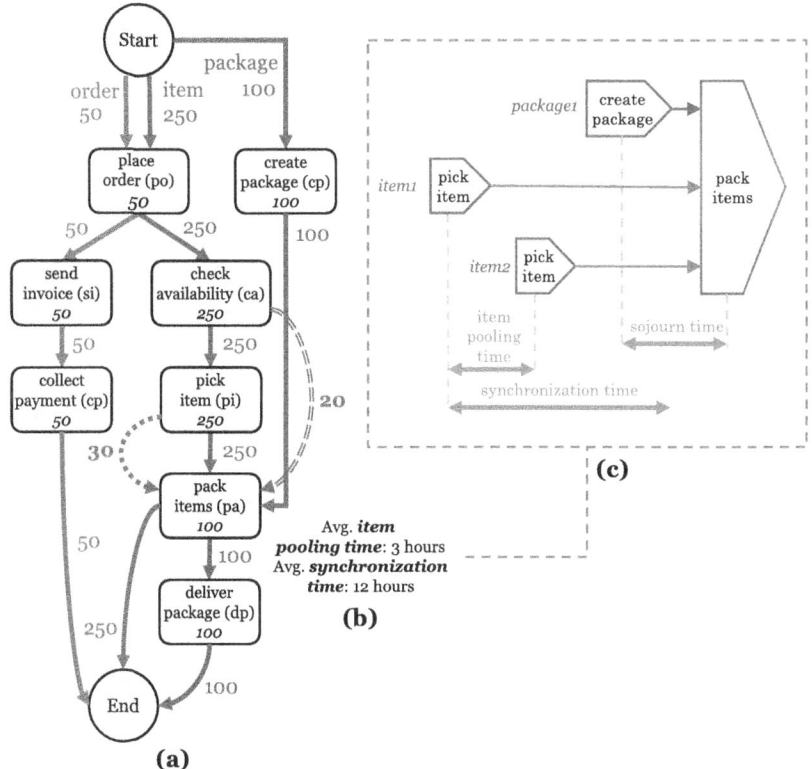

Fig. 1. A motivating example: (a) an example of OC-DFGs, (b) object-centric performance metrics for activity *pack items*, (c) object-centric performance metrics for a *pack items* event.

which compares observed behavior (event log) with modeled behavior (process model), aids in identifying discrepancies between actual process execution and the process model. Moreover, performance indicators can be visualized and analyzed within the process model's context, offering clear insights into process delays or inefficiencies.

A variety of modeling formalisms is employed to represent processes in object-centric process mining, including Proclets [14], Object-Centric Behavioral Constraint (OCBC) models [17], and Object-Centric Petri Nets (OCPNs) [4]. Techniques for conformance checking [5,7,15,20] and performance analysis [21,22] have been developed using the formal semantics of these advanced modeling formalisms.

Despite the prevalent adoption of directly-follows-based process maps in traditional process mining tools, attributed to their simplicity and interactive filtering capabilities, their object-centric equivalents, i.e., Object-Centric DFGs (OC-DFGs) [1], have not been widely adopted in practice due to the lack of capabilities for conformance checking and performance analysis. To address this gap, our work presents novel methods for conformance checking and performance analysis using OC-DFGs.

In more detail, this work presents a conformance checking technique using an OC-DFG and object-centric event logs. To that end, we translate each directly-follows graph constituting an OC-DFG to a workflow net and compute *alignments* of the corresponding projected event log on the workflow net. Next, we visualize the alignments using the OC-DFG. For example, Fig. 1(a) shows an example of an OC-DFG that visualizes alignments. The process model describes an order-to-cash process with three object types: *order*, *item*, and *package*. *Log moves* of 30 items, i.e., the actual behavior between *pick item* and *pack items* that is missing in the model, are described as a dotted line. *Model moves* of 20 items, i.e., 20 items that skip *pick item*, are described as a double line.

Next, we introduce a performance analysis technique by projecting the alignments on the process model and computing object-centric performance metrics such as synchronization time and pooling time. For example, a synchronization time for execution of *pack items* with *package1*, *item1*, and *item2*, described in Fig. 1(b), refers to the time taken from the picking of the last item to the creation of packages, as shown in Fig. 1(c). A pooling time of items in the execution refers to the time taken from the first picking to the last picking, as depicted in Fig. 1(c).

The proposed approach is fully implemented as a web application with a dedicated user interface. Furthermore, we evaluate the approach by applying it in a real-life loan application process using the implementation.

The remainder of this paper is organized as follows. We discuss the related work in Sect. 2. Next, we introduce preliminaries on event data and OC-DFGs in Sect. 3. Afterward, we explain an approach to conformance checking using OC-DFGs and alignments in Sect. 4 and performance analysis in Sect. 5. Then, Sect. 6 introduces the implementation of the proposed approach, presents a use case study on a real-life loan application process using the implementation, and discusses the utility and limitations of the proposed approach. Finally, Sect. 7 concludes the paper.

2   Related Work

This section introduces related work on object-centric process mining and process mining using directly-follows graphs.

2.1   Object-Centric Process Mining

This research aligns with recent developments in object-centric process mining [1]. Traditional process mining methods assume each event is associated with a single case, viewing the event log as a collection of isolated event sequences. Object-centric process mining deviates from this assumption, permitting an event to associate with multiple cases, leading to shared events between sequences, i.e., a graph of events.

A range of process modeling formalisms has been proposed to address the complexities of multiple case notions. Proclets [14], the first modeling technique designed to describe interacting workflow processes, was later extended by artifact-centric modeling [11]. DB nets [19], a modeling technique built on colored Petri nets, were also introduced. Li et al. [17] proposed Object-Centric Behavioral Constraint (OCBC) models, merging data models with process models. In addition, Object-Centric Petri Nets (OCPNs), a restricted variant of colored Petri nets, were suggested, where places are typed, tokens refer to objects, and transitions correspond to activities [4]. Moreover, typed Jackson nets, a subclass of typed Petri nets, were recently suggested along with a unique reconstructability property [9]. Alongside these advanced modeling languages, OC-DFGs have been presented as a simpler solution to address the inherent complexity in advanced modeling languages [1].

Numerous techniques have been proposed to facilitate object-centric process mining using these advanced modeling formalisms. Initially, process discovery techniques were suggested to automatically discover process models in Proclets [18], OCBC [17], and OCPNs [4] from event data. Furthermore, conformance checking techniques were developed to assess the discrepancies between modeled behavior and actual behavior in event data using Proclets [15], OCBC [5], and OCPNs [4,7]. Moreover, techniques for performance analysis have been introduced. For instance, an approach to object-centric performance analysis based on OCPNs was proposed in [21].

While several process mining approaches utilizing OC-DFGs exist, such as the baseline approach for discovering OC-DFGs from object-centric event logs [1] and the technique by Berti and van der Aalst for discovering Multiple View Point (MVP) models from databases [10], there remains a gap in the area of conformance checking and performance analysis using OC-DFGs. To our knowledge, there is no existing work that tackles these two aspects with OC-DFGs. This paper, therefore, strives to fill this gap by developing a technique for conformance checking and performance analysis using OC-DFGs.

2.2   Process Mining Using Directly-Follows Graphs

Directly-Follows Graphs (DFGs) are advantageous in traditional process mining due to their simplicity, allowance for vagueness, and scalability [6]. First, these models are easier to understand for end-users, as they circumvent the complexity often associated with formal models like Petri nets and BPMN. Secondly, they accommodate vagueness, allowing them to represent the majority of process behaviors without having to account for every outlier or deviation. Finally, directly-follows graphs are highly scalable. They can handle logs with millions of events efficiently, which is a crucial requirement for commercial process mining tools.

Many techniques have been developed to enhance process mining using DFGs. First, various discovery techniques aim to lessen the complexity of DFGs. For instance, Heuristic Miner [23] excludes infrequent behavioral relations based on the frequency of occurrences. Leemans [16] present a technique for conformance checking and performance analysis with DFGs. They propose a DFG-to-Petri-net conversion to align the model with a log. This alignment serves as a foundation for analyzing discrepancies between the log and the model. Subsequently, they determine performance metrics by projecting alignments onto DFGs. In our work, we employ the method proposed by Leemans [16] to calculate alignments for each DFG that forms a part of an OC-DFG. The alignment is then projected onto the DFG to compute object-centric performance metrics.

3   Preliminaries

In this section, we introduce object-centric event logs, object-centric directly-follows graphs, and alignments.

3.1   Object-Centric Event Logs (OCELs)

An object-centric event log overcomes the limitation of traditional event logs by relating events to multiple interacting objects [1]. To define object-centric event logs, we first introduce several universes.

Definition 1 (Universes). $\mathbb{U}_{ei}$ *is the universe of event identifiers,* $\mathbb{U}_{oi}$ *is the universe of object identifiers,* $\mathbb{U}_{act}$ *is the universe of activity names,* $\mathbb{U}_{time}$ *is the universe of timestamps,* $\mathbb{U}_{ot}$ *is the universe of object types,* $\mathbb{U}_{attr}$ *is the universe of attributes,* $\mathbb{U}_{val}$ *is the universe of values, and* $\mathbb{U}_{map} = \mathbb{U}_{attr} \nrightarrow \mathbb{U}_{val}$ *is the universe of attribute-value mappings. For any* $f \in \mathbb{U}_{map}$ *and* $x \notin dom(f)$, $f(x) = \bot$.

Using the universes, we define an object-centric event log.

Definition 2 (Object-Centric Event Log). *An object-centric event log is a tuple* $L = (E, O, \mu, R)$, *where* $E \subseteq \mathbb{U}_{ei}$ *is a set of events,* $O \subseteq \mathbb{U}_{oi}$ *is a set of objects,* $\mu \in (E \to \mathbb{U}_{map}) \cup (O \to \mathbb{U}_{map})$ *is a mapping, and* $R \subseteq E \times O$ *is a relation, s.t. for any* $e \in E$, $\mu(e)(act) \in \mathbb{U}_{act}$ *and* $\mu(e)(time) \in \mathbb{U}_{time}$, *and for any* $o \in O$, $\mu(o)(type) \in \mathbb{U}_{ot}$. $\prec_L \subset E \times E$ *is a total order such that for any pair of events* $e_1, e_2 \in E : e_1 \prec_L e_2$ *implies* $e_1 \neq e_2$ *and* $\mu(e_1)(time) \leq \mu(e_2)(time)$.

For the sake of brevity, we denote $\mu(e)(x)$ as $\mu_x(e)$ and $\mu(o)(x)$ as $\mu_x(o)$. Table 1 describes a fragment of a simple event log $L_1 = (E_1, O_1, \mu_1, R_1)$ with $E_1 = \{e_1, e_2, \dots\}$, $O_1 = \{o_1, , i_1, i_2, \dots\}$, $R_1 = \{(e_1, o_1), (e_1, i_1), \dots\}$, $\mu_{act}(e_1) = po$, $\mu_{time}(e_1) = $ 25-11-2023:09.35, $\mu_{type}(o_1) = order$, and $\mu_{type}(i_1) = item$.

The event sequence and trace of an object are defined as follows.

Table 1. A fragment of an event log

event id	activity	timestamp	order	item
e_1	place order (po)	25-11-2023:09.35	$\{o_1\}$	$\{i_1, i_2, i_3\}$
e_2	check availability (ca)	25-11-2023:13.35	$\emptyset$	$\{i_1\}$
e_3	check availability (ca)	25-11-2023:18.35	$\emptyset$	$\{i_2\}$
e_4	send invoice (si)	26-11-2023:15.35	$\{o_1\}$	$\emptyset$
...	...	...	...	...

Definition 3 (Event Sequences and Traces). *Let $L = (E, O, \mu, R)$ be an object-centric event log. For object $o \in O$, $seq(o) = \langle e_1, e_2, \ldots, e_n \rangle$ s.t. $\{e_1, e_2, \ldots, e_n\} = \{e \in E \mid (e, o) \in R\}$ and $e_i \prec_L e_j$ for any $1 \leq i < j \leq n$ is the sequence of all events where the object is involved in. For object $o \in O$, $trace(o) = \langle a_1, a_2, \ldots, a_n \rangle$ s.t. $seq(o) = \langle e_1, e_2, \ldots, e_n \rangle$ and $a_i = \mu_{act}(e_i)$ for any $1 \leq i \leq n$ is the trace of the object.*

Given L_1, $seq(o_1) = \langle e_1, e_4 \rangle$ and $trace(o_1) = \langle po, si \rangle$.

3.2   Object-Centric Directly-Follows Graphs (OC-DFGs)

An OC-DFG can be conceptualized as a layered arrangement of Directly-Follows Graphs (DFGs), where each DFG corresponds to a distinct object type within a business process. In this layered structure, the arcs, as well as the start and end activities, are always associated with a specific object type. The interlinking within this multilayered construct is achieved through shared activities which, acting as interconnecting nodes, bridge different object types.

Definition 4 (OC-DFGs). *An OC-DFG $M = (OT, A, A_{start}, A_{end}, R)$ consists of*

- *$OT \subseteq \mathbb{U}_{ot}$ is a set of object types,*
- *$A \subseteq \mathbb{U}_{act}$ is a set of activities,*
- *$A_{start} = \{start_{ot} \mid ot \in OT\}$ is a set of start activities and $A_{end} = \{end_{ot} \mid ot \in OT\}$ is a set of end activities such that $(A_{start} \cup A_{end}) \cap A = \emptyset$, and*
- *$R \subseteq \{(ot, a_1, a_2) \in OT \times (A \cup A_{start}) \times (A \cup A_{end}) \mid (a_1 \in A_{start} \implies a_1 = start_{ot}) \wedge (a_2 \in A_{end} \implies a_2 = end_{ot})\}$ is a set of edges labeled with the corresponding object type.*

Figure 1 shows an OC-DFG with *order*, *item*, and *package* as object types. An OC-DFG can be discovered from an object-centric event log.

Definition 5 (Discovering OC-DFGs). *Let $L = (E, O, \mu, R)$ be an object-centric event log. $disc(L) = (OT, A, A_{start}, A_{end}, R)$ is the corresponding OC-DFG where*

- *$OT = \{\mu_{type}(o) \mid o \in O\}$ is the set of object types,*
- *$A = \{\mu_{act}(e) \mid e \in E\}$ is the set of activities,*
- *$R = \{(ot, a_i, a_{i+1}) \mid o \in O \wedge ot = \pi_{type}(o) \wedge trace(o) = \langle a_1, a_2, \ldots, a_n \rangle \wedge a_0 = start_{ot} \wedge a_{n+1} = end_{ot} \wedge 0 \leq i \leq n\}$ is the set of edges with the corresponding object type.*

4   Conformance Checking

Conformance checking techniques compare an event log with a process model, which can be constructed either manually or by process discovery techniques. The commonly used technique for conformance checking involves calculating alignments on Petri nets. In this section, we describe how alignments can be applied to OC-DFGs and visualized accordingly.

Figure 2 shows an overview of the approach. First, an OC-DFG is translated into multiple workflow nets, each of which represents a unique object type. The OCEL is then converted into a set of event logs projected on different object types. Subsequently, alignments are calculated for each workflow net and event log pair. Finally, these individual alignments are visualized within the OC-DFG.

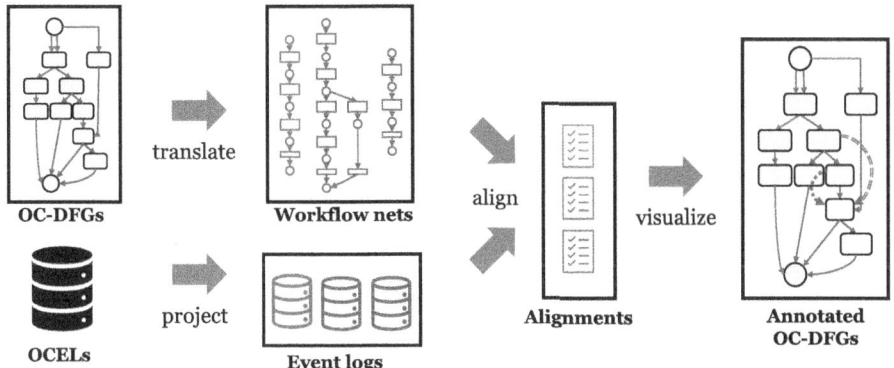

Fig. 2. An overview of conformance checking using OC-DFGs

4.1   Translating an OC-DFG to Workflow Nets

First, we translate an OC-DFG, i.e., $M = (OT, A, A_{start}, A_{end}, R)$, into a set of DFGs projected on object types OT. For $ot \in OT$, a projected DFG ($DFG_{ot} = (A_{ot}, start_{ot}, end_{ot}, R_{ot})$) consists of

- a set of activity nodes, i.e., $A_{ot} = \{a \in A \mid \exists_{a' \in A} (ot, a, a') \in R \vee (ot, a', a) \in R\}$,
- a start node, i.e., $start_{ot} \in A_{start}$,
- an end node, i.e., $end_{ot} \in A_{end}$, and
- a set of edges, i.e., $R_{ot} = \{(s, t) \in (A_{ot} \cup \{start_{ot}\}) \times (A_{ot} \cup \{end_{ot}\}) \mid (ot, s, t) \in R\}$.

Figure 3(a) shows an example of projecting M_1 to DFG_{item}.

Next, we translate each DFG to a workflow net. A translated workflow net ($PN_{ot} = (P, T, F, l)$) from a DFG ($DFG_{ot} = (A_{ot}, start_{ot}, end_{ot}, R_{ot})$) consists of

- a set of places that correspond to the nodes, i.e., $P = A_{ot} \cup \{start_{ot}, end_{ot}\}$,
- a set of transitions correspond to the edges, i.e., $T = R_{ot}$,
- a set of arcs that connect the transitions (i.e., the edges) to the corresponding places (i.e., the source and target nodes), i.e., $F = \bigcup_{(s,t) \in R_{ot}} \{(s, (s, t)), ((s, t), t)\}$, and
- a labeling function l such that, for $(s, t) \in T$, $l(s, t) = t$ if $t \neq end_{ot}$. $l(s, t) = \tau$ otherwise.

Figure 3(b) shows translating DFG_{item} to PN_{item}.

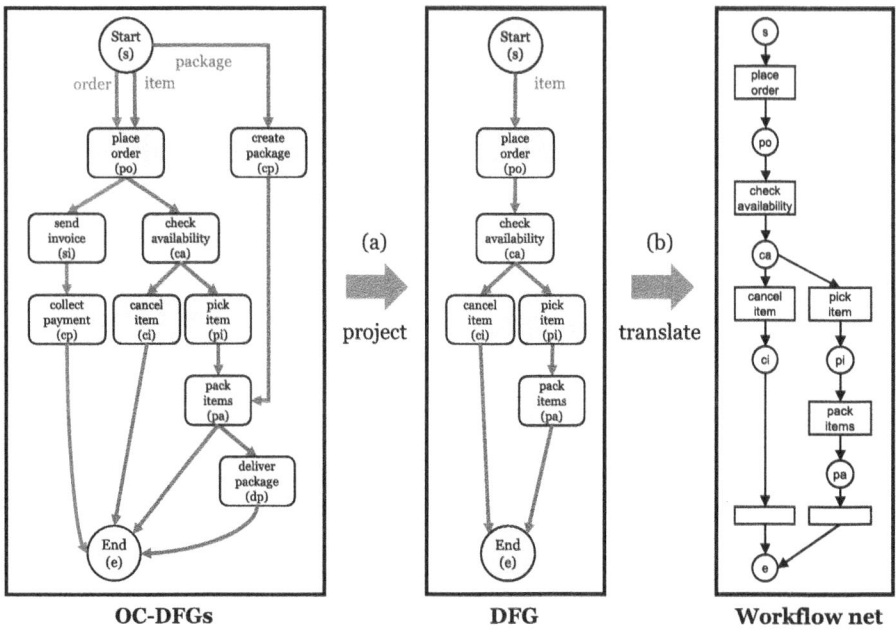

OC-DFGs   **DFG**   **Workflow net**

Fig. 3. (a) A OC-DFG is projected on *item* object type, resulting in a DFG representing the process model of items. (b) A projected DFG is translated to a workflow net.

4.2 Projecting an OCEL to Event Logs

We convert an object-centric event log $L = (E, O, \mu, R)$ into projected event logs, i.e., $\{L_{ot_1}, L_{ot_2}, \dots\}$, each corresponding to an object type of the OC-DFG. For any object type ot, we create a projected event log $L_{ot} = [seq(o) \mid o \in O \land \mu_{type}(o) = ot]$. Given the object-centric event log L described in Table 2, we derive $L_{item} = [\langle e_{11}, e_{12}, e_{13}, e_{21} \rangle, \dots]$ and $L_{package} = [\langle e_{20}, e_{21}, e_{22} \rangle, \dots]$.

4.3 Aligning Event Logs with Workflow Nets

Optimal alignments are then computed for each object type ot, utilizing the corresponding projected event log L_{ot} and the translated workflow net PN_{ot}. Several techniques have been developed to calculate optimal alignments. The most recognized approach employs the A^* algorithm to identify the shortest path in the reachability graph of a synchronous product net [3], using the marking equation of workflow nets. Figure 4 showcases optimal alignments of the package and item object types, respectively. For example, *item2* has a log move at e_{17} (*pick item*), e_{18} (*inspect item*), and e_{19} (*report item*), whereas *item3* has a model move at *pick item*.

package1's alignment				item1's alignment				item2's alignment							item3's alignment			
log	e_{20}	e_{21}	e_{22}	e_{11}	e_{12}	e_{13}	e_{21}	e_{11}	e_{14}	e_{15}	e_{17}	e_{18}	e_{19}	e_{21}	e_{11}	e_{16}	$\gg$	e_{21}
model	cp	pa	dp	po	ca	pi	pa	po	ca	pi	$\gg$	$\gg$	$\gg$	pa	po	ca	pi	pa

... (to the right)

Fig. 4. Optimal alignments for the package and item object types

4.4 Visualizing Alignments

Alignments result in two types of deviations: log moves and model moves. Log moves suggest that an event appeared in the log that the model did not permit. These are visualized using a dotted edge connecting model nodes, indicating additional events executed between the source and target activities in the model. There are five distinct situations in which log moves may occur, and each is visualized in a different way. Assuming that Fig. 5(a) presents a DFG for a specific object type, the situations can be explained as follows:

1. A log move at the beginning of an alignment, e.g., $\langle(e_1, \gg), (e_2, a), (e_3, b)\rangle$, is visualized as a dotted edge connecting the start node to the activity of the first synchronous move (cf. Fig. 5(b)).
2. A log move in the middle of an alignment, e.g., $\langle(e_1, a), (e_2, \gg), (e_3, b)\rangle$, is visualized as a dotted edge connecting the activities of the preceding and following synchronous moves (cf. Fig. 5(c)). Additionally, we handle two special cases:
   (a) If $\mu_{act}(e_2) = a$, the log move is visualized as a self-loop to the activity of the preceding synchronous move (cf. Fig. 5(d)).
   (b) If $\mu_{act}(e_2) = b$, the log move is visualized as a self-loop to the activity of the following synchronous move (cf. Fig. 5(e)).

Table 2. A fragment of an event log.

event id	activity	timestamp	order	item	package
...	...	...	...	...	...
e_{11}	place order (po)	27-11-2023:09.35	$\{o_1\}$	$\{i_1, i_2, i_3\}$	$\emptyset$
e_{12}	check availability (ca)	27-11-2023:13.35	$\emptyset$	$\{i1\}$	$\emptyset$
e_{13}	pick item (pi)	28-11-2023:11.35	$\emptyset$	$\{i1\}$	$\emptyset$
e_{14}	check availability (ca)	28-11-2023:12.35	$\emptyset$	$\{i2\}$	$\emptyset$
e_{15}	pick item (pi)	28-11-2023:18.35	$\emptyset$	$\{i2\}$	$\emptyset$
e_{16}	check availability (ca)	29-11-2023:09.35	$\emptyset$	$\{i3\}$	$\emptyset$
e_{17}	pick item (pi)	29-11-2023:10.35	$\emptyset$	$\{i2\}$	$\emptyset$
e_{18}	inspect item (ii)	29-11-2023:15.35	$\emptyset$	$\{i2\}$	$\emptyset$
e_{19}	report item (ri)	29-11-2023:17.35	$\emptyset$	$\{i2\}$	$\emptyset$
e_{20}	create package (cp)	30-11-2023:11.35	$\emptyset$	$\emptyset$	$\{p1\}$
e_{21}	pack items (pa)	30-11-2023:12.35	$\emptyset$	$\{i1, i2, i3\}$	$\{p1\}$
e_{22}	deliver package (dp)	02-12-2023:17.35	$\emptyset$	$\emptyset$	$\{p1\}$
...	...	...	...	...	...

3. A log move at the end of an alignment, e.g., $\langle (e_1, a), (e_2, b), (e_3, \gg) \rangle$, is visualized as a dotted edge connecting the activity of the last synchronous move to the end node (cf. Fig. 5(f)).

Conversely, model moves, indicated by a double-line edge bypassing a model node, suggest that an event that the model dictates should have occurred was missing from an event log trace.

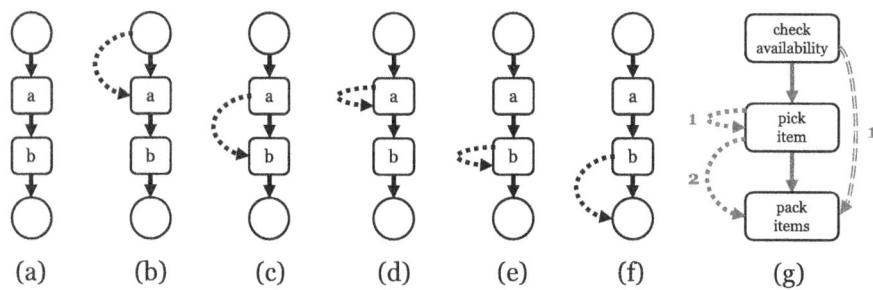

Fig. 5. (a) A DFG, (b–f) Five different types of log moves, and (g) Log moves (dotted lines) and model moves (double line) visualized on an OC-DFG.

Figure 5(g) visualizes the alignments described in Fig. 4. For example, the dotted edge connecting *pick item* node shows the log move of *pick item* (with 1 indicating *item2*'s log move at e_{17}, i.e., *pick item*), while the dotted edge connecting *pick item* to *pack items* indicates the log move between *pick item* and *pack items* (with 2 denoting *item2*'s log moves at e_{18}, i.e., *inspect item*, and e_{19}, i.e., *report item*, respectively). Moreover, the double-line edge in Fig. 5(g) demonstrates the model move at *pick item* (with 1 indicating *item3*'s model move at *pick item*).

5   Performance Analysis

Using the alignments computed in the previous section, we calculate various object-centric performance metrics in relation to the process model. Firstly, the alignments are projected onto the process model. An *event relationship* is then established by linking each event for which performance metrics need to be computed with the preceding events. Lastly, object-centric performance measures for an event are computed using its associated event relationship.

5.1   Projecting Alignments

First, we project alignments on the process model. The synchronous moves of an alignment are linked to the corresponding activities in the process model. For instance, synchronous moves of *item1*, such as (e_{11}, po), (e_{12}, ca), (e_{13}, pi), and (e_{21}, pa), are linked to respective activities *place order*, *check availability*, *pick item*, and *pack items*, as depicted in Fig. 6(a).

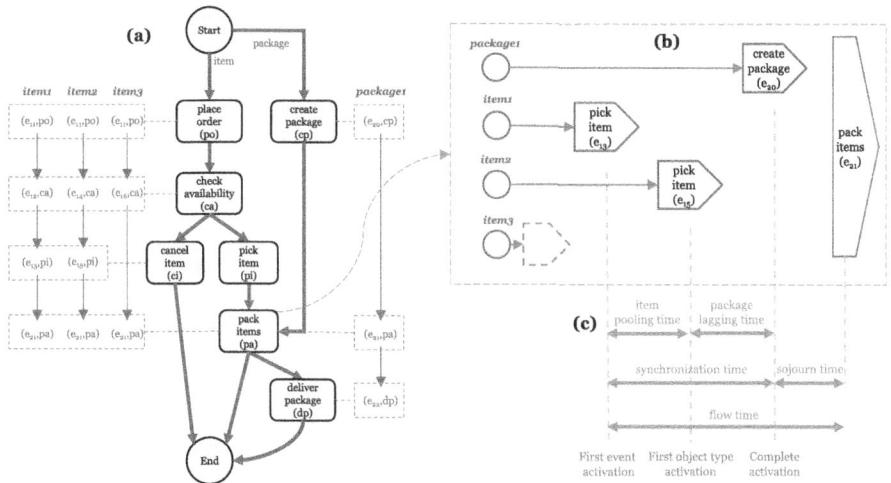

Fig. 6. (a) A OC-DFG projected by alignments. (b) A event relationship of e_{21} (*pack items*). (c) Object-centric performance metrics of e_{21} (*pack items*).

5.2 Establishing Event Relationship

Then, we define an event relationship for each event by associating the event with its preceding events. Figure 6(b) demonstrates an event relationship of e_{21}, where *item1*'s e_{13}, *item2*'s e_{15}, and *package1*'s e_{20} are the preceding events. This is done by first identifying the objects involved in the event by collecting the synchronous moves for the event. For instance, the synchronous moves for e_{21} include *item1*'s (e_{21}, pa), *item2*'s (e_{21}, pa), *item3*'s (e_{21}, pa), and *package1*'s (e_{21}, pa). From this, it is deduced that *item1*, *item2*, *item3*, and *package1* are involved in e_{21}.

Next, for each object, we identify the event that corresponds to the preceding activity of the target activity in the model, e.g., the preceding activity of *pack items* is *pick item* and *create package*. For example, *item1*'s e_{13} and *item2*'s e_{15} correspond to the activity *pick item*, while *package1*'s e_{20} corresponds to activity *create package*. Note that *item3* does not have an event that corresponds to the preceding activity of *pack items*.

5.3 Object-Centric Performance Measures

With the event relationship of an event, we can calculate various object-centric performance measures. To achieve this, we introduce the concepts of *first event activation*, *first object type activation*, and *complete activation*. The first event activation refers to the first event in the event relationship. For instance, e_{13} is the first event activation for the event relationship described in Fig. 6(b).

Next, the first object type activation refers to the last event of the object type that occurs the earliest among other object types. For example, e_{15} is the first object type activation as it is the last event of the *item* object type, which occurs earlier than the last event of the *package* object type. Finally, the complete activation refers to the last event among the preceding events, i.e., e_{20}.

Using these concepts, we explain each object-centric performance measure as follows. Figure 6(c) shows an example of object-centric performance measures related to e_{21} (*pack items*).

- *Sojourn time* is the time difference between the target event and the complete activation.
- *Synchronization time* is the time difference between the complete activation and the first event activation.
- *Flow time* is the time difference between the target event and the first event activation.
- *Lagging time for an object type* is the time difference between the last event of the object type and the first object type activation.
- *Pooling time for an object type* is the interval between the last event of the object type and the first event of the object type.

6   Application and Discussion

This section presents a tool that implements the proposed conformance checking and performance analysis. Next, we conduct a use case study on a real-life process using the tool. Finally, we discuss the utility and limitations of the proposed method.

6.1   Implementation: *Explori*

The approach described in this paper has been implemented as an open-source web application named *Explori*, based on the Python library OCPA [8]. The source code and manual are available at https://github.com/gyunamister/Explori. The tool consists of four functional components: *event log management, process discovery, conformance checking*, and *performance analysis*.

Event Log Management. This component aims to support users to manage and interact with Object-Centric Event Logs (OCELs). It allows users to view available OCELs, delete existing OCELs, upload new OCELs, and select an OCEL to start a new analysis. Users can upload an OCEL in either CSV or JSONOCEL formats. After the file is uploaded, it will be displayed in the list of available OCELs. The deletion of an OCEL will remove all associated information, such as the cached analysis results.

Process Discovery. This component aims to discover an OC-DFG from a selected OCEL. Figure 7(a) is the OC-DFG that is discovered from an uploaded OCEL. Each node in the OC-DFG represents an activity that occurred in the

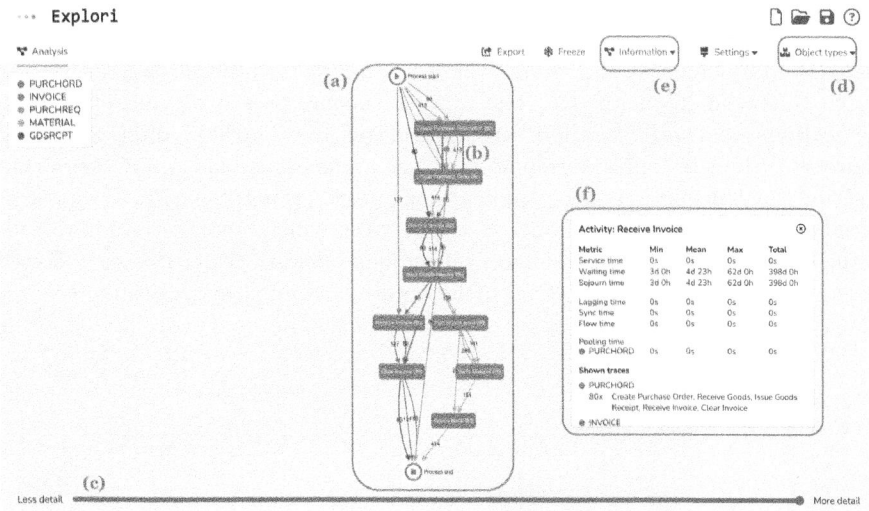

Fig. 7. A screenshot of *Explori* (Color figure online)

event log, while the edges indicate the directly-follows relation between the source activity and target activity. Multiple edges between two nodes signify the involvement of multiple object types in both activities, with each edge color representing a specific object type. For instance, Fig. 7(b) indicates the involvement of *MATERIAL* (represented as the yellow-colored edge) and *PURCHREQ* (represented as the green-colored edge) in *Create Purchase Requisition* and *Create Purchase Order*. The graph can be panned, zoomed, and nodes can be repositioned for better visualization.

OCELs often contain a mix of typical process behavior and outlier cases, which can complicate the model. To filter the graph and display only the most frequent behavior, users can adjust *threshold slider* at the bottom of the page (cf. Fig. 7(c)). The threshold represents the proportion of objects shown in the model. Moreover, users can also filter the graph by selecting specific object types of interest through the "Object Types" button in the top right corner shown in Fig. 7(d).

Conformance Checking. This component aims to compute alignments. The information dropdown in the navigation bar (cf. Fig. 7(e)) allows users to display alignment information within the process model. Log moves are visualized as dotted edges, while model moves are visualized as double-line edges.

Performance Analysis. This component aims to compute various performance measures. Users can click on nodes and edges to display the information box that contains object-centric performance metrics for the selected element, as shown in Fig. 7(f). The performance metrics include the flow, sojourn, synchronization, pooling, and lagging time.

6.2   Real-Life Use Case: Loan Application Process

Using the implementation, we analyze a real-life loan application process of a Dutch Financial Institute [13]. The process encompasses two types of objects: *applications* and *offers*, where an application may include multiple offers. Figure 8(a) depicts a process model of 1,317 applications and 4,457 offers that describe the cancellation of the applications under various scenarios. The process initiates when a customer submits an application, which, upon acceptance, leads to the bank generating and communicating loan offers to the customer, followed by the eventual cancellation of both the application and any associated offers.

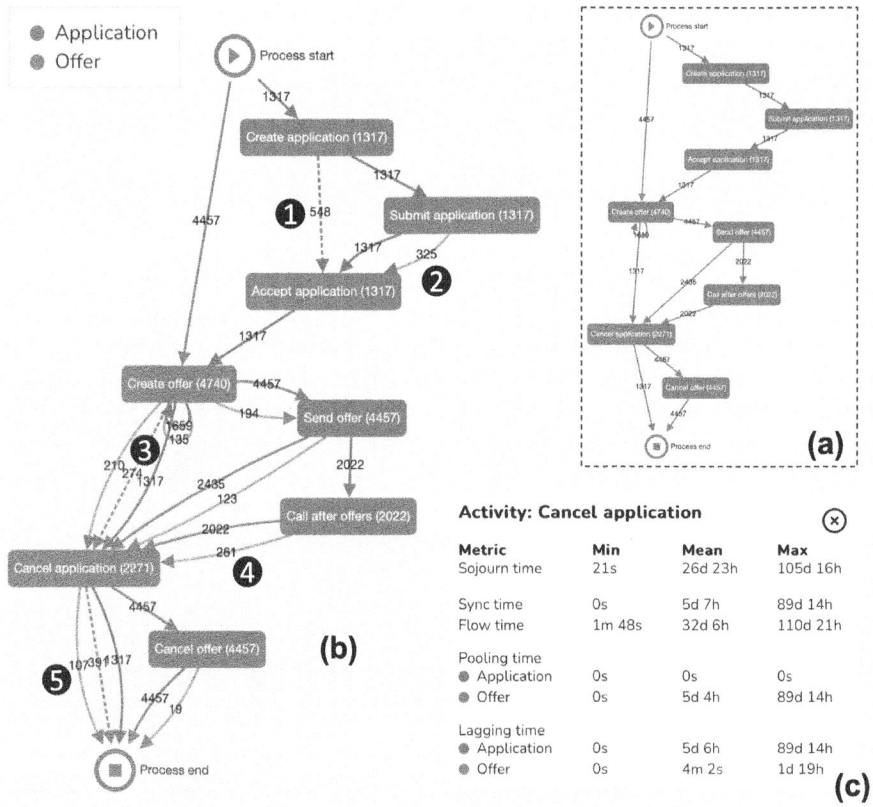

Fig. 8. (a) A OC-DFG representing the loan application process, (b) conformance checking visualization, and (c) performance analysis of *cancel application* activity

Conformance Checking. A conformance check was performed using an event log comprising 31,010 events related to the 2,302 applications and 5,128 offers. Figure 8(b) visualizes the conformance checking outcomes. Model moves at ① indicate the omission of the *submit application* activity for 548 applications, attributed to in-person customer visits negating the need for this activity. Log moves at ② highlight 325 applications where an additional *handle leads* activity was necessary prior to acceptance. Model moves at ③ show 274 offers being canceled immediately after creation, bypassing the *send offer* and *call after offers* activities. Log moves at ④ and ⑤ illustrate additional activities required for 261 offers (i.e., *return offer*) and 107 applications (i.e., *call incomplete files*), respectively, following cancellation.

Performance Analysis. The cancellation process for applications engages a variable number of offers along with their corresponding applications. As per the process model, three activities precede the action of canceling applications. Within the lifecycle of applications, the preceding activity is consistently *create offer*, whereas the preceding activities for offers encompass *create offer*, *send offer* and *call after offers* activities.

The average sojourn time, i.e., the duration from the initiation of the preceding activities until the cancellation of an application, is approximately 26 days and 23 h. Thus, it suggests that, on average, about 26 days elapse to cancel applications following the execution of *create offer*, *send offer*, and *call after offers*. The synchronization time, i.e., the duration between completing the first and last of these activities before cancelling an application, averages at 5 days and 7 h, suggesting a coordination time taken for completing all necessary prerequisites.

Furthermore, the average pooling time for offers, defined as the period from when the first to the last offer is prepared for cancellation, is 5 days and 6 h. The average lagging time of applications at 5 days and 6 h, contrasts with a notably shorter lagging time of offers, merely 4 min. This indicates that applications are typically cancelled only after consolidating all related offers.

6.3   Discussion

The use case study conducted offers valuable insights into the practical application of OC-DFGs for conformance checking and performance analysis in a real-life loan application process. The application of our approach has demonstrated its utility in simplifying complex process models and making them more accessible for analysis.

However, the proposed approach has limitations that stem from the "decompose and recombine" paradigm, particularly in the recombination phase, where a coherent global alignment is needed. Currently, the approach lacks a systematic way to ensure that shared transitions across local alignments are treated consistently. This can lead to incoherencies in the global view of the process when different object views suggest different local alignments of the same transition. For instance, consider the transition *cancel application*, which occurs in the life cycle of both offers and applications. In the local alignment for the *offer*, we might decide to model skip the *cancel application*. However, this same transition, when viewed from the *application* perspective, could be critical and thus included in the alignment. This discrepancy leads to an incoherent global view where the offer is canceled, but the application is not, resulting in a conflicting representation of the process.

Such inconsistencies are particularly problematic in performance analysis. For example, if the *cancel application* is skipped in one object view but not in another, it would be unclear how to calculate metrics such as synchronization time related to *cancel application* activity within the broader context of the application process.

Additionally, the simplification of synchronization semantics in the translation from DFGs to workflow nets is a limitation. This overlooks the dynamics of concurrent processes and may lead to a misrepresentation of performance metrics that are subject to the sequencing of interdependent activities. For instance, in our case study involving the loan application process, activities such as *send* and *call* may occur in parallel. However, if we were to represent this scenario in a simplified DFG, the concurrency will be depicted as a series of sequential steps.

This sequential representation could significantly skew performance analysis. Suppose *call* typically takes longer than *send*. In that case, the time between the start of *send* and its completion is not just the execution time of that single activity but also includes the waiting time until *call* is completed.

To address these challenges, a systematic approach should be developed for the recombination phase, which can reconcile the local alignments into a coherent global alignment, thereby aligning shared transitions consistently across all object views. Moreover, future work should explore simple, but richer process modeling formalisms that can capture the concurrent nature of business processes. This could involve incorporating elements from other process modeling languages that include explicit constructs for concurrency and synchronization.

7   Conclusion

In this paper, we address the gap in the literature concerning OC-DFGs for conformance checking and performance analysis in object-centric process mining. OC-DFGs, while noted for their simplicity and practical applications, have previously lacked comprehensive support for these essential tasks. We developed and presented an approach that utilizes OC-DFGs for both conformance checking and performance analysis. By translating each directly-follows graph in an OC-DFG to a workflow net, we compute alignments for improved conformance checking. Additionally, we utilize these alignments for performance analysis, computing object-centric performance metrics. The proposed approach is fully implemented as a web application and is further validated through a case study on a real-life loan application process.

References

1. van der Aalst, W.M.P.: Object-centric process mining: dealing with divergence and convergence in event data. In: Ölveczky, P.C., Salaün, G. (eds.) SEFM 2019. LNCS, vol. 11724, pp. 3–25. Springer, Cham (2019). https://doi.org/10.1007/978-3-030-30446-1_1

2. van der Aalst, W.M.P.: Object-centric process mining: the next frontier in business performance. White paper, Celonis (2023)

3. van der Aalst, W.M.P., Adriansyah, A., van Dongen, B.F.: Replaying history on process models for conformance checking and performance analysis. WIREs Data Min. Knowl. Discov. **2**(2), 182–192 (2012)

4. van der Aalst, W.M.P., Berti, A.: Discovering object-centric Petri nets. Fundam. Informaticae **175**(1–4), 1–40 (2020)

5. van der Aalst, W.M.P., Li, G., Montali, M.: Object-centric behavioral constraints. CoRR abs/1703.05740 (2017)

6. van der Aalst, W.M.P., De Masellis, R., Di Francescomarino, C., Ghidini, C.: Learning hybrid process models from events - process discovery without faking confidence. In: Carmona, J., Engels, G., Kumar, A. (eds.) BPM 2017. LNCS, vol. 10445, pp. 59–76. Springer, Cham (2017). https://doi.org/10.1007/978-3-319-65000-5_4

7. Adams, J.N., van der Aalst, W.M.P.: Precision and fitness in object-centric process mining. In: ICPM 2021, pp. 128–135 (2021)

8. Adams, J.N., Park, G., van der Aalst, W.M.P.: ocpa: A Python library for object-centric process analysis. Softw. Impacts **14**, 100438 (2022)

9. Barenholz, D., Montali, M., Polyvyanyy, A., Reijers, H.A., Rivkin, A., van der Werf, J.M.E.M.: There and back again - on the reconstructability and rediscoverability of typed Jackson nets. In: Gomes, L., Lorenz, R. (eds.) PETRI NETS 2023. LNCS, vol. 13929, pp. 37–58. Springer, Cham (2023). https://doi.org/10.1007/978-3-031-33620-1_3

10. Berti, A., van der Aalst, W.: Extracting multiple viewpoint models from relational databases. In: Ceravolo, P., van Keulen, M., Gómez-López, M.T. (eds.) SIMPDA 2018-2019. LNBIP, vol. 379, pp. 24–51. Springer, Cham (2020). https://doi.org/10.1007/978-3-030-46633-6_2

11. Cohn, D., Hull, R.: Business artifacts: a data-centric approach to modeling business operations and processes. IEEE Data Eng. Bull. **32**(3), 3–9 (2009)

12. Diba, K., Batoulis, K., Weidlich, M., Weske, M.: Extraction, correlation, and abstraction of event data for process mining. WIREs Data Min. Knowl. Discov. **10**(3), e1346 (2020)

13. van Dongen, B.: BPI challenge 2017 (2017). https://doi.org/10.4121/UUID: 5F3067DF-F10B-45DA-B98B-86AE4C7A310B.   https://data.4tu.nl/articles/_/ 12696884/1

14. Fahland, D.: Describing behavior of processes with many-to-many interactions. In: Donatelli, S., Haar, S. (eds.) PETRI NETS 2019. LNCS, vol. 11522, pp. 3–24. Springer, Cham (2019). https://doi.org/10.1007/978-3-030-21571-2_1

15. Fahland, D., de Leoni, M., van Dongen, B.F., van der Aalst, W.M.P.: Conformance checking of interacting processes with overlapping instances. In: Rinderle-Ma, S., Toumani, F., Wolf, K. (eds.) BPM 2011. LNCS, vol. 6896, pp. 345–361. Springer, Heidelberg (2011). https://doi.org/10.1007/978-3-642-23059-2_26

16. Leemans, S.J.J., Poppe, E., Wynn, M.T.: Directly follows-based process mining: exploration & a case study. In: ICPM 2019, pp. 25–32. IEEE (2019)

17. Li, G., de Carvalho, R.M., van der Aalst, W.M.P.: Automatic discovery of object-centric behavioral constraint models. In: Abramowicz, W. (ed.) BIS 2017. LNBIP, vol. 288, pp. 43–58. Springer, Cham (2017). https://doi.org/10.1007/978-3-319-59336-4_4

18. Lu, X.: Artifact-centric log extraction and process discovery. Unpublished Master's thesis, Eindhoven University of Technology (2013)

19. Montali, M., Rivkin, A.: DB-Nets: on the marriage of colored Petri nets and relational databases. In: Koutny, M., Kleijn, J., Penczek, W. (eds.) Transactions on Petri Nets and Other Models of Concurrency XII. LNCS, vol. 10470, pp. 91–118. Springer, Heidelberg (2017). https://doi.org/10.1007/978-3-662-55862-1_5

20. Park, G., van der Aalst, W.M.P.: Monitoring constraints in business processes using object-centric constraint graphs. In: Montali, M., Senderovich, A., Weidlich, M. (eds.) ICPM 2022. LNBIP, vol. 468, pp. 479–492. Springer, Cham (2022). https://doi.org/10.1007/978-3-031-27815-0_35

21. Park, G., Adams, J.N., van der Aalst, W.M.P.: OPerA: object-centric performance analysis. In: Ralyté, J., Chakravarthy, S., Mohania, M., Jeusfeld, M.A., Karlapalem, K. (eds.) ER 2022. LNCS, vol. 13607, pp. 281–292. Springer, Cham (2022). https://doi.org/10.1007/978-3-031-17995-2_20

22. Park, G., Comuzzi, M., van der Aalst, W.M.P.: Analyzing process-aware information system updates using digital twins of organizations. In: Guizzardi, R., Ralyté, J., Franch, X. (eds.) RCIS 2022. LNBIP, vol. 446, pp. 159–176. Springer, Cham (2022). https://doi.org/10.1007/978-3-031-05760-1_10

23. Weijters, A., van der Aalst, W., Alves De Medeiros, A.: Process mining with the HeuristicsMiner algorithm. BETA publicatie: working papers (2006)

A Universal Approach to Feature Representation in Dynamic Task Assignment Problems

Riccardo Lo Bianco[1,2]([⊠]), Remco Dijkman[1,2], Wim Nuijten[1,2], and Willem van Jaarsveld[1,2]

[1] Eindhoven University of Technology, Eindhoven, The Netherlands
{r.lo.bianco,r.m.dijkman,w.p.m.nuijten,w.l.v.jaarsveld}@tue.nl
[2] Eindhoven Artificial Intelligence Systems Institute, Eindhoven, The Netherlands

Abstract. Dynamic task assignment concerns the optimal assignment of resources to tasks in a business process. Recently, Deep Reinforcement Learning (DRL) has been proposed as the state of the art for solving assignment problems. DRL methods usually employ a neural network (NN) as an approximator for the policy function, which ingests the state of the process and outputs a valuation of the possible assignments. However, representing the state and the possible assignments so that they can serve as inputs and outputs for a policy NN remains an open challenge, especially when tasks or resources have features with an infinite number of possible values. To solve this problem, this paper proposes a method for representing and solving assignment problems with infinite state and action spaces. In doing so, it provides three contributions: (I) A graph-based feature representation of assignment problems, which we call assignment graph; (II) A mapping from marked Colored Petri Nets to assignment graphs; (III) An adaptation of the Proximal Policy Optimization algorithm that can learn to solve assignment problems represented through assignment graphs. To evaluate the proposed representation method, we model three archetypal assignment problems ranging from finite to infinite state and action space dimensionalities. The experiments show that the method is suitable for representing and learning close-to-optimal task assignment policies regardless of the state and action space dimensionalities.

Keywords: Business Process Optimization · Assignment Problem · Markov Decision Process · Deep Reinforcement Learning · Graph Neural Networks

1 Introduction

Dynamic task assignment problems, referred to in this work as assignment problems, involve optimizing the allocation of resources to tasks in a process, where new tasks (and possibly new resources) become available at runtime. In their broad context, assignment problems encompass diverse resource and task types, compatibility constraints on the possible assignments, time-related variables such

© The Author(s), under exclusive license to Springer Nature Switzerland AG 2024
A. Marrella et al. (Eds.): BPM 2024 Forum, LNBIP 526, pp. 197–213, 2024.
https://doi.org/10.1007/978-3-031-70418-5_12

as resource schedules, time-related variables, and stochastic variables. In recent years, Deep Reinforcement Learning (DRL) was proposed as the state-of-the-art solution approach for assignment problems [1–3]. Most DRL methods rely on a neural network (NN) to encode a policy function. This function takes a representation of the state of a process, called an observation, as input and produces a valuation over the possible actions, the assignments, as output.

The ideal way to tackle an assignment problem would be to model it using widely recognized notations, like BPMN or Petri Nets (PNs), and let an algorithm learn a suitable policy using observations expressed through the same notation without additional feature engineering. This is the direction taken in [1], where a PN variant suitable to encode assignment problems is proposed, together with an algorithmic approach to train DRL models on the problems expressed through in PN form. However, in current DRL approaches, multi-layer perceptrons (MLPs) remain the most commonly used neural networks for policy approximation, and this is the case for [1] as well. MLPs require fixed-size observation vectors, which makes modeling the features of observation and action vectors challenging. In particular, the feature representation problem is challenging when observation and action vectors have unknown sizes during the design phase and could potentially grow to infinity.

Figure 1 exemplifies this problem. On the left, it depicts a process model representing a consulting firm with a single employee, r_1, who can perform projects. Projects are characterized by a type, a categorical value that identifies the project type, and a budget, a value in euros representing the project's profitability. The optimization objective is to maximize profits, given by the sum of projects' budgets. On the right, the figure presents the direct translation of the process state into the observation vectors that existing DRL solution approaches use [1–3]. The elements of the input vectors are the number of projects of each possible type, while those of the output vector are the possible assignments of resource to project type (if multiple projects of the same type are queued, a FIFO strategy is applied). Since the projects are characterized by an unbounded value (the budget), the input and output sizes grow to infinity. Consequently, feature representation in vectorial form is not suitable for this problem, nor for any problem having tasks or resources characterized by continuous or unbounded variables, which encompasses many real-world problems. A prominent example in this sense is given by problems where the features depend on time, which is naturally an unbounded metric. It is possible that suitable observation vectors can be devised through feature engineering for single problem instances, but this is a tedious and error-prone task. Moreover, it may lead to observations that do not encode all information in the state space, which results in suboptimal policies being learned. For example, in the problem presented in Fig. 1, the budget could be limited to a maximum value, but this choice would make the algorithm incapable of handling cases with a budget higher than the maximum. Finally, imposing a maximum budget would make the state and action spaces finite, but their dimensionalities would still grow exponentially with the number of available resources and task types in the system, which poses a significant limitation to the applicability of DRL as a solving method [4].

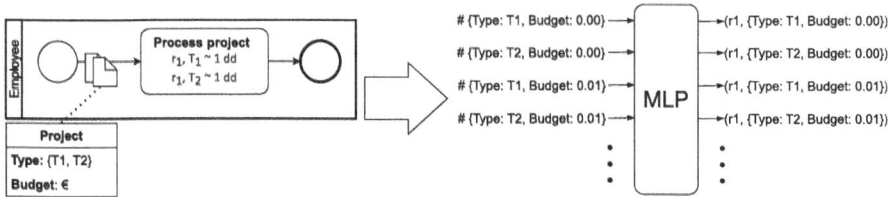

Fig. 1. Feature representation problem

In the general case, the representation problem is twofold: on the one hand, the observation vector grows to infinity whenever continuous or unbounded variables are to be taken into account, and on the other, the action vector also grows to infinity as the result of the need to enumerate all possible assignments.

In this work, we refine the methods presented in [1] by providing three contributions:

1. A graph-based feature representation approach for assignment problems, which we call assignment graph.
2. A mapping from marked Colored Petri Nets (CPN) to assignment graphs.
3. An adaptation of the Proximal Policy Optimization (PPO) algorithm that can learn to solve assignment problems expressed as assignment graphs.

To evaluate the proposed method, we provide three instances of the assignment problem with increasing state and action space dimensionality. We show empirically that the novel feature representation method allows us to seamlessly learn a close-to-optimal policy on assignment problems with finite and infinite state and action spaces.

Against this background, the remainder of this paper is structured as follows. Section 2 outlines the relevant literature. Section 3 introduces the necessary background knowledge and notation. Section 4 outlines the notion of assignment graph and the Petri Net to assignment graph mapping algorithm. Section 5 provides an overview of a DRL algorithm's design capable of handling dynamic assignment graphs. In Sect. 6, three assignment problem instances with increasing state and action space dimensionalities are presented, and the new feature representation method is applied to learn a close-to-optimal policy in all three cases. Section 7 discusses the proposed method's merits and limitations and reflects on possible directions of future research.

2 Related Work

The problem of encoding large or infinite state spaces for task assignment problems is well established in the operation research literature. Early works in this field highlighted limitations in solving variants of the fleet dispatching problem with infinite state spaces [5]. The most widely adopted technique used to handle such problems has been aggregation. This procedure involves projecting the original problem to a state space with low dimensionality, solving the aggregated problem, and disaggregating to find the solution to the original problem [6]. Multiple aggregation techniques have been proposed [7,8], but all entail arbitrary choices from the modeler or the costly tuning of ad-hoc hyperparameters. In the field of business process optimization, assignment problems were studied through predictive process monitoring [9,10] and prescriptive process monitoring [11,12]. However, these methods assume predefined task and resource types, which limits their applicability to finite state spaces. In contrast, this work proposes a seamless feature representation method suitable for infinite and state and action spaces.

We will frame assignment problems as a Markov Decision Process (MDP) to solve them through DRL. In this work, we will rely on the Action-Evolution Petri Net (A-E PN) framework [1], an extension of CPN suitable to model and solve MDPs. In their original definition, A-E PNs were limited to representing finite state spaces in vectorial form as shown in Fig. 3. In this work, we lift this limitation by defining a translation from a marked A-E PN to a graph, called assignment graph, that can be used as observation for a Graph Neural Network (GNN) to learn effective assignment policies.

In recent years, the use of graph-based observations has been explored as a feature representation tool for DRL applied to specific dynamic assignment problem classes, in particular in the field of transportation [13–15]. However, these works focus on a single problem instance and do not provide a reusable approach to feature representation for assignment problems.

The use of GNNs is increasingly being explored in predictive process monitoring [16,17]. Still, the predictive setting differs significantly from decision-making, and no direct translation is possible.

Some works employed GNNs to analyze Petri Nets [18,19]. In [18], the close relationship between graphs and PNs is exploited to address the problem of process discovery as a supervised learning problem. In the paper, the authors encode a set of execution traces as a graph, then extract an unmarked Petri Net from the graph and train a GNN to operate deletions on the Petri Net to improve the fitness of the process model against the traces in the dataset. The paper is one of the first to employ GNNs on Petri Nets, but the algorithm is only suitable for unmarked PNs. By contrast, the technique proposed in this work enables the representation of marked A-E PNs. In [19], the authors propose a novel GNN layer that handles a specific Petri Net variant combining timed-place Petri nets and a system of simple sequential processes with resources to optimize the dynamic scheduling problem of flexible manufacturing systems. In their work, the authors focus on defining the novel GNN layer considering a single flexible manufacturing system scheduling problem, thus tackling a smaller class of problems than ours. Moreover, the system considered does not model resources with different properties (e.g., colors) and does not deal with dynamic graphs. In the current work, we employ a general-purpose GNN layer to solve assignment problems, focusing on defining an easily adaptable translation technique from marked A-E PN to assignment graph.

3   Background

Since we aim to provide a universal approach to feature representation for task assignment problems, we need to introduce a suitable modeling method and a suitable solution method. This section provides the background regarding these two components: we present a CPN variant from the literature that can be used to model assignment problems with finite state spaces. We also delineate the main steps of the PPO algorithm and the DRL algorithm we employ as a solution method.

This work expresses assignment problems using Petri Nets as a modeling language. Basic PNs are suitable to model and simulate processes of arbitrary complexity, but in their basic form, they lack the expressive power to model decision processes. In particular, PNs are not equipped with constructs to define which actions can be taken, nor the goodness of the actions taken. To bridge this gap, Action-Evolution Petri Nets were proposed [1]. A-E PNs are an extension of Timed-Arc Colored Petri Nets (T-A CPN) that serve as a modeling and solving framework for assignment problems. This definition means any assignment problem modeled through A-E PN can be directly fed to a RL algorithm to learn a close-to-optimal policy without performing feature engineering. A-E PNs distinguish two types of transitions: action transitions represent events that require the agent to make an assignment by choosing the tokens used to fire them; evolution transitions represent events that happen in the environment independently of the agent's decisions and they are fired non-deterministically. Only one type of transition can fire at a given time, based on a network tag that is possibly updated every time a transition is fired. Transitions are associated with a reward signal to account for the actions' goodness.

A reduced version of A-E PN's formal definition is reported in Definition 1. For the complete definition and the firing rules, we refer the reader to [1].

Definition 1 (Action-Evolution Petri Net). *An Action-Evolution Petri Net (A-E PN) is a tuple $AEPN = (\mathcal{E}, P, T, F, C, G, E, I, L, l_0, \mathcal{R}, \rho_0)$, where $\mathcal{E}$ is a finite set of types called color sets, P is a finite set of places, T is a finite set of transitions, F is a finite set of arcs, $C : P \rightarrow \mathcal{E}$ is a color function that maps each place into a set of possible token colors, G is a guard function, E is an arc expression function, L is a transition tag function, I is an initialization function, l_0 is the network's initial tag, which can be one of 'a-transition' representing that the next step is a decision (or action) step or 'e-transition', representing that the next step is a regular firing of transitions, $\mathcal{R}$ is the transition reward function, ρ_0 is the initial network reward.*

The two elements that limit applicability of AE P-N to finite state space assignment problems are $\mathcal{E}$ and C. Consequently, we will adapt those in the next section.

Figure 2, taken from [1], shows how A-E PN is embedded in the RL cycle.

In the basic A-E PN framework, the observation manager is responsible for presenting a vectorized representation of the A-E PN marking to the agent, a

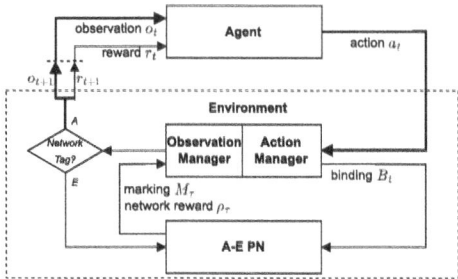

Fig. 2. The reinforcement learning cycle with A-E PN

vector containing, for each place, the number of tokens of each color in the place's color set. Similarly, the action manager is responsible for ingesting a vectorized representation of the action chosen by the agent, which is a vector having one element for each possible assignment. Both observation and action vectors are treated statically, and it is assumed that the tokens in the network fall into a limited set of types whose attributes' possible values are pre-defined by the PN color sets. The problem of state representation in assignment problems through A-E PN arises in the same modalities presented in Fig. 1.

In this work, we use a modified version of the observation manager and the action manager to handle assignment graphs of varying sizes instead of static vectors. Assignment Graphs are then fed to a specialized implementation of the PPO [20] algorithm to learn close-to-optimal assignment policies.

PPO is one of the most widely adopted DRL algorithms. It is an on-policy method, which means it learns the policy used to make environmental decisions using the policy itself. We can summarize PPO's functioning mechanism as the repetition of three steps:

1. **Samples collection:** the agent interacts with the environment following the current policy and collects a set of trajectories. Each trajectory contains a sequence of states, actions, rewards, and new states.
2. **Advantage computation:** the agent uses the generalized advantage estimation technique [21] to compute an estimate of how much better or worse each action was compared to the average action the current policy would have taken in that state.
3. **Policy and value function optimization:** the agent optimizes the policy to make the actions with higher advantages more likely and those with lower advantages less likely. This is done using a surrogate objective function that forces the new policy to stay close to the old policy. Parallelly, the agent updates the weights of a value network, an estimator of the total value of each encountered state based on the observed total discounted rewards at the end of trajectories to refine the computation of the advantage terms in the next steps of the optimization process.

4   Feature Representation Through Assignment Graphs

This section defines a feature representation approach, namely assignment graphs, suitable for representing assignment problems with both finite and infinite state spaces, such that they can serve as input to a Neural Network. To be able to model and simulate the desired assignment problem, we first introduce a suitable PN variant. We then provide the formal definition of assignment graph and its properties. Lastly, we present a two-step mapping from a marked PN to the corresponding assignment graph.

To provide a universal approach to feature representation for assignment problems, we first need to adopt a modeling technique that is suitable to represent assignment problems with finite or infinite state space. To this end, we provide a modified version of Definition 1, which we call attributed A-E PN.

Definition 2 (Attributed Action-Evolution Petri Net). *An Attributed Action-Evolution Petri Net (A-E PN) is a tuple $AEPN = (A, P, T, F, \mathcal{A}, G, E, I, L, l_0, \mathcal{R}, \rho_0)$, where $(P, T, F, G, E, I, L, l_0, \mathcal{R}, \rho_0)$ follow Definition 1, and:*

- *A is a finite set of types called attributes.*
- *$\mathcal{A}$ is an attribute function that maps each place p into a set of attributes. Each token on p must have a color that is composed of the token's time and a value for each of the attributes in $\mathcal{A}(p)$.*

The main difference between A-E PN (and accordingly CPN) and attributed A-E PN is that the former expects all the places' color sets to be finite and defined at the initialization. In contrast, the latter only expects the attributes of tokens that flow in each place to be given at the initialization, while the values associated with each attribute in a given marking do not need to be pre-defined. This work will consider cases where attributes represent values in $\mathbb{R}$. From this point onward, we will use the term A-E PN to refer to attributed A-E PNs.

In Definition 3, we introduce assignment graphs as a representation of assignment problems that can serve as input to a Neural Network to learn effective assignment policies.

Definition 3 *(Assignment Graph).* *An assignment graph G is a tuple $G = (V, D, Y, A, \phi, \theta)$ where:*

- *V is a finite set of nodes.*
- *$D \subseteq V \times V$ is a set of ordered pairs of vertices, known as (directed) edges.*
- *$Y = \{A\_Transition, E\_Transition, \dots\}$ is a finite set of node types, where $A\_Transition$ represents the action type and $E\_Transition$ represents the evolution type. Other node types represent the different places in the PN, discriminated on the base of their attributes.*
- *A is a finite set of attributes.*
- *$\phi : V \to Y$ is a function assigning a type to each node.*
- *$\theta : Y \to 2^A$ is a function assigning a set of attributes to each type.*

An assignment graph respects the following two properties:

- For any two nodes $v_1, v_2 \in V$ where either $\phi(v_1) = \phi(v_2) = $ A_Transition, $\phi(v_1) = \phi(v_2) = $ E_Transition, or $\phi(v_1) = $ A_Transition and $\phi(v_2) = $ E_Transition, there does not exist a directed edge $d \in D$ such that $d = (v_1, v_2)$ or $d = (v_2, v_1)$.
- For any node $v \in V$ with $\phi(v) = $ A_Transition $\vee$ $\phi(v) = $ E_Transition, the set of attributes assigned to it by θ is empty, i.e., $\theta(v) = \emptyset$.

Having provided the necessary definitions, we can describe the two-step algorithm to translate a marked A-E PN to the corresponding assignment graph. To clarify the algorithm's steps, we will refer to the problem of a consulting firm with a single employee in Fig. 3. Tasks arrive at a constant frequency of two per time unit, characterized by a type and a budget. The resource must be allocated to tasks to maximize the total budget, considering that the completion time of an assignment is always equal to one time unit. The problem is modeled through one evolution transition ($Arrive$) and one action transition ($Start$), which awards a reward every time an assignment is made. The top left triple presents, in order, the network tag (which has value A, meaning that only action transitions are enabled), the current time, and the cumulative reward.

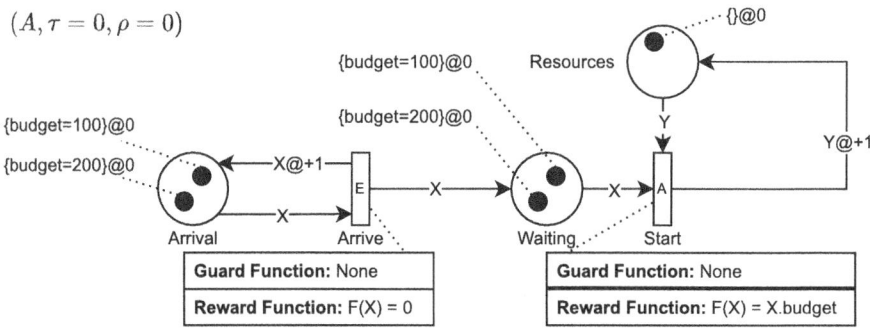

Fig. 3. Marked A-E PN for a simple consulting firm.

It may seem tempting to translate a marked A-E PN directly to a heterogeneous graph with places and transitions as (heterogeneous) nodes and edges corresponding to the A-E PN arcs. However, this approach proves problematic whenever a place contains more than one token since the corresponding node in the assignment graph should hold an embedding of the tokens, which is infeasible as their number is variable. Moreover, there would be no obvious way to encode the possible assignments. For the reasons above, we operate an intermediate step, namely, the expansion, to bring a given marked A-E PN to an equivalent representation suitable for translation to an assignment graph.

Definition 4 (Expansion). *An expansion is a function from a source marked A-E PN S to a target marked A-E PN T that is functionally equivalent to S but contains at most one token in every place.*

In this work, we do not discuss the equivalence between source A-E PN and target A-E PN in terms of PN properties such as reachability or safety, since such properties are not relevant for the translation of the target A-E PN to the assignment graph.

At a high level, the expansion function operates in three main steps:

1. **Place expansion:** each place in the source network, for each token in its marking, is expanded to a new place in the target network marked with that token. Empty places are expanded to a single empty place.
2. **Transition expansion:** every evolution transition in the source network is expanded to a new evolution transition in the target network. Every action transition is expanded to a new action transition in the target network for each timed binding (a set of tokens enabling the original transition).
3. **Arc expansion:** new arcs are created in the target network connecting expanded places to expanded transitions, and vice-versa, obtained from places and transitions connected by an arc in the source network.

The expansion function is described in Algorithm 1. In the algorithm, we use the subscript notation to refer to the individual elements of a given A-E PN. For example, if S is a marked A-E PN, we refer to the set of places in S as P_S. We will also use the notation M_p to refer to the timed marking of place p. we use the notation $'$ to refer to the creation of new PN elements. We denote with $p(t)$ a place p marked with a single token t. M_p represents the marking of a place p. b_{tr} is the timed binding of a transition tr. Moreover, we introduce an auxiliary multi-set H to map places of S into places of T.

The output of the expansion function applied on the A-E PN in Fig. 3 is reported in Fig. 4.

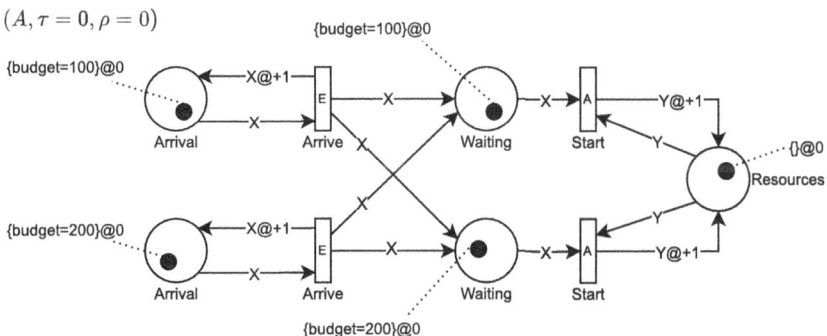

Fig. 4. A-E PN expanded from the A-E PN in Fig. 3.

Algorithm 1. A-E PN Expansion

1: **procedure** EXPAND($\mathcal{S}$)
2:     $A_T \leftarrow A_{\mathcal{S}}, P_T \leftarrow \emptyset, T_T \leftarrow \emptyset, \mathcal{A}_T \leftarrow \emptyset, H \leftarrow \emptyset$
3:     **for** $p \in P_{\mathcal{S}}$ **do**
4:         **if** $M_p \neq \emptyset$ **then**
5:             **for** $t \in M_p$ **do**
6:                 $P_T \leftarrow P_T \cup \{p'(t)\}, H \leftarrow H \cup (p, p'(t))$
7:         **else**
8:             $P_T \leftarrow P_T \cup \{p\}$
9:     **for** $tr \in T_{\mathcal{S}}$ **do**
10:         **if** $L(tr) =$ 'E' **then**
11:             $T_T \leftarrow T_T \cup \{tr\}$
12:             **for** $a \in F_{\mathcal{S}}$ **do**
13:                 **if** $a[0] = tr$ **then**
14:                     $F_T \leftarrow F_T \cup \{a\}$
15:         **else**
16:             **for** $b \in b_{tr}$ **do**
17:                 $tr' \leftarrow tr, T_T \leftarrow T_T \cup \{tr'\}$
18:                 **for** $a \in F_{\mathcal{S}}$ **do**
19:                     $H' \leftarrow (p, p') \in H$ s.t. $p = a[1], P' \leftarrow \{e[1] \forall e \in H'\}$
20:                     **if** $a[0] = tr$ **then**
21:                         $F_T \leftarrow F_T \cup tr' \times P'$
22:                     **else**
23:                         **if** $a[1] = tr$ **then**
24:                             $F_T \leftarrow F_T \cup P' \times tr'$
25:     **return** T

The target A-E PN is easily translatable to a decision graph since every place now contains a single token, and every action transition in T corresponds to one enabling timed binding of the original action transition in $\mathcal{S}$. To operate the translation, we define the Petri Net to assignment graph mapping function.

Definition 5 (Mapping from A-E PN to assignment graph). *A mapping from (expanded) A-E PN to assignment graph is a function from a marked A-E PN T to an assignment graph G having one node for each place or transition in T and one (directed) edge for each arc in T to the corresponding source and destination nodes in G.*

At a high level, the mapping function operates in three main steps:

1. **Place mapping:** every non-empty place is mapped to a node with the same attributes as the single token in the original place, plus its time.
2. **Transition mapping:** every transition is mapped to a node without attributes.
3. **Arc mapping:** every arc not connected to empty places is mapped to an edge connecting nodes corresponding to the arc's source and destination.

The mapping algorithm is described in Algorithm 2. In the algorithm, we use the notation $v(t)$ to refer to the value of attributes of token t, $\tau(t)$ to refer to its time, and $v(n')$ to refer to the values associated with attributes of the type of the newly created node n'. Finally, we introduce two support multi-sets, H_p and H_t, used to store tuples of nodes and transitions in the A-E PN and the corresponding edges in the assignment graph.

Algorithm 2. Petri Net to Assignment Graph Mapping

1: **procedure** MAP($\mathcal{T}$)
2:    $V_G \leftarrow \emptyset, D_G \leftarrow \emptyset, Y_G \leftarrow \emptyset, D_G \leftarrow \emptyset, A_G \leftarrow A_T, H_p \leftarrow \emptyset, H_t \leftarrow \emptyset$
3:    **for** $\forall p \in P_T$ **do**
4:       **if** $\exists p$ s.t. $p(t)$ **then**
5:          $V_G \leftarrow V_G \cup \{n'$ s.t. $\phi(n') = \{\text{TIME}\} \cup \mathcal{A}(p) \wedge v(n') = \{\tau(t)\} \cup v(t)\}$
6:          $H_p \leftarrow H_p \cup (p, n')$
7:    **for** $\forall tr \in T_T$ **do**
8:       **if** $L_T(tr) = $ 'E' **then**
9:          $V_G \leftarrow V_G \cup n'$ s.t. $\phi(n') = \text{E\_Transition} \wedge v(n') = \emptyset$
10:       **else**
11:          $V_G \leftarrow V_G \cup n'$ s.t. $\phi(n') = \text{A\_Transition} \wedge v(n') = \emptyset$
12:       $H_{tr} \leftarrow H_{tr} \cup (tr, n')$
13:    **for** $\forall a \in F_T$ **do**
14:       **if** $a[0] \in P_T$ **then**
15:          $D_G \leftarrow D_G \cup \{(H_p[a[0]], H_{tr}[a[1]])\}$
16:       **else**
17:          $D_G \leftarrow D_G \cup \{(H_{tr}[a[0]], H_p[a[1]])\}$
18:    **return** G

The mapping function's output applied on the A-E PN in Fig. 4 is reported in Fig. 5.

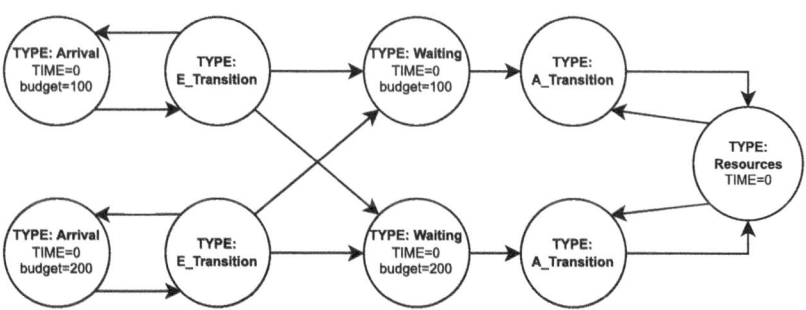

Fig. 5. Decision graph mapped from the A-E PN in Fig. 4.

In an assignment graph, selecting a node corresponding to an action transition directly translates to choosing a binding to fire the transition. By following the expansion-mapping procedure, we obtain an assignment graph where each binding is represented as a node, effectively casting the task assignment problem to a node selection problem on the assignment graph. Moreover, the feature representation technique based on assignment graphs produces graph observations that encode the entirety of the information contained in a given PN marking.

5   Learning Policies on Graphs

This section provides an overview of the steps necessary to train PPO using assignment graphs as observations. To the authors' knowledge, this work presents the first application that leverages fully dynamic graphs in PPO.

Decision graphs represent a flexible feature representation approach, but their shapes vary significantly from sample to sample. For this reason, we need NN layers capable of handling heterogeneous graphs of varying sizes seamlessly, such as the Heterogeneous Graph Attention Layer (HANConv) [22]. In HANConv layers, information is shared across nodes of the same type and then across nodes of different types, effectively bringing each node to a fixed-size representation known as embedding. As shown in Fig. 6, both the policy and the value networks employ the HANConv layer as the encoder. The two networks, however, differ in the decoding phase:

- **Policy network decoder:** encoded action nodes are passed through a two-layer MLP with RELU activation function one by one, producing a value per each node as output. The outputs of the MLP are stacked in a vector that is then passed through a softmax layer that ensures they collectively sum up to 1, effectively turning the MLP outputs into the probability each node has to be selected.
- **Value network decoder:** all encoded nodes are passed through an aggregation layer that computes their mean, and the resulting vector is passed to a two-layer MLP with RELU activation function, producing a single value as output.

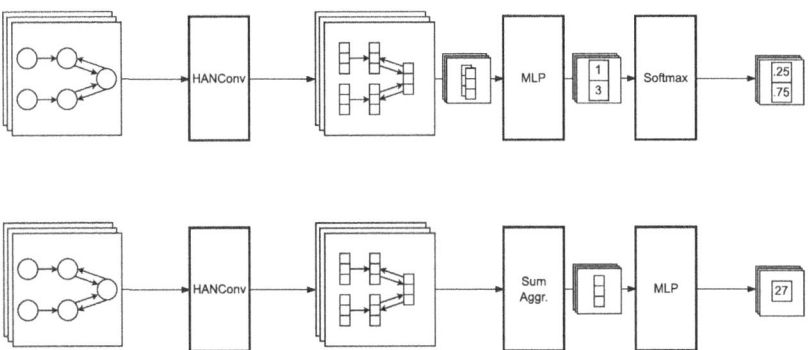

Fig. 6. Policy and Value Networks

However, choosing a suitable network architecture is not enough to handle assignment graphs in PPO. In fact, the dynamic nature of assignment graphs introduces two main challenges when performing batch learning:

1. **Batches contain samples of different sizes:** every assignment graph can be of a different size, making the batching procedure complex. Instead of setting a maximum graph size m and padding every observation to size m, we adopt the approach used in the *Pytorch Geometric* library [23], based on storing a large (sparse) adjacency matrix by concatenating those of single observations along the main diagonal. This approach is much more memory-efficient and allows for modeling assignment graphs of arbitrary sizes.

2. **The number of actions is variable:** in classic RL applications, all possible actions are listed at initialization. In the proposed method, actions correspond to v_a nodes in the assignment graphs, and their number is variable from sample to sample. The use of HanConv ensures that the order of actions is irrelevant (permutation invariance), and the variable amount of actions can be handled using indexed softmax to ensure the action probabilities computed at the output of the actor are correctly associated with the observation that generated them.

Having provided an overview of the PPO implementation used to handle assignment graphs, we can proceed with evaluating the proposed method.

6   Evaluation

This section is dedicated to the definition of a set of three archetypal problem instances of increasing complexity: one with finite and small state/action space, easily representable in vectorial form; one with finite but large state/action space, which proves hard to solve by resorting on vectorial observations; one with infinite state/action space, not representable in vectorial form. We show that PPO can learn close-to-optimal assignment policies in all three cases.

In Fig. 7, we report the initial configuration of a problem entailing a consulting company with a single employee deciding on which projects to take on, given that each project has a budget and the employee's objective is the maximization of the collected budget. In this case, the projects can only have a budget of 100 or 200, based on their type.

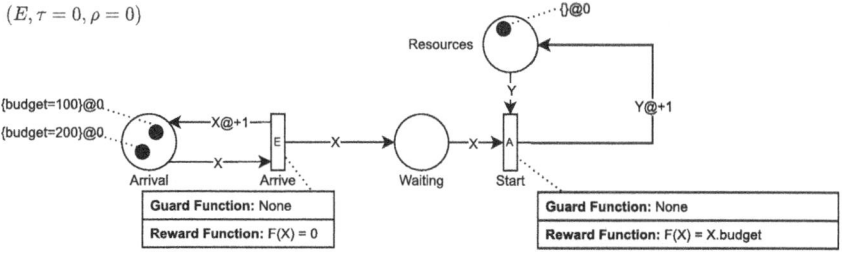

Fig. 7. Initial marking of the A-E PN representation of the first problem.

In Fig. 8, we report the initial configuration of a more complex problem where project budgets are values in euros sampled from uniform distributions between 70 and 130 for projects of type 0 and between 170 and 230 for projects of type 1, rounded to the second decimal value. In this case, the state and action spaces are still finite, but their dimensionality is very large.

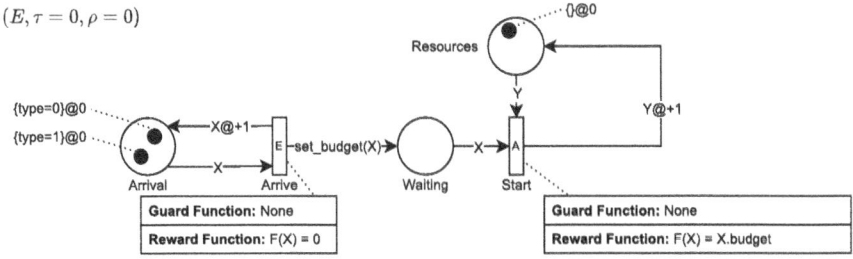

Fig. 8. Initial marking of the A-E PN representation of the second problem.

In Fig. 9, we report the initial configuration of a problem where the task budgets are sampled from a Gaussian distribution with mean respectively 100 and 200, depending on the task type, and standard deviation 10. Tasks enter the system according to exponential interarrival times with rate 1. New projects can only be accepted during a finite time window, which is a feature of the project and is sampled from an exponential distribution with rate $\frac{1}{2}$. Additionally, we consider two resources having different processing times. The Gaussian budgets and the time window features, set for each task through the *set_*attrs arc function, make the vectorial state and action spaces infinite.

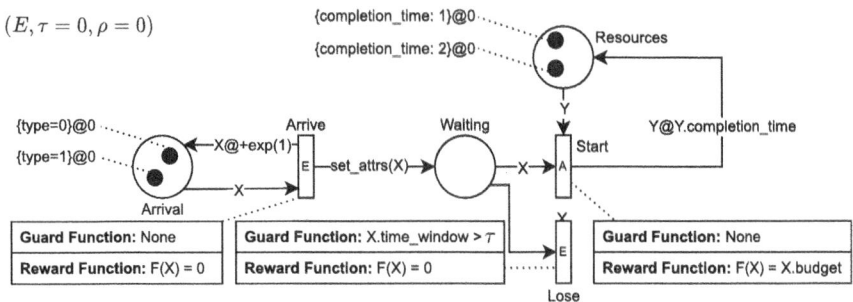

Fig. 9. Initial marking of the A-E PN representation of the third problem.

The three problems were modeled and, where possible, solved with a PPO instance that uses vector observations (*PPO Vector*) and one that uses assignment graphs (*PPO Graph*). As a baseline, we provide the random policy (*Random*) and a simple but effective greedy heuristic (*Greedy*) that always prioritizes the tasks with the highest budget. Experiments were conducted on 10^5 independent episodes, truncated after 10 time units. The results are reported in Table 1 with the mean and variance of the total reward accumulated in each episode. We include an indication of the order of magnitude of state and action space in the vectorial form.

Table 1. Results for the presented problem instances.

Problem		Results			
Instance	Size	Random	Greedy	PPO Vector	PPO Graph
Figure 7	10^0	1349 ± 95	2000 ± 0	2000 ± 0	2000 ± 0
Figure 8	10^4	1348 ± 109	1998 ± 53	1625 ± 130	1999 ± 54
Figure 9	∞	2096 ± 231	2266 ± 316	Not representable	2341 ± 329

The experimental results outline how the assignment graph provides a suitable approach to feature representation regardless of the sizes of the action and state spaces. In the first problem instance, all solving methods reach perfect convergence. However, in the second instance, the PPO instance trained with vector observations struggles to converge, whereas the instance trained on assignment graphs performs the same as the heuristic. Even in the last problem, where the effect of random events is substantial, the PPO instance trained through assignment graphs matches the performance of the heuristic. All experiments are available in a proof-of-concept Python package[1].

7   Conclusions and Future Work

This paper proposed a universal feature representation approach for assignment problems, providing three contributions: the introduction of the assignment graph as a graph-based feature representation for assignment problems, the establishment of a mapping from marked CPN to assignment graphs, and the adaptation of the PPO algorithm to handle assignment graphs as observations.

These contributions were evaluated through their application to three archetypal assignment problems of increasing complexity. The results demonstrated that our method can learn close-to-optimal task assignment policies, regardless of the dimensionalities of the state and action spaces, where existing approaches were limited to state and action spaces of finite size.

Looking forward, the inherent scalability of the proposed universal feature representation approach shows promise. However, this work only focused on small-scale use cases, while its possible limitations are still to be fully explored. For this reason, future work will focus on applying the proposed methods to model and solve real-world problems.

---

[1] The code is publicly available at https://github.com/lobiaminor/BPM2024.

References

1. Lo Bianco, R., Dijkman, R., Nuijten, W., van Jaarsveld, W.: Action-evolution Petri nets: a framework for modeling and solving dynamic task assignment problems. In: Di Francescomarino, C., Burattin, A., Janiesch, C., Sadiq, S. (eds.) BPM 2023. LNCS, vol. 14159, pp. 216–231. Springer, Cham (2023). https://doi.org/10.1007/978-3-031-41620-0_13

2. Żbikowski, K., Ostapowicz, M., Gawrysiak, P.: Deep reinforcement learning for resource allocation in business processes. In: Montali, M., Senderovich, A., Weidlich, M. (eds.) ICPM 2022. LNBIP, vol. 468, pp. 177–189. Springer, Cham (2023). https://doi.org/10.1007/978-3-031-27815-0_13

3. Middelhuis, J., Lo Bianco, R., Scherzer, E., Bukhsh, Z., Adan, I., Dijkman, R.: Learning policies for resource allocation in business processes (2024)

4. Tavares, A.R., Anbalagan, S., Marcolino, L.S., Chaimowicz, L.: Algorithms or actions? A study in large-scale reinforcement learning. In IJCAI, pp. 2717–2723 (2018)

5. Psaraftis, H.N.: A dynamic programming solution to the single vehicle many-to-many immediate request dial-a-ride problem. Transp. Sci. **14**(2), 130–154 (1980)

6. Rogers, D.F., Plante, R.D., Wong, R.T., Evans, J.R.: Aggregation and disaggregation techniques and methodology in optimization. Oper. Res. **39**(4), 553–582 (1991)

7. Cheung, R., Powell, W.B.: Shape - a stochastic hybrid approximation procedure for two-stage stochastic programs. Oper. Res. **48**(1), 73–79 (2000)

8. Spivey, M.Z., Powell, W.B.: The dynamic assignment problem. Transp. Sci. **38**(4), 399–419 (2004)

9. Park, G., Song, M.: Prediction-based resource allocation using LSTM and minimum cost and maximum flow algorithm. In: 2019 International Conference on Process Mining (ICPM), pp. 121–128 (2019)

10. Park, G., Song, M.: Optimizing resource allocation based on predictive process monitoring. IEEE Access **11**, 38309–38323 (2023)

11. Van Der Aalst, W.M.P., Kumar, A., Verbeek, E.M.W.: Dynamic work distribution in workflow management systems: how to balance quality and performance. J. Manag. Inf. Syst. **18**(3), 157–193 (2002)

12. Dasht Bozorgi, Z., Teinemaa, I., Dumas, M., La Rosa, M., Polyvyanyy, A.: Prescriptive process monitoring based on causal effect estimation. Inf. Syst. **116**, 102198 (2023)

13. Gammelli, D., Yang, K., Harrison, J., Rodrigues, F., Pereira, F.C., Pavone, M.: Graph neural network reinforcement learning for autonomous mobility-on-demand systems. In: 2021 60th IEEE Conference on Decision and Control (CDC), pp. 2996–3003 (2021)

14. Yan, Y., Deng, Y., Cui, S., Kuo, Y., Chow, A., Ying, C.: A policy gradient approach to solving dynamic assignment problem for on-site service delivery. Transp. Res. Part E: Logist. Transp. Rev. **178**, 103260 (2023)

15. Begnardi, L., Baier, H., van Jaarsveld, W., Zhang, Y.: Deep reinforcement learning for two-sided online bipartite matching in collaborative order picking. In: Proceedings of the 15th Asian Conference on Machine Learning (ACML2023). Proceedings of Machine Learning Research (2023)

16. Chiorrini, A., Diamantini, C., Mircoli, A., Potena, D.: Exploiting instance graphs and graph neural networks for next activity prediction. In: Munoz-Gama, J., Lu, X. (eds.) ICPM 2021. LNBIP, vol. 433, pp. 115–126. Springer, Cham (2022). https://doi.org/10.1007/978-3-030-98581-3_9

17. Weinzierl, S.: Exploring gated graph sequence neural networks for predicting next process activities. In: Marrella, A., Weber, B. (eds.) BPM 2021. LNBIP, vol. 436, pp. 30–42. Springer, Cham (2022). https://doi.org/10.1007/978-3-030-94343-1_3
18. Sommers, D., Menkovski, V., Fahland, D.: Process discovery using graph neural networks (2021)
19. Hu, L., Liu, Z., Hu, W., Wang, Y., Tan, J., Wu, F.: Petri-net-based dynamic scheduling of flexible manufacturing system via deep reinforcement learning with graph convolutional network. J. Manuf. Syst. **55**, 1–14 (2020)
20. Schulman, J., Wolski, F., Dhariwal, P., Radford, A., Klimov, O.: Proximal policy optimization algorithms (2017)
21. Schulman, J., Moritz, P., Levine, S., Jordan, M., Abbeel, P.: High-dimensional continuous control using generalized advantage estimation (2018)
22. Wang, X., et al.: Heterogeneous graph attention network (2021)
23. Fey, M., Lenssen, J.E.: Fast graph representation learning with PyTorch geometric. CoRR, abs/1903.02428 (2019)

Towards Immersive Environments for Declarative Process Models

Simon James Jensen and Hugo A. López$^{(\boxtimes)}$

DTU Compute, Technical University of Denmark,
Kongens Lyngby, Denmark
hulo@dtu.dk

Abstract. Declarative Process Models (DPMs) provide a visual representation of processes based on rules, yet novice users often struggle to understand them, associating representations with imperative flows. This study explores a novel approach by investigating the potential of leveraging 3D gaming environments to represent and execute processes. We conducted semi-structured interviews with experts in 3D modeling, game development, visual arts, and process management to elicit metaphors explaining process behaviors. These metaphors informed the creation of an artifact representing DPMs in 3D worlds in canonical and domain-specific representations. Our validation phase involved think-aloud sessions and semi-structured interviews with novice users to assess their perception and understanding of processes depicted in 3D environments. Results suggest a positive impact of 3D process representations on users' understanding of declarative process models, with a preference for domain-specific 3D representations. This research contributes to advancing process modeling by addressing challenges faced by novice users and offering innovative solutions within the realm of 3D representations.

Keywords: Declarative Process Models · Process Simulation · Human-Computer Interaction · Serious Games

1 Introduction

Process models are a central artifact in the business process modeling lifecycle. Either for communication, enactment, or discovery, they help human practitioners understand, control and optimize processes. Our focus is the communicative and cognitive aspects behind models. In the end, a process model is an abstract representation of a process, and it provides value to an organization if the stakeholders involved in the process can understand it.

The most common way of representing process models is via two-dimensional representations. Notations such as BPMN, Petri Nets, or EPCs lie in this set. These notations describe the flow of the process, and while the notational set of elements can be extensive, it is still well-understood how sequences of actions can represent a process flow. This is not the case for *declarative process modeling notations* (DPMN). The set of constructs of DPMN like Declare [22] or

A. Marrella et al. (Eds.): BPM 2024 Forum, LNBIP 526, pp. 214–231, 2024.
https://doi.org/10.1007/978-3-031-70418-5_13

DCR graphs [10] describes causal relations between events. These notations are ideal for expressing processes with a large degree of variability or rework. By expressing the flows implicitly, declarative models tend to have fewer edges than their imperative counterparts, where the density of models is correlated to comprehension [16]. However, density alone does not make declarative models more understandable [6], and some works suggest that the choice edges as the representation of constraints may be semantically perverse [7,13,27], that is, the representation used conveys the opposite meaning of the constraint at stake.

This paper challenges the idea that two-dimensional representations are the only (and best) representation for processes. In particular, our models are abstractions of real scenarios where multiple agents interact in spatio-temporal settings. For this, our objective is *to explore the affordances of virtual worlds in the comprehension of knowledge-intensive processes*, the same typology of processes that declarative processes excel in representing. Virtual worlds allow us to explore novel ways of navigating and interacting with the process. They map seamlessly the open-world assumption used in declarative process models, where every action is allowed unless explicitly forbidden. Moreover, the inclusion of space and time allow for novel representations of constraints as changes in the surrounding world. If realized, the impact of this work will potentiate the creation of novel learning environments for teaching, simulation, and upskilling of process participants in knowledge-intensive processes.

To reap the benefits of an effective virtual learning environment, this paper has the following objectives. First, we need to design a virtual world environment that considers both the semantics of declarative process models and the ways novice users interact with the environment. Second, it needs to create an artifact that shows that such an environment is feasible, and finally, it needs to evaluate the perception of users when interacting with the environment. These tasks were realized in the following manner. To elicit the requirements for creating immersive virtual environments for DPMs, we assembled a cohort of experts in process modeling, game design, virtual reality, and arts. Second, we use the metaphors to implement a 3D simulator in Unity3D. Third, we validated the usability of the artifact by executing a pilot study with domain specialists in knowledge-intensive processes at a hospital in Denmark. The results of our pilot study suggest that there is a user preference towards 3D representations in aspects such as usability, learnability, and engagement, but the choice between 2D and 3D representations for future use may depend on the intended use.

Paper Structure: Section 2 discusses related work. Section 3 covers the necessary background on the selected DPM notation: DCR graphs. Section 4 documents the process followed to elicit requirements for the representation of DCR graphs in virtual environments. Section 5 introduces the design choices in the creation of the virtual world. Section 6 documents the validation process followed. Section 7 presents the analysis of the results. Section 8 describe provides further details on the reproducibility of the experiments. Finally, Sect. 10 concludes and outlines future work.

2   Related Work

Previous studies have explored augmenting the representation of imperative models like BPMN with 3D environments. For instance, works such as [1,11,21,24], highlight the positive impacts of 3D visualizations on learning and stakeholder engagement. However, by being based on imperative notations and standard representations, the 3D representations limit the interactions of the user to fixed paths with predefined shapes, failing to explore the dynamic potential of virtual environments. Moreover, virtual reality (VR) environments for collaborative and immersive exploration of business processes such as [2,19] are challenged by the loss of the "big picture", VR sickness, and the cumbersome nature of VR gear have been noted [20]. Contrary to these approaches, our virtual environment proposes leveraging an open-world 3D game environment based on the semantics of DPMNs, offering a level of interactivity and dynamism not previously explored. Moreover, it departs from the idea that representations of process concepts in 3D worlds may differ from their 2D representations, thus allowing the opportunity for new (and even, domain-specific) process representations. Moreover, our work addresses the drawbacks of VR by eliminating the need for cumbersome equipment and reducing the risk of VR sickness.

3   Background

3.1   DCR Graphs

We take DCR graphs [10] as a representative declarative notation. We selected DCR among other declarative notations due to its application in industry and adoption by the public sector in several countries [9]. While multiple extensions of DCR graphs exist including control flow, time [12], subprocesses [4], data [26], and inter-process communication [5], we use the classical definition considering control-flow constraints [10]. An example of a classical process model representation for DCR graphs is shown in Fig. 1. We provide the intuition behind DCR graphs, and we direct the reader to [8] for its formalization.

A DCR Graph is a multi-directed graph whose nodes represent *activities* and its edges represent *rules* binding activities. Each activity is mapped to a textual *label*. The rules define the restrictions and govern the behavior of the activities. There is no lower or upper bound to how often an activity is *executed*, as long as its restrictions are respected.

The *state* of an activity depends on three *markings*. An activity can be either *included* (i.e.: *Go shopping*) or *excluded* (i.e.: *Prepare sandwich*). An excluded activity cannot be executed, nor can it impose rules on other activities. Once an activity is executed it changes marked its *executed* marking (i.e.: *Eat sandwich*) An activity may have a *pending* marking, denoting that it must eventually be executed (or excluded). A graph also contains runtime and global properties. An activity is *enabled* if it is not excluded and does not have previous conditions or milestones pending for execution. Also, a graph is in *accepting state* if it has no included and pending activity. Moreover, the basic relations found in

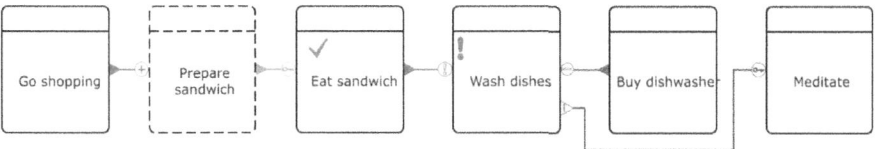

Fig. 1. The DCR graph process of sandwich making, using Trinh's notation [27] (Color figure online)

DCR Graphs include *condition, response, milestone, include,* and *exclude.* Rules connect a *source* activity to a *target* activity and the target marking.

At the start of the process in Fig. 1, the activities of *meditate, wash dishes, buy dishwashers* and *go shopping* can be executed freely, as often as desired, and in any order. This is because they do not include incoming conditions (i.e. yellow relations) or milestones (i.e. violet arrow). *Prepare Sandwich* cannot be executed, since it is currently excluded, and *Eat Sandwich* can be executed even if it is conditioned, because *Prepare sandwich* is excluded. The *condition rule* specifies that the target activity may not execute so long as the source activity is included and has yet to execute. The *include rule* specifies that executing the source activity will mark the target activity as included. This behavior can be seen in the *Go shopping* activity. If *Prepare sandwich* is executed, it will fulfill the condition rule placed on *Eat sandwich,* enabling this activity for execution. The *response* rule (in blue) specifies that if the source activity is executed, the target activity becomes pending. Therefore, if *Eat sandwich* is executed at any point after it was enabled, the process will require that *Wash dishes* is eventually executed to consider the trace accepting. Finally, the milestone rule (in violet) specifies that as long as the source activity is pending, the target activity is disabled and cannot be executed. Thus the *Meditate* activity cannot be currently executed until the *Wash dishes* activity is executed.

4   Eliciting Requirements from Experts

This study's first part focused on identifying how to represent the concepts existing in DCR graphs in a virtual world representation. Compared to related work that overlays 2D representations in 3D world, we wanted to explore the design space in search of representations that could fit better the concepts. Our team had set an original requirement that the virtual representation of activities should be easy to specialize for the domain. This follows recommendations for semantic transparency [17] that favours icons over abstract geometrical representations. As the transparency of the symbols used for activities will depend on the problem domain, we will leave this requirement aside, and focus on identifying how to best represent the remaining DCR concepts including markings, enabledness and the effect of relations. To do so, we conducted semi-structured interviews with experts in visual arts, digital games, UX, and process modeling to identify requirements for the 3D visualization of DCR graphs. In particular, the cohort included the following profiles: i. a UX designer with experience in game development and process models, ii. a visual artist and designer, iii. an associate professor in software engineering with expertise in BPM, iv. an associate professor in digital design with expertise in game design. The selection of the cohort aimed a balancing considerations between process modeling experts and HCI/game development experts. Each expert was previously briefed on the purpose of the research. They provided consent to use their information for data collection. Each participant was presented with a familiarization phase where the concepts of DCR graphs were introduced without introducing their existing graphical representation, and comprehension tests were executed to ensure there were no ambiguities. After this phase, each expert underwent a semi-structured interview to generate requirements for each concept in DCR graphs. The interviews focused on visualizing the basic elements of a DCR graph in a game environment. They were later transcribed and coded using grounded-theory principles [3]. We executed three rounds of coding: In the initial phase, we reviewed the interview transcripts to identify comments relevant to the study, anonymized data, and corrected transcriptions. An initial set of codes was generated including navigability and game design. Then a coding phase was executed. Initial codes were categorized to fit the conceptual model of DCR graphs [13]. Finally, each of the codes was grouped into major categories covering the semantics of the model and 3D-specific features affecting the artifact such as animation, effects, domain, etc. This set of requirements was used to derive the initial features of the artifact, selecting those with agreement among interviewees and it was feasible to implement for this proof of concept. Recommendations that contradict each other (e.g. choose a fantastical setting instead of a domain-specific one) were combined and discussed.

Requirements: We report the most salient requirements to describe DCR concepts according to the coding. Concerning the representation of **activities**, its representation should convey the idea of a spatial placeholder that can be refined according to the domain of the process. Its representation requires an interactive element identifiable from the rest and used consistently in all activities. It was suggested to define markings as effects that can be applied to the visual representation of activities. For **markings**, it was suggested that the user should be able to identify *excluded* activities and they should not draw attention until they are included. Activities that are included and do not happen should be easy to identify, and visual clues drawing attention should disappear upon activity execution, transitioning to a modifier denoting exhaustion. Finally, *pending* and included activities should draw the most attention to the map. As the **enabled** state of an activity depends on its marking and its neighbors, the user must differentiate enabled from disabled activities. To preserve DCR's open-world assumption it was suggested to render disabled activities visible to users allowing the user to relate the action's effects at runtime. Finally, enabled activities should draw attention from the user.

For **relations**, it was suggested to depart from the description of the behavior of each type, and that their visual representation might not be necessarily be

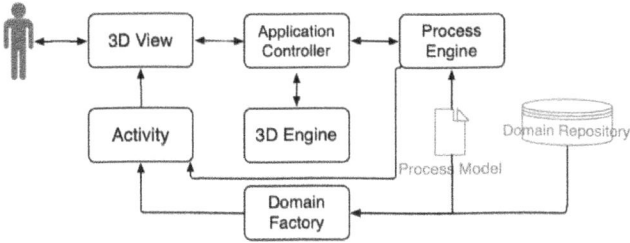

Fig. 2. Application architecture

explicit (drawn), but instead queried by demand. Rather than the relations, what you care is on the effects that the application of such an event has in the environment (e.g. *"If we if we want to keep an overview, we can show how the game will change if you execute an activity. If it's something that's critical to know beforehand we could have a small indicator when were looking at the map, just to show you that we aren't there yet, but then when you tab you get more details on why aren't we there yet."*). For **conditions and milestones**, the focus lies on source activities. Target activities from a condition or a milestone whose source has not been executed cannot be enabled, and sources requiring execution should be identifiable upon the user's query (e.g.: *"You need to do something before and then also you will use light to show what is the other thing you have to do it before or and then this is"*). For **responses** it was suggested to make explicit the effects of the relation by ambient modifiers (for instance, a light). No particular suggestions were provided on the representation of the **include and exclude** relations, thus we will use the intuition behind it. If an activity becomes pending due to the firing of a source from an include/exclude relation, its modifiers for inclusion are applied. For all relations, experts suggested that there should be a specific animation for each type of relation. Relations can be represented as direct paths, and one-to-many/many-to-one/many-to-many rules can be showcased with different representations. For *process-related* properties, it was suggested that if a process is in an **accepting** state, a global property affecting the world (e.g. a light) should be switched.

Concerning *environmental variables*, there was an agreement that the navigation should emulate an open world where the only non-interactive elements should be non-enabled activities. Users should be able to navigate through their execution history and view upcoming steps to plan the next execution. Moreover, it was suggested to use a composition of different views to support comprehension: a first-person view for interaction, a global view for panning, and a timeline view for replication.

5   Artifact Design

A virtual reality environment was designed based on the requirements in Sect. 4. Figure 2 describes the architecture of the main components. We extend a classi-

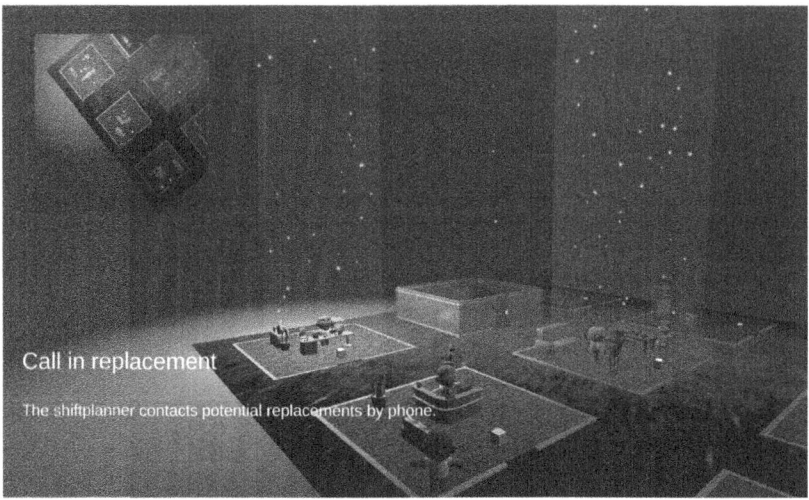

Fig. 3. A DCR graph in a domain-specific representation

cal MVC pattern where the controller interacts with a process and 3D engines. The *application controller* initializes the view according to initial markings, processes user inputs, updates the process model based on user actions, and handles synchronization between the game view and the process model. The *3D View* presents the game's user interface, including activities, visual effects, and camera views. This component also shows diverse camera perspectives to enhance interaction and comprehension of the game's spatial environment. The *activity* component renders DCR activities in the 3D view, integrating a standard (canonical) view of activities with their marking. It is in charge of adapting the visual representation to specific domains. The *Domain Factory* works as an abstract factory pattern: it provides a catalog of coherent visual representations for each domain, where the assets used for each activity are stored in a *domain repository*. The *process engine* governs the execution of the process, parsing the process model and updating it according to the semantics in [10]. Finally, the *3D Engine* enables the rendering of the 3D environment. It provides functionalities for updating the 3D views according to user interactions.

In the following lines, we discuss how we implemented expert's requirements.

We decided to represent activities as cells in a map (e.g. Fig. 4). In their canonical representation, they only contain interactive elements used to execute activities (e.g. Fig. 4a), but they can be specialized according to the domain (e.g. Fig. 4b). All activities are prone to change in their representation according to the markings, as exhibited for disabled activities in Fig. 4c. We implemented markings as overlays applied to activities. For pending markings we used a pulsating animation to attract an action. For excluded markings we applied transparencies to the representation of the activity (e.g. 5c). We apply a color modifier to executed activities to differentiate them with the non-executed ones.

(a)   Enabled   activity, canonical

(b)   Enabled   activity, domain-specific

(c)   Disabled   activity, canonical

Fig. 4. DCR activities: canonical vs. domain-specific representations

(a) Included, not exe-cuted, pending

(b) Included, executed, not pending

(c) Excluded, not exe-cuted, not pending

Fig. 5. Activity markings (selected)

To denote the event state we represent disabled events enclosed in walls (e.g. Fig. 4c), and no modifiers are applied to enabled events. Concerning relations, we represented condition and milestone relations using visual cues. A condition (resp. milestone) from A to B will enrich the activity representation of both with surrounding matching colors. When a queried activity is the target of multiple unmet unfulfilled conditions (resp. milestones), it will identify each of them via an unique color, that will be removed after their execution. Both source and target activities are marked with symbols to denote the reading order of the relation. To represent the process state, we used an environmental variable: ambient light. Daylight was used for accepting graphs, while non-accepting graphs will be represented as a dark space, highlighting the activities that will render the graph accepting (e.g. Fig. 5c).

Finally, we cover two game-specific features: navigability and views. For navigation, we designed the 3D environment as an open-world map where the user can explore the world according to the open-world assumptions for DPMNs. Moreover, the artifact implements three views: First-person, third-person and a global view The first-person view (e.g.: Fig. 3) positions the player directly within the world and it is expected to aid the sense of presence. The global view (e.g.: Fig. 6) offers a broader vantage point for the user and bears familiarity with 2D representations of models, which may support the transition for expert users. Moreover, this view allows the user to perform what-if analysis, by simu-

 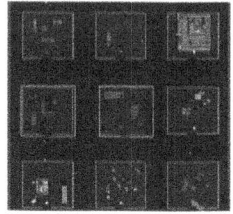 

(a) First person view   (b) Global view: what-   (c) Third person view
                          if scenario

Fig. 6. Integration of views

lating the effects of the events selected by the user. Moreover, the inclusion of a global map view is expected to alleviate navigational challenges, allowing players navigate over large diagrams in a quick fashion. These view are interchangeable, and in some cases (e.g. Fig. 3) integrated to support cognitive integration.

6   Artifact Validation

We tested the artifact with potential end users to understand how the new 3D representations contribute to their process understanding or preferences. We performed two pilot evaluations with small cohorts ($n_1 = 4, n_2 = 4$), with the takeaways from the first cycle used to improve the artifact before the second cycle. At the same time, we collected data regarding comprehension and preferences. The two rounds included different types of participants. In the first round, we tested with four public health professionals specializing in roster planning between the mid-30s and mid-50s. This cohort was new to process modeling but had expertise in the domain. The second round included three novice users and one expert in DCR graphs between the late 20 s and early 50 s. A register sickness process was presented in both rounds, and no participant took part in both experiments to avoid learning bias. Figure 7 illustrates the process in its 2D representation. Each participant was briefed on the scope of the research project and provided consent to participate.

Our validation phase assessed the participants' perceptions of the usefulness, learnability, commitment, preference, and comprehension of DCR graphs via two experiments: The first experiment compared the comprehension of a process model in the classical 2D DCR representation (used in tools like [15]) and its corresponding canonical 3D representation. The second experiment compared a process representation in the canonical 3D representation against a 3D representation where the representation of the activities was adapted to the domain.

We ran the experiments using a within-subject design. The order of experiments was randomized externally (which experiment was conducted first) and internally (the sequence of representations tested). Participants were asked to navigate the 3D environment and articulate their thoughts in the spirit of concurrent think-aloud methods [25], complemented with semi-structured interviews

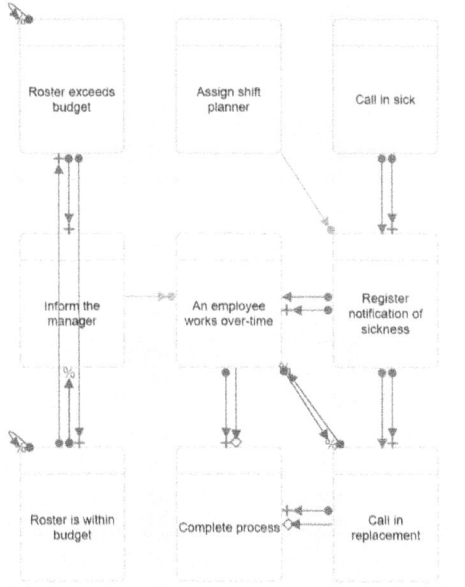

Fig. 7. Register sickness process: 2D representation of the process in Fig. 3

to ensure insights regarding the understandability of the representations. During the think-aloud session, the interviewer would remind them to keep speaking and ask open-ended follow-up questions. If a particular screen feature went undocumented, the researcher would ask the user if they noticed it and followed the protocol. The interviewer was careful to avoid asking leading questions and to every possible extent let the information flow only from the participant. Once then, interviews were transcribed and coded. We removed personal information and corrected discrepancies against voice recordings during the first coding phase. Each data fragment was timestamped and grouped according to the representation discussed. Then a second coding phase was executed, assigning to each fragment tags the representation commented on, its testing order, and the DCR concept. Lastly, the data was clustered per type of representation. The results of the coding and transcripts are available in [14].

6.1   Discussion of Validation Results

We will discuss the codes providing insights about the affordances of the 3D representations compared to 2D versions.

For **markings**, we observed that participants generally understood the representations used to denote completed, pending, enabled, or excluded markings, but interpretations varied significantly between the 2D and 3D representations. In the 2D representation, the use of color to denote enabled activities was well understood, however, the shapes used to describe markings were not always understood. The comprehension of the semantics behind the symbols used to describe executed and pending was not always perceived by participants. The 3D representation introduced interactive elements and novel representations for markings with visual cues like fog and light effects. These features added a layer of engagement but also confused some participants. At the activity level, the animations used to represent executed markings were sometimes misunderstood. However, the visual cues given to describe changes in the process state such as the removal of fog or the use of light sources guided participants toward pending activities.

For the representation of **relations**, participants often interpreted the arrows used by the 2D representation terms of execution order or activity status, reflecting a wrong, linear perspective not existing in DCR graphs. In the 3D representation, the use of interactive elements and environmental cues offered more engagement but also introduced certain challenges. The representation of include and exclude relations was correctly understood as the ability to enable or disable events. The representation of the condition relations was changed between the first and the second experiment as the initial design used colored particles to indicate conditions were not always effective and participants overlooked the cues. For the second iteration conditions were augmented with lock-key symbols and the understanding improved significantly. These metaphors, along with more explicit visual and auditory feedback on successful or failed interactions, clarified the effects of actions within the process. The representation of responses was perceived as activities that needed to be executed linearly. The understanding regarding milestones was challenging in both the 2D and 3D representation, but participants recognized the dependency relations in the 3D version thanks to key-lock relations.

Regarding the understandability of **process variables**, the change in lighting conditions denotes that the acceptance introduced dual results. On the one hand, the change in lighting successfully drew attention to the execution of pending activities. However, the intended meaning (to denote that a process instance is not accepting) was mostly missed by the participants who understood it as time passing or a cue to focus on specific tasks.

Regarding the use of **labels** to describe activities was divisive: some participants relied heavily on them for logical progression, while others prioritized visual cues over textual information.

On the perception of the virtual environment, we expected to see an open-world navigation from the participants. However, participants often approached the exploration of the world linearly. This behavior was constant for both the 2D and 3D representations. Repeated interactions with the 3D model gradually led some participants to notice the DCR graph's dynamic capabilities, but it was not intuitive in early interactions. Moreover, we could observe that the interactive elements and spatial layout introduced in the 3D representations encouraged exploration but also reinforced linear progression perceptions.

Overall, the general qualitative evaluation of the 3D environment was mixed, with some participants finding navigation less intuitive than expected and expressing a preference for a more linear navigation. However, these results are contrasted with the results participant preference survey executed later.

Summary of 2D vs Canonical 3D Representation: Participants' experiences between a 2D DCR graph representation and a canonical 3D representation revealed several reactions primarily centered around comprehension, engagement, and navigation. In the 2D representation users appreciated the clear visual cues like transparency for disabled activities and green check-marks for executed activities. In the canonical 3D representation the 3D environment's attempt to signify executed activities with animated geometric objects was met with initial confusion, with some participants mistaking these for obstacles. However, this confusion is typically resolved with further interaction. The 3D representation struggled to communicate the executed status of activities as clearly as the 2D, with varied interpretations of the animated objects' meaning.

The navigation and overall interaction with the process model changed significantly in the 3D environment. The sense of exploration was heightened, but the linear execution present in the interpretation of 2D DCR graphs persisted, with participants often looking for a sequence to follow. The 3D environment's spatial layout and visual effects offered new ways to understand activity states but also introduced complexities in grasping the semantics of DCR graphs. The use of familiar, contextual visual cues and metaphors, such as locked doors to signify disabled activities and keys for enabling conditions, provided a more relatable and understandable experience. This made the process more engaging but also helped in drawing parallels to their real-world applications, potentially aiding in quicker comprehension and more meaningful interaction with the graph. In summary, while the 2D representation provided clarity and simplicity in understanding activity states, the canonical 3D representation offered a more immersive and exploratory experience. However, with this immersion came challenges in intuitively understanding the semantics of DCR graphs, indicating a need for more intuitive visual cues and guidance within the 3D environment.

Summary of Canonical and Domain-Specific 3D Representations: Comparing users' experiences between the canonical 3D representation and the domain-specific 3D representation of DCR graphs reveals nuanced differences in engagement, comprehension, and usability. The addition of domain-specific elements aimed to enhance intuitive understanding and relate more closely to users' professional contexts. The domain-specific 3D representation did enhance the users' ability to relate the DCR graph to practical scenarios, as seen in their improved navigation and decision-making within the graph. However, the same challenges in the navigation of declarative processes persisted, similar to the transition from 2D to canonical 3D representations.

To conclude, domain-specific 3D representations improved engagement and contextual understanding over the canonical 3D model, yet both shared challenges in communicating the dynamic, non-linear nature of DCR graphs.

6.2   Participant Preferences

After the think-aloud tests were performed, the 8 participants were invited to fill out surveys regarding their preferences for a visual representation. We collected data in four usability dimensions:

- Usability: what is the preferred process representation for ease of use,
- Learnability: what is the process representation that helps to understand the underlying process the most
- Engagement: which process representation motivated the user to explore the process more
- Future preference: which process representation will be preferred for future use

Three comparisons were performed: the first trial compared the 2D vs. the canonical 3D representation, the second compared the canonical 3D vs domain-specific 3D representations, and the final one compared they compared all representations together. We report the results from the 5 participants who filled out the surveys, summarized in Table 1.

Usability: In the three comparisons there was a preference for 3D representations for ease of use, with a higher preference for domain-specific environments. The utterances suggest a memory effect not existing in 2D representations: *"it was easier (with a domain-specific 3D model) to remember the relationships between the elements and their location"*, however, participants also commented on the advantage of having a model as an abstraction: *"It was easier to form an overview (when speaking about the 2D representation)"*.

Learnability: Overall, there was a preference for 3D representations as enablers for learning processes over 2D representations, with a preference for domain-specific representations. At the same time, suggestions for a combination of 2D views to guide the navigation of 3D environments were posed: *"one could see both processes[1] simultaneously"*.

Engagement: For the sample considered, there was a marked tendency to consider 3D environments more engaging than their 2D counterparts, where participants commented on how more details engage exploration: *"it is even more fun since there was more to look at"* (when describing the preference for a domain-specific 3D representation).

Preference for Future Use: Our results are inconclusive for such a small sample, and we can see a division in the preference between the domain-specific 3D representation and a 2D one. The utterances suggest two different types of uses in the minds of the participants. The first one, as an overview: *"it is faster to see the relations between the elements"* (when choosing 2D representations), and the second, as a game *"it is the most fun"* (when preferring domain-specific 3D representations).

---

[1] Referring to views.

Table 1. Results of user survey on usability, learnability, engagement, and user preferences. Text in bold denotes the highest preference.

Trial	Preference	Usability		Learnability		Engagement		User Preference	
		N	%	N	%	N	%	N	%
1	2D	2	40%	2	40%	1	20%	**3**	**60%**
1	Canonical 3D	**3**	**60%**	**3**	**60%**	**4**	**80%**	2	40%
2	Canonical 3D	1	20%	2	40%	2	40%	2	40%
2	Domain-specific 3D	**4**	**80%**	**3**	**60%**	**3**	**60%**	**3**	**60%**
3	2D	1	20%	1	20%	0	0%	2	40%
3	Canonical 3D	1	20%	1	20%	1	20%	1	20%
3	Domain-specific 3D	**3**	**60%**	**3**	**60%**	**4**	**80%**	2	40%

7   Analysis of the Results

Overall, there appears to be a preference towards the 3D representation in usefulness, learnability, and engagement. The preference for the domain-specific 3D representation goes in line with previous research that demonstrates preferences for icon-rich notations compared with notations based on abstract symbols [11,23,28]. However, it is interesting that some participants preferred the 2D representation for future use. The pragmatics of the model may play a role here, where 3D models could be helpful for simulation and teaching scenarios, and modeling or analysis could benefit from 2D models.

Across both 2D and 3D representations of processes, navigating declarative models was perceived as challenging. In this dimension there was no difference between the domain-specific and the canonical representations. Despite briefings about DCR's dynamic capabilities, participants frequently treated the process as a sequence of steps to be completed in order. This was particularly evident in the 3D versions, where interactive elements and spatial layout encouraged exploration and reinforced linear progression perceptions. We could attribute this to a higher intrinsic cognitive load in understanding declarative models. Moreover, noticing the two techniques used to parse the models (label-first or model-first) underscores the importance of balancing text and visual elements in process visualization tools, and the need to keep both labels and activity representations aligned. In addition, other factors such as the layout of the process or the visual cues selected may also contribute to a linear navigation from participants. This indicates that while domain-specific elements can aid in comprehension and engagement, further research into how to support the navigation of declarative processes is still required.

Overall, both the 2D and 3D representations each had strengths and weaknesses in illustrating certain aspects of DCR graphs and each representation presented challenges in comprehension and engagement. In particular, the 3D representations added a further specialization for the representation of activities. Also, it showcased that causal relations may very well be represented with animations and not only with a standard graph-based representation. This additional levels may increase visual expressiveness [17], however, it also opens the room for semantic perversity by choosing visual representations that obfuscate their intended meaning. In some concepts both the 3D and 2D representations struggled to provide an intuitive semantics. Concepts such as the milestone relation was commonly misunderstood, and we believe the explanation might lie in a higher intrinsic complexity concept rather than of its representation. The study highlights the importance of clear, intuitive design and the potential for interactive, game-like elements to enhance user understanding and interaction with complex process models. However, the sample size of the experiments does not allow us to deduce definite conclusions, and we would like to perform a larger validation phase with more participants to understand the effects of 3D representations of models in future work.

8   Reproducibility

We adhere to the open science principles. In particular, our virtual world can be explored freely at https://bit.ly/3DDCRG. The GitHub repository https://github.com/GloriousHypnotoad/DCR-Graph-artifact contains the Unity3D code. Interview protocols used for the requirements elicitation phase the validation phase, and the interview transcripts can be found in [14].

9   Threats to Validity

Our main threat to validity is the size of the samples used in the validation experiments, too small to be generalizable. While lacking statistical significance, the sample size is similar to the understandability studies for VR tools in BPMN [2,20] and on par with standard system usability tests [18]. Moreover, our coding sessions revealed perceptions that could be difficult to identify in quantitative studies and suggest possible hypotheses that need to be tested in further empirical studies with larger samples.

10   Conclusion and Future Work

This paper aimed at answer what are the affordances provided by immersive environments in the representation of declarative processes. To do this we collected requirements regarding how to make immersive environments from a multi-disciplinary cohort in process modelling, UX, game design and arts. These requirements were then studied for feasibility, and they were integrated in a prototypical 3D representation of a DCR graph. The results of the pilot studies suggest that not only declarative models can be represented without the need of edges and still be understandable, but they also suggest a stronger preference in usefulness, learnability and engagement against classical 2D representations.

For future work, we would like to explore the cognitive integration between the 2D and 3D representations of declarative process model. These can be in the shape of overlay process maps that help the navigation of the 3D environment. In addition, we believe that finding good heuristics for declarative process modelling layout will help in the understandability of process models. Finally, we would like to extend the pilot study presented here towards a large-scale experiment with statistical significance.

Acknowledgements. This work was supported by the researchgrant "Center for Digital ComplianceE (DICE)" (VIL57420) from VILLUM FONDEN.

References

1. Abdul, B.M., Corradini, F., Re, B., Rossi, L., Tiezzi, F.: UBBA: unity based BPMN animator. In: Cappiello, C., Ruiz, M. (eds.) CAiSE 2019. LNBIP, vol. 350, pp. 1–9. Springer, Cham (2019). https://doi.org/10.1007/978-3-030-21297-1_1
2. Brown, R., Recker, J., West, S.: Using virtual worlds for collaborative business process modeling. Bus. Process. Manag. J. **17**(3), 546–564 (2011)
3. Charmaz, K.: Constructing Grounded Theory. Sage (2014)
4. Debois, S., Hildebrandt, T., Slaats, T.: Hierarchical declarative modelling with refinement and sub-processes. In: Sadiq, S., Soffer, P., Völzer, H. (eds.) BPM 2014. LNCS, vol. 8659, pp. 18–33. Springer, Cham (2014). https://doi.org/10.1007/978-3-319-10172-9_2
5. Debois, S., López, H.A., Slaats, T., Andaloussi, A.A., Hildebrandt, T.T.: Chain of events: modular process models for the law. In: Dongol, B., Troubitsyna, E. (eds.) IFM 2020. LNCS, vol. 12546, pp. 368–386. Springer, Cham (2020). https://doi.org/10.1007/978-3-030-63461-2_20
6. Fahland, D., et al.: Declarative versus imperative process modeling languages: the issue of understandability. In: Halpin, T., et al. (eds.) BPMDS/EMMSAD -2009. LNBIP, vol. 29, pp. 353–366. Springer, Heidelberg (2009). https://doi.org/10.1007/978-3-642-01862-6_29
7. Figl, K., Di Ciccio, C., Reijers, H.A.: Do declarative process models help to reduce cognitive biases related to business rules? In: Dobbie, G., Frank, U., Kappel, G., Liddle, S.W., Mayr, H.C. (eds.) ER 2020. LNCS, vol. 12400, pp. 119–133. Springer, Cham (2020). https://doi.org/10.1007/978-3-030-62522-1_9
8. Hildebrandt, T., Mukkamala, R.R., Slaats, T.: Nested dynamic condition response graphs. In: Arbab, F., Sirjani, M. (eds.) FSEN 2011. LNCS, vol. 7141, pp. 343–350. Springer, Heidelberg (2012). https://doi.org/10.1007/978-3-642-29320-7_23
9. Hildebrandt, T.T., et al.: EcoKnow: engineering effective, co-created and compliant adaptive case management systems for knowledge workers. In: ICSSP, pp. 155–164 (2020)
10. Hildebrandt, T.T., Mukkamala, R.R.: Declarative event-based workflow as distributed dynamic condition response graphs. arXiv preprint arXiv:1110.4161 (2011)
11. Leyer, M., Brown, R., Aysolmaz, B., Vanderfeesten, I., Turetken, O.: 3D virtual world BPM training systems: process gateway experimental results. In: Giorgini, P., Weber, B. (eds.) CAiSE 2019. LNCS, vol. 11483, pp. 415–429. Springer, Cham (2019). https://doi.org/10.1007/978-3-030-21290-2_26

12. López, H.A., Debois, S., Slaats, T., Hildebrandt, T.T.: Business process compliance using reference models of law. In: FASE 2020. LNCS, vol. 12076, pp. 378–399. Springer, Cham (2020). https://doi.org/10.1007/978-3-030-45234-6_19
13. López, H.A., Simon, V.D.: How to (re) design declarative process notations? A view from the lens of cognitive effectiveness frameworks. In: 15th IFIP Working Conference on the Practice of Enterprise Modeling 2022. CEUR-WS (2022)
14. López, H.A., Jensen, S.J.: Towards Immersive Process Simulation for Declarative Models - Accompanying Material (2024). https://doi.org/10.5281/zenodo.11425176
15. Marquard, M., Shahzad, M., Slaats, T.: Web-based modelling and collaborative simulation of declarative processes. In: Motahari-Nezhad, H.R., Recker, J., Weidlich, M. (eds.) BPM 2015. LNCS, vol. 9253, pp. 209–225. Springer, Cham (2015). https://doi.org/10.1007/978-3-319-23063-4_15
16. Mendling, J., Reijers, H.A., Cardoso, J.: What makes process models understandable? In: Alonso, G., Dadam, P., Rosemann, M. (eds.) BPM 2007. LNCS, vol. 4714, pp. 48–63. Springer, Heidelberg (2007). https://doi.org/10.1007/978-3-540-75183-0_4
17. Moody, D.: The "physics" of notations: toward a scientific basis for constructing visual notations in software engineering. IEEE Trans. Softw. Eng. **35**(6), 756–779 (2009)
18. Nielsen, J., Landauer, T.K.: A mathematical model of the finding of usability problems. In: Proceedings of the INTERACT'93 and CHI'93 Conference on Human Factors in Computing Systems, pp. 206–213 (1993)
19. Oberhauser, R., Pogolski, C.: VR-EA: virtual reality visualization of enterprise architecture models with ArchiMate and BPMN. In: Shishkov, B. (ed.) BMSD 2019. LNBIP, vol. 356, pp. 170–187. Springer, Cham (2019). https://doi.org/10.1007/978-3-030-24854-3_11
20. Oberhauser, R., Pogolski, C., Matic, A.: VR-BPMN: visualizing BPMN models in virtual reality. In: Shishkov, B. (ed.) BMSD 2018. LNBIP, vol. 319, pp. 83–97. Springer, Cham (2018). https://doi.org/10.1007/978-3-319-94214-8_6
21. Panzoli, D., Lelardeux, C.P., Galaup, M., Lagarrigue, P., Minville, V., Lubrano, V.: Interaction and communication in an immersive learning game: the challenges of modelling real-time collaboration in a virtual operating room. In: Ma, M., Oikonomou, A. (eds.) Serious Games and Edutainment Applications, pp. 147–186. Springer, Cham (2017). https://doi.org/10.1007/978-3-319-51645-5_7
22. Pesic, M., Schonenberg, H., Van der Aalst, W.M.: DECLARE: full support for loosely-structured processes. In: 11th IEEE International Enterprise Distributed Object Computing Conference (EDOC 2007), pp. 287–287. IEEE (2007)
23. Petre, M.: Why looking isn't always seeing: readership skills and graphical programming. Commun. ACM **38**(6), 33–44 (1995)
24. Saito, S.: ProcessCity: visualizing business processes as city metaphor. In: Cappiello, C., Ruiz, M. (eds.) CAiSE 2019. LNBIP, vol. 350, pp. 207–214. Springer, Cham (2019). https://doi.org/10.1007/978-3-030-21297-1_18
25. Kuusela, H., Paul, P.: A comparison of concurrent and retrospective verbal protocol analysis. Am. J. Psychol. **113**, 387–404 (2000)
26. Strømsted, R., López, H.A., Debois, S., Marquard, M.: Dynamic evaluation forms using declarative modeling. In: 16th International Conference on Business Process Management (BPM'18), pp. 172–179. CEUR Workshop Proceedings (2018)

27. Trinh, D.M.T., Abbad-Andaloussi, A., López, H.A.: On the semantic transparency of declarative process models: the case of constraints. In: Sellami, M., Vidal, M.E., van Dongen, B., Gaaloul, W., Panetto, H. (eds.) CoopIS 2023. LNCS, vol. 14353, pp. 217–236. Springer, Cham (2023). https://doi.org/10.1007/978-3-031-46846-9_12

28. Zenner, A., Makhsadov, A., Klingner, S., Liebemann, D., Krüger, A.: Immersive process model exploration in virtual reality. IEEE Trans. Vis. Comput. Graph. **26**(5), 2104–2114 (2020)

Inferring Missing Event Log Data from IoT Sensor Data - A Case Study in Manufacturing

Alexander Seeliger[1,2]([✉]) [iD], Markus Schreiber[3], Florian Giger[2],
Joachim Metternich[3], and Max Mühlhäuser[2] [iD]

[1] Celonis Labs GmbH, Munich, Germany
a.seeliger@celonis.com
[2] Telecooperation Lab, Technical University of Darmstadt, Darmstadt, Germany
{giger,max}@tk.tu-darmstadt.de
[3] Institute for Production Management, Technology and Machine Tools,
Technical University of Darmstadt, Darmstadt, Germany
{m.schreiber,j.metternich}@ptw.tu-darmstadt.de

Abstract. Market forces such as rising amounts of product variants and decreasing batch sizes lead to higher complexity in manufacturing processes. Therefore, production management's demand for data-based process transparency is growing continuously as well as the number of companies turning to process mining to address these challenges. Information systems in production usually do not provide readily available event log data for the analysis. This paper investigates several techniques for inferring missing event log data in production processes by extracting events with timestamps from sensor data from machines and link them to process instances. We demonstrate the effectiveness of our approach in a real-world manufacturing environment. The evaluation of the resulting event logs revealed that the quality of the timestamps and the assignment of the actual process instances is sufficient to apply process mining techniques that would have required both greater effort and higher cost intensity if a traceability system had been implemented.

Keywords: Process mining · Internet of Things · Event logs · Manufacturing

1 Introduction

Industrial companies are challenged by rising complexity in their manufacturing processes, for example due to constantly rising demands for customized products. In response to competitors, companies create more product variants to stay attractive on their markets [9]. In the context digitizing the manufacturing sector, also known as Industry 4.0 (I4.0), data-based transparency is needed to tackle complexity and support the effective decision-making of managers [22].

A. Marrella et al. (Eds.): BPM 2024 Forum, LNBIP 526, pp. 232–248, 2024.
https://doi.org/10.1007/978-3-031-70418-5_14

Out of 104 interviewed companies, 40% have stated to use process mining in production, aiming for process transparency and improvements as their top two reasons [8, 19].

However, manufacturing processes are often supported by non-process aware and/or proprietary software systems [23], hindering the use of process mining techniques. As a result, process mining projects in production tend to fail due to missing and unreliable data [18]. In particular, data acquisition is considered a key factor in improving availability and reliability of process data and hence an event log's quality [6, 11]. In this paper, we propose a framework that aims to reconstruct process data, i.e., event logs, solely based on input from *Internet of Things* (IoT) sensors that organizations equip their machines with. These sensors measure various metrics to accurately monitor the condition of a machine or production line. However, the sensor data is not inherently linked to process activities, instances, or models, which makes their use for process mining a major challenge [12].

We demonstrate the effectiveness of our approach by applying it to the manufacturing process of the *TU Darmstadt ETA-Fabrik* and *Prozesslernfabrik CiP* [26] of an entire week with the goal to infer event logs from IoT sensor data for further analysis with process mining. The process consists of several real-life manufacturing steps like sawing, lathing, milling, cleaning, laser marking, and hardening. The production site is equally equipped with IoT sensors at each machine as well as a traceability system [19] which is used for component tracking. Traceability data is collected from scanning sets of optical codes (data matrix code and QR-Code) of the components and the machines by which they are processed. The recorded scanning timestamps act as the ground truth for evaluating the obtained results. Simultaneously different sensor data from each machine is collected and centrally stored. By using these sensor data, we aim to reconstruct the trace of each individual part at the floor shop without relying on the expensive traceability system. The results and lessons learned of the study under real-life conditions are reported in this paper.

The rest of the paper is organized as follows: First, we introduce related work (Sect. 2). Then, we introduce our framework using IoT sensor data to reconstruct events logs for the use in process mining techniques (Sect. 3). In Sect. 4, we present our case study setup and show the results. Afterwards, we discuss the results and present our findings (Sect. 5). Finally, we conclude the paper and give an outlook for future work (Sect. 6).

2   Related Work

Traceability systems are integral to the Industry 4.0 vision of smart factories, providing insights into the precise location of individual objects within the production line. These systems have a great potential to enhance the quality and availability of data for effective Process Mining analyses by capturing, storing, and utilizing data such as product IDs, locations, and timestamps in production processes [20]. Typically, traceability systems use *automatic identification* (autoID) technologies such as optical codes or *Radio Frequency Identification*

(RFID) to generate process data. In the investigated real-life production network a traceability system is used to provide process data for evaluating its performance. The implementation of a traceability system requires not only the initial investment of the corresponding hardware, but also the effort to configure the traceability system adequately generating discrete data points in production needed to derive process traces. Therefore, a necessary data acquisition concept including the corresponding configuration of the traceability system is developed [20,21]. The concept ensures that every product manufactured is tracked through an individual optical data matrix code when arriving and before leaving any station. In order to avoid the financial impact and mentioned configuration effort when implementing a traceability system, this paper presents an alternative approach using data generated by IoT sensors instead of generated trace data to create valid process traces.

With the increasing equipment of production machines with low-cost IoT sensors to collect metrics in periodical time intervals, more and more IoT data is available for analysis. However, this data is disconnected from processes [12] because there is typically no case identifier. Van Eck et al. [5] segment sensor data to discover user interaction models using bottom-up change point detection, then identified segments are assigned to individual cases, providing start and end timestamps. Real-time locating systems to reconstruct process models are investigated by Senderovich et al. [24]. Based on prior knowledge about time and space, sensor data is mapped to event logs by relating identified user interactions to cases. Similarly, Dogan et al. [3,4] use indoor localization systems in a shopping mall environment to track customer journeys with Bluetooth beacons and infrared technology. Vitali et al. [27] correlate sensor data to event log to identify interconnections between different processes. A *complex event processing*-based approach which correlates and composites sensor data with event logs is introduced by Seiger et al. [23].

The analysis of IoT sensor data itself is another wide area of research: For instance, there exist different types of *change point detection* to identify points in time where the behavior of a time series changes, such as sliding window approaches [1], binary segmentation [15], or bottom-up segmentation [13]. Another technique is *pattern matching* which specifically searches for repeating behavior in a time series. Approaches range from template and rule-based methods [7] over hybrid methods that combine correlation coefficient, rule-sets, and sliding window approaches [29], to similarity search-based methods [17]. A common problem in pattern matching is that the template may not be exactly time-aligned, which is why Dynamic Time Warping (DTW) [2] was introduced, which tries to compute the lowest distance by matching each point between two time series. *Motif discovery* [10,14,25] aims to obtain templates capturing repeating sequences in a time series for pattern matching, also called motifs. Due to the computation of such a motif being rather expensive, the concept of a *matrix profile* [28] is introduced to overcome computational issues. The matrix profile is an efficient data structure to compute distances between subsequences within a time series.

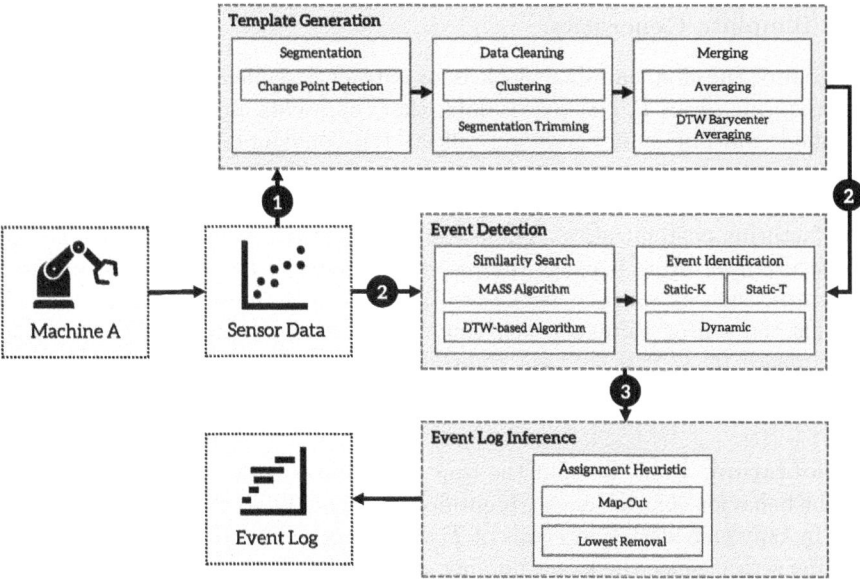

Fig. 1. Overview of the framework, showing the three main building (blue boxes), their relationships (arrows), and the introduced methods (orange boxes). (Color figure online)

3 Methodology

In this section, we introduce our framework for inferring event logs from IoT sensor data in manufacturing processes. The goal of the framework is to define concepts and methods that bridge the gap between IoT sensor data and processes, linking machine data to process instances. As a result, the output of the framework is an inferred event log that accurately represents the events with start and end timestamps of each machine and process instance. Our framework consists of three main building blocks (see Fig. 1):

1. **Template generation:** Extract a template pattern from IoT raw sensor data to determine the repeating active working cycle of a machine processing a component within the process (Sect. 3.1).
2. **Event detection:** Determine accurate start and end timestamps for each machine representing the active working cycle (Sect. 3.2).
3. **Event log inference:** Assigning the detected events to the corresponding process instance to infer the final event log (Sect. 3.3).

Next, we provide a detailed description of the three building blocks and introduce methods for implementation.

3.1   Template Generation

The goal of the first building block is to extract a sequence template $R \in \mathbb{R}^n$ from the time series T, i.e., sensor data, that represents an active working cycle of a single machine and, therefore, reflects the behavioral sensor value pattern for which we want to infer an event. We aim for a template that is specific to represent start and end points of a cycle but is also generic to deal with potential manufacturing customization. These start and end timestamps of machines are specifically interesting because the respective material has additional practical key states, such as waiting, transportation, and idle times. The time series T contains the sensor value at each time in the analyzed interval. To generate the template R, we first segment the time series, perform data cleaning of each identified segment, and finally merge all identified repeating segments.

Segmentation. First, we split the time series into smaller segments where significant behavior changes occur. Identified segments build the basis for the template by only considering intervals of T which represent active working cycles of the time series. Since the exact number and the structure of the working cycles is unknown, we identified *approximate change point detection* approaches being a good fit for segmentation. Specifically, we found sliding window [1], binary segmentation [15], and bottom up segmentation [13] approaches provide most accurate results in our experiments (see Sect. 4). The time series is split into smaller segments by providing change points that determine significant behavior changes. The sensitivity of the methods can be determined by a threshold parameter. Using a low threshold value avoids missing the split of passive and active segments but may introduce more segments within the passive interval of the time series. As a result, we obtain segments of the original time series that are candidates for creating the template R.

Data Cleaning. Next, we classify the identified segments into the *active* and *passive* states using clustering (see A2). Specifically, we use k-means clustering with $k = 2$. The segments of the cluster with the lower average values are classified as passive and, thus, removed. As a result, we obtain only segments of T which are good candidates to represent the active working cycles of the machine. In preliminary experiments with the real-life dataset, we found that an additional step is needed to trim trailing low sensor values of each segment. This trailing values are typically not part of the active working cycle but have been detected by error. We only consider the trailing third of the segment $T_{i,j}$ to find a cutoff point x such that it maximizes the difference between the subsegments it separates.

Merging. Lastly, we generate a template R for each machine from all segments that were classified as active. The basic idea is to merge all segments into a single template, representing all characteristics of a single working cycle of a machine including potential manufacturing customizations. A naive approach is to just averaging all segments to retrieve the template. However, in order to retain the specific characteristics, i.e., avoid the smoothing out effect due to averaging, of the segments, we propose the use of *Dynamic Time Warping Barycenter Averaging* (DBA) [16] to merge the patterns. DBA aligns all identified active working cycle segments such that different lengths are adjusted and characteristic values, like spikes and plateaus, remain in the resulting template.

3.2   Event Detection

The goal of the second building block is to determine the exact events, i.e., the start and end timestamps of the active working cycles of a machine for each process instance, by matching the template R in the time series T.

Similarity Search. The first step is to determine all occurrences of the template R in the time series T. We identified two methods for computing the distance to obtain the positions where the template occurs in T: the MASS algorithm [14] and Dynamic Time Warping (DTW) [2].

The MASS algorithm is an efficient method to compute a distance profile D between a template R and all subsequences of a time series T with the length of R using the Euclidean distance. The distance profile represents the distance of the template at each point of the T which allows to determine at which points in T the template matches. However, using the Euclidean distance has the disadvantages that it may inaccurately find matches due to distortion, misalignment, and different sequence lengths.

We propose a customized algorithm based on the MASS algorithm with the DTW distance function to overcome these issues. Similar to the MASS algorithm, a matrix is computed by measuring the distance between all subsequences of T and R. Each row in the matrix maintains the distance between the template and the subsequence but with different lengths to find the best match using the DTW distance function. From each row, we then take the lowest distance along with the ending index of the subsequence in T to which this distance was computed. An additional vector is stored to capture the respective length of the subsequences at which the distance is the lowest. The total lowest distance values then build the matrix profile (see Fig. 2).

Since the computational complexity of the algorithm lies in $O(n^3)$ with n being the length of the template, two simplifications are considered:

(1) **Limit window size.** The DTW distance function is limited by a window size that determines how far two values of the compared sequences can be apart to be considered as similar.
(2) **Limit subsequence size.** The length of the subsequences is limited to reduce the number of comparisons.

Event Identification. The last step is to obtain the start and end timestamp for each event by computing the local minima, i.e., the lowest distance between

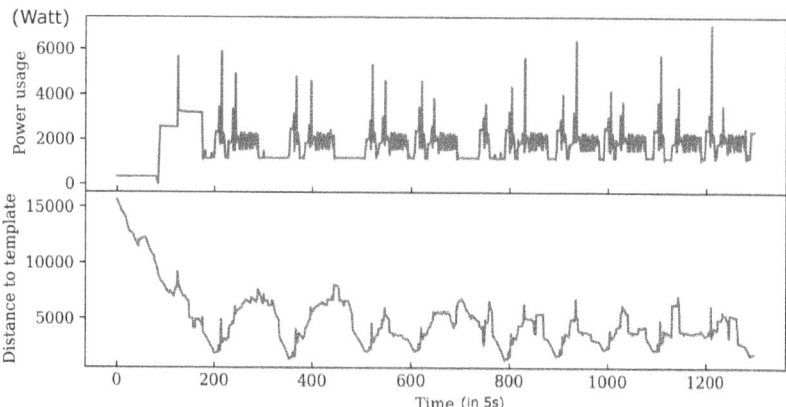

Fig. 2. The original time series T and the distance profile D of a template R using the DTW-based similarity search algorithm of machine HaasST10 (y-axis = watt; x-axis = time in 5 s).

the template and the time series, from the distance profile (see bottom part of Fig. 2). We compute the moving average within a specific time window of the distance profile and obtain the points where the gradient is zero. These points are then sorted to determine the best local minima. We propose three heuristics to determine the final events:

- **Static-K** - Only the top-K event candidates with the lowest distance are considered as active working cycles. This strategy can be used if the number of events is known beforehand.
- **Static-T** - Only the events below a predefined threshold ϕ are used to represent active working cycles. This strategy can be used if the number of events is unknown.
- **Dynamic** - k-means clustering determines two clusters to discriminate between relevant and irrelevant events. The cluster with the lowest average distance and its indices are used.

Finally, the events are generated by turning the final set into tuples with start and end timestamps by adding the lengths to the start index.

3.3   Event Log Inference

In the third and final building block, we assign each event to a process instance to connect IoT related data and process data together under the assumption that the order and number of process instances is known beforehand (see A4).

It is straight forward to assign events to the corresponding process instances if the number of identified events is equal to the number of instances, which is always the case for the *static-K* heuristic. However, other inference heuristics may lead to a negative or positive difference which does not allow for exact matching. In case we have identified too few events, the thresholds for *static-T* or the parameters for *dynamic* must be adjusted towards producing more events.

In order to adequately assigning events to process instances when there are more events than instances, we propose two assignment heuristics. These heuristics assume that at least some one-to-one mappings exists, e.g., through timestamps or partial order:

- **Map-Out** makes use of the template length and tries to map instances to events that are within the interval of a template without overlaps. Events that are outside the intervals are removed.
- **Lowest Removal** removes the events with the lowest similarity value in between known one-to-one mappings, until the number of required events is attained.

As a final result, we get a set of start and end events, each assigned to a specific process instance to build the inferred event log which can be used to apply any process mining technique.

4   A Case Study in Manufacturing

In this section, we present the results and lessons learned of our case study applying the proposed framework, which infers event log data for process mining from IoT sensors data recorded in a production process. We compare different methods for each proposed building block and apply them to a real-life production environment to show their respective performance.

We implemented our framework[1] using Python with the use of different existing libraries such as *"ruptures"*[2] (bottom up, binary, and sliding window segmentation), *"scikit-learn"* (k-means clustering), *"tslearn"* (DTW barycenter averaging), *"dtaidistance"*[3] (dynamic time warping), and *"STUMPY"*[4] (matrix profiles).

4.1   Use Case

Process Description. We applied our approach to a real manufacturing process using data from TU Darmstadt ETA-Fabrik and Prozesslernfabrik CiP [26]. The manufacturing process consists of several steps that are interconnected with each other in two separate factories (see Fig. 3). The process begins with sawing the raw material with a Kasto machine and produces cylindrical metal disks. In the next step, the metal disk is machined using a lathing and milling machine. The produced piece is then cleaned using several machines. Finally, to harden the metal disk it gets into a gas nitriding oven.

---

[1] https://github.com/alexsee/self-templating-event-detection.

[2] https://centre-borelli.github.io/ruptures-docs/.

[3] https://dtaidistance.readthedocs.io/en/latest/.

[4] https://stumpy.readthedocs.io/en/latest/.

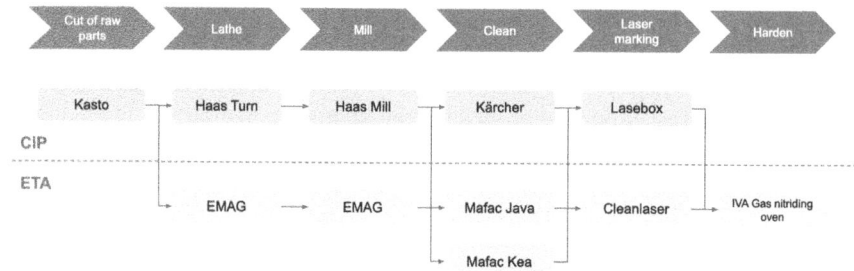

Fig. 3. Overview of the manufacturing process in ETA and CiP manufacturing factory at TU Darmstadt.

Dataset. The dataset consists of data from two main data sources over a period of 5 days. First, the entire process was tracked by a traceability system, recording the exact timestamps of every individual component being processed by each machine. Traceability data is collected from scanning sets of optical codes (data matrix code and QR-Code) of the components and the machines by which they are processed. The recorded scanning timestamps act as the ground truth for evaluating the obtained results. Second, simultaneously sensor data from each machine was collected and centrally stored with a data point sampled every 5 s.

In our evaluation, we only use the power usage sensors as the input for our framework. We exclude time spans where faulty and missing values occurred during the data collection.

4.2   Experimental Setup

We evaluate three aspects of our approach and compare the results with appropriate baselines and the ground truth data collected by the traceability system:

1. **Segmentation accuracy.** We compare three different segmentation methods and evaluate the quality with the ground truth segments.
2. **Event detection accuracy.** We compare our DTW-based similarity search on different sensor data patterns with the Euclidean-based similarity search used by the MASS algorithm.
3. **Event log inference accuracy.** We evaluate the event log inference and report the quality of the event to process instance assignment.

We measure the following two aspects:

(i) For segmentation and pattern matching, we measure the quality by computing *precision*, *recall*, and *F1-score* of the identified change points compared to the ground truth. We allow a small window (of 60 s) within we consider an identified change-point as a true positive.
(ii) For event log accuracy, we measure the component assignments of each identified pattern and compare it with the ground truth.

In the following, we report the results of our experiments.

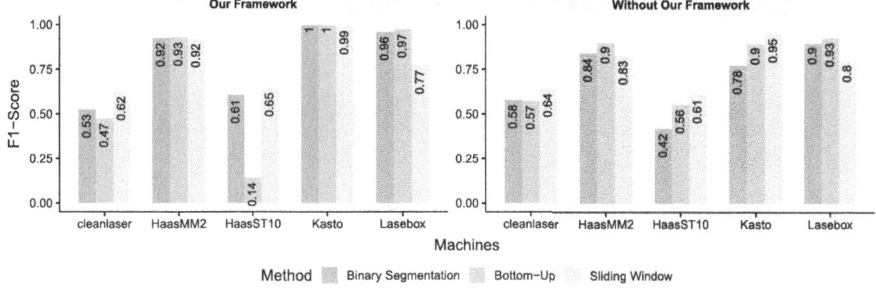

Fig. 4. Average F1-scores across all conventional change-point detection methods and the event identification grouped by the different machines.

Table 1. Average F1-scores of the time series segmentation grouped by method and event identification heuristic (static-T, static-K, and dynamic).

Mode/Method	Static-T	Static-K	Dynamic	Our Framework
Bottom-up	0.687	0.689	0.657	0.771
Binary Segmentation	0.758	0.739	0.690	0.703
Sliding Window	0.778	0.770	0.761	0.767

4.3 Segmentation Accuracy

We compare three change point detection algorithms (sliding window, bottom-up, and binary segmentation) with our approach. We compute the optimal penalty value of each method and proceed with the best results to determine the final events for comparison. Our approach is configured with the DTW-based similarity search and *static-T* as the event identification heuristic. We report the average F1-scores over all manufacturing days for each machine type individually, reflecting correctly detected change points. The results are depicted in Fig. 4.

On average the F1-scores show an improvement over the conventional approaches (see Table 1). However, we also observed substantially lower results for the machines *Haas Turn* with bottom-up and *Cleanlaser* across all methods. For *Haas Turn* we found that even with a high penalty value, many change points are detected that are close to each other and, therefore, do not accurately represent the segments. In particular, high value spikes are separately identified as a segment although these are part of a larger actual segment.

We also measured the offset between the identified change points and the ground truth (see Fig. 5). The results show that across all methods change points are detected on average a few seconds earlier with a standard deviation of 20 s.

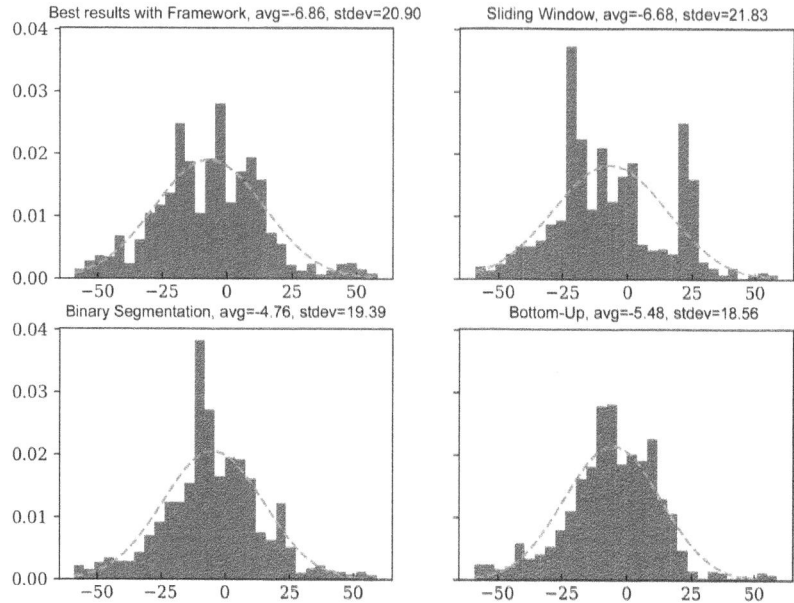

Fig. 5. Average offset of detected true positive change-points compared to the ground truth in seconds.

4.4 Event Detection Accuracy

The second part of our evaluation deals with the pattern matching accuracy, comparing MASS and our DTW-based similarity search. Based on the results of the previous section, we use *static-T* for the comparison. Figure 6 shows the average F1-scores, precision, and recall results for MASS and DTW-based similarity search. We evaluated start and end change-points individually to investigate if either can be detected more accurately. The results show significantly better results for our custom DTW-based similarity search which better handles distortion, misalignment, and different sequence lengths than the Euclidean distance used by the MASS algorithm. Better results are mainly influenced by the lower recall. Consequently, the MASS algorithm produces more false negatives related to the number of true positives than the DTW-based similarity search.

4.5 Event Log Inference Accuracy

Lastly, we evaluate the inferred event log and compare it with the ground truth. Based on the previous results, we selected the sliding window approach with DTW-based similarity search as the setup for evaluating the event log inference.

Table 2 shows the accuracy of the event log inference for the event identification heuristics (*static-T*, *static-K*, and *dynamic*) and the assignment heuristics (*map-out* and *lowest removal*). For both heuristics, *static-K* achieved the best

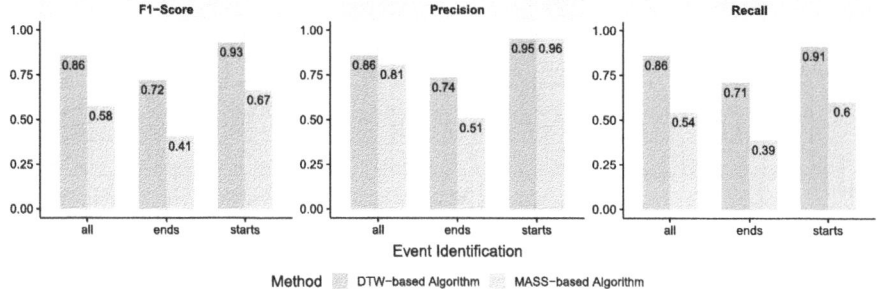

Fig. 6. Average F1-scores, precision, and recall for the MASS- and DTW-based similarity search, showing the values for all, only start, and only end change-points.

Table 2. Accuracy of event identification combined with assignment heuristics.

Heuristics	Static-T	Static-K	Dynamic
Map-Out	0.694	0.858	0.658
Lowest Removal	0.793	0.853	0.758

accuracy with 85.8% correctly matched events. The heuristic *lowest removal* achieved a generally better accuracy for *dynamic* and *static-K* event detection. We found that *lowest removal* and *static-K* decreases the risk of misalignment of events due to the lower number of detected events in general. Contrary, *map-out* with *dynamic* tend to detect a higher number of events and increasing the risk of misalignment. Interestingly, *static-T* leads to lower accuracy results although achieving the highest F1-scores for detecting change points.

Figure 7 shows the result of the event log inference of the machine *Cleanlaser* on a single day. We can see that the detection of the actual start and end events is mostly accurate. However, we also observed that in some cases the actual end event is not detected correctly.

5   Discussion

5.1   Results

The results of our evaluation suggest that for the initial segmentation the *sliding window* approach performed best. In particular, it is more suitable than the *bottom-up* method which achieved similar results as standalone but tends to detect end events within the active sequence. For the sliding window approach, it is essential to select a window size that is similar to the expected size of the active working cycle. Selecting an appropriate size leads to more accurate detection results around active sequences at the cost of change points amidst inactive sequences. In general, we found that the *static-T* strategy provides best results for change-point detection, however, *static-K* delivered better results in the context of event inference and assignment.

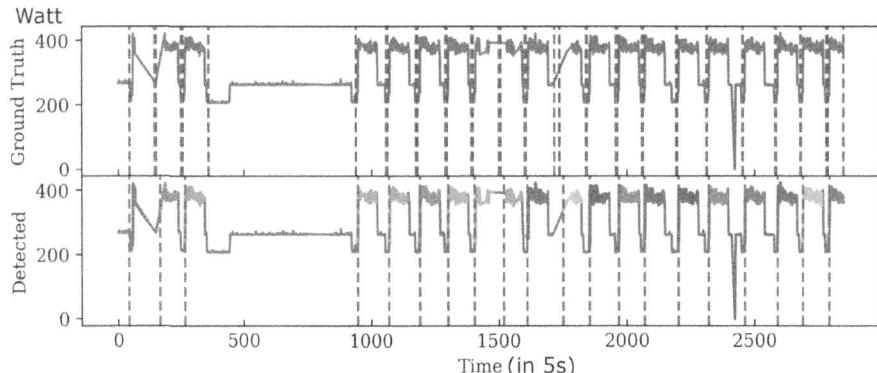

Fig. 7. Time series excerpt for machine *Cleanlaser* on 2020-06-26 with event identification heuristic *static T* of one day compared to the ground truth. Green dotted lines indicate a start of an event whereas a red dotted line indicates the end of the event (y-axis = watt; x-axis = time in 5 s). (Color figure online)

Our presented approach improves F1-score performance in average compared to conventional change point detection when applied with optimal settings. This is mainly achieved by considering the specific characteristics of a manufacturing process, such as repeated working cycles, ramp-up, and cool-down phases of machines. Furthermore, our approach explicitly classifies events into start and end events to improve event log inference performance, which is not possible with conventional change point detection methods.

Instead of using Euclidean similarity search, such as used in MASS, we introduced a customized DTW-based similarity search that provides better working pattern detection performance of machines. The algorithm explicitly considers time shifting, differences in amplitudes, and differences in sequence length of working cycles. The evaluation results show an average accuracy of 85.8% regarding event log inference when using our proposed framework on this specific dataset.

5.2   Limitations

The reality shows that manufacturing environments vary significantly and share different characteristics. In our case study, we only investigated the applicability to a single manufacturing process and, thus, considered the following limitations:

A1 **Single process observation:** Machines only perform activities related to the process being investigated and analyzed.

A2 **Two machine states:** Machines only have the states *active*, i.e., a component is currently processed, and *inactive*, i.e., no component is processed. However, we don't assume that these states have a constant value or are detectable by a fixed threshold. Instead, we detect patterns that represent an active work cycle of the machine (see Fig. 2).

A3 **Repeating pattern:** Machines perform a single task which is repeated for each component being processed. However, the task may be distorted or may be of different length.

A4 **Known order of component:** The order and identifier of each component entering the process is known.

We acknowledge that these assumptions may limit the selected methods we applied but the core concepts of our framework are still applicable in a modified fashion. Still, we found that our framework works under real-life conditions and provides accurate timestamps to reconstruct manufacturing processes.

Another aspect we would like to mention is that we only considered the use of the power consumption to detect machine patterns in our case study. However, there may be other sensor values that are better indicators for detecting working cycles but these are the most easiest to install and equip machines with. We also observed that certain machines (e.g., Kärcher) consume constant amounts of power during the production and, therefore, these sensors cannot be used.

Lastly, our framework does not consider storage places or queues between machines which would allow analyzing the intralogistics. However, due to the detection of start and end timestamps of machine working cycles, we can precisely identify breaks and waiting times of machines within the entire process. Our approach is capable of inferring event logs on the level of start and end events of each component and machine which follow the first-out strategy without being able to identify the individual component. Other strategies are currently not supported.

6   Conclusion

In this paper, we introduced a novel framework for inferring event log data from IoT sensor data in the context of production processes. Our framework consists of three main building blocks to assist the process of connecting machine-centric sensor data to process instances. For each building block, we provided methods for implementation and evaluated them within the context of a real-life manufacturing process. The conducted experimental evaluation of our approach shows the general applicability of the framework compared to the used traceability system. Based on the available traceability events, the proposed methods provide a high accuracy of the identified events solely using the machines' sensor data.

From a practical point of view, companies have to consider some requirements to apply our alternative approach. Most importantly, the application of the approach depend on employed staff with competencies in the field of data analytics and algorithm development. Further important requirements are a fully equipped production line with IoT sensors that allow data collection and the inference to events as well as the configuration of various method parameters and the validation of the working cycle template. In comparison, the usage of a traceability system as data supplier appears to be more obvious. Traceability systems are typically available in manufacturing companies (see Sect. 2). In order to use traceability systems for systematic data generation, production managers need develop a data acquisition strategy and adapt the hardware set-up accordingly.

From a scientific perspective, an important application field of our alternative approach can be found in the process industry characterized by continuous production processes where physical components may not be tracked via autoID technologies. Also, the alternative approach can be applied in discrete production processes. On the one hand, based on the high average accuracy of the inferred events shown in our results, the approach may be very useful substituting a lacking traceability system that generates start and end timestamps of individual machining processes to determine the performance of a production network. On the other hand, it may complement a detailed data acquisition strategy in an entire value stream, firstly by capturing intralogistic timestamps based on a traceability system and secondly using machine timestamps based on IoT sensor data. In contrast to a traceability system, the used IoT data in our approach facilitates the distinction of different operating states of production machines such as warm-up phase, operation phase, or cool-down phase.

As for future work, it would be interesting to compliment the use of traceability systems with our alternative approach. Firstly, research needs to provide a systematic way to use traceability systems as data supplier. Secondly, it needs to be investigated how the alternative approach can support the systematic generation of event data indicating different operating states of a single machine. Currently, our framework assumes that a machine only performs one single task on components and that this repeating task has the same pattern. However, it may be the case that different products lead to slightly different operating states and, therefore, the pattern may be different. Additionally, we have not considered explicit dependencies between machines which could improve and simplify the assignment of process instances to events. If we know that certain tasks depend on each other, this order of tasks can be used to improve the assignment of each event to process instances.

References

1. Ahmed, E., Clark, A., Mohay, G.: A novel sliding window based change detection algorithm for asymmetric traffic. In: International Conference on Network and Parallel Computing, pp. 168–175. IEEE (2008)
2. Berndt, D.J., Clifford, J.: Using dynamic time warping to find patterns in time series. In: KDD Workshop, vol. 10, pp. 359–370 (1994)
3. Dogan, O., Bayo-Monton, J.L., Fernandez-Llatas, C., Oztaysi, B.: Analyzing of gender behaviors from paths using process mining: a shopping mall application. Sensors 19(3), 557 (2019)
4. Dogan, O., et al.: Individual behavior modeling with sensors using process mining. Electronics 8(7), 766 (2019)
5. van Eck, M.L., Sidorova, N., van der Aalst, W.M.P.: Enabling process mining on sensor data from smart products. In: International Conference on Research Challenges in Information Science (RCIS). IEEE (2016)

6. Flack, C., Dreher, S., Birk, A., Wilhelm, Y.: Process Mining in der Produktion: Spezifische Herausforderungen bei der Anwendung. Zeitschrift für wirtschaftlichen Fabrikbetrieb **115**(11), 829–833 (2020)

7. Fu, T., Chung, F., Luk, R., Ng, C.: Stock time series pattern matching: template-based vs. rule-based approaches. Eng. Appl. Artif. Intell. **20**(3), 347–364 (2007)

8. Galic, G., Wolf, M.: Delivering Value with Process Analytics - Adoption and Success Factors of Process Mining (2020)

9. Gottmann, J.: Produktionscontrolling. Springer, Wiesbaden (2016)

10. Hu, J., Li, B., Kihara, D.: Limitations and potentials of current motif discovery algorithms. Nucleic Acids Res. **33**(15), 4899–4913 (2005)

11. Jahn, M.: Industrie 4.0 Konkret. Springer, Wiesbaden (2017)

12. Janiesch, C., et al.: The Internet of Things meets business process management: a manifesto. IEEE Syst. Man Cybern. Mag. **6**(4), 34–44 (2020)

13. Keogh, E., Chu, S., Hart, D., Pazzani, M.: An online algorithm for segmenting time series. In: International Conference on Data Mining, pp. 289–296. IEEE (2001)

14. Mueen, A.: Time series motif discovery: dimensions and applications. Wiley Interdisc. Rev. Data Min. Knowl. Discov. **4**(2), 152–159 (2014)

15. Olshen, A.B., Venkatraman, E.S., Lucito, R., Wigler, M.: Circular binary segmentation for the analysis of array-based DNA copy number data. Biostatistics **5**(4), 557–572 (2004)

16. Petitjean, F., Ketterlin, A., Gançarski, P.: A global averaging method for dynamic time warping, with applications to clustering. Pattern Recogn. **44**(3), 678–693 (2011)

17. Rakthanmanon, T., et al.: Searching and mining trillions of time series subsequences under dynamic time warping. In: International Conference on Knowledge Discovery and Data Mining. ACM Press (2012)

18. Reinkemeyer, L.: Process Mining in Action. Principles, Use Cases and Outlook, Santa Barbara (2020)

19. Schreiber, M., Bausch, P., Windecker, M., Metternich, J.: Traceability-Systeme als Enabler für Process Mining Eine Interview-Studie zur industriellen Praxis. ZWF Zeitschrift fuer Wirtschaftlichen Fabrikbetrieb **116**(10), 722–727 (2021)

20. Schreiber, M., Metternich, J.: Data value chains in manufacturing: data-based process transparency through traceability and process mining. Procedia CIRP **107**, 629–634 (2022)

21. Schreiber, M., Weisbrod, N., Ziegenbein, A., Metternich, J.: Tool management optimisation through traceability and tool wear prediction in the aviation industry. Prod. Eng. Res. Devel. **17**(2), 185–195 (2023)

22. Schuh, G., Anderl, R., Dumitrescu, R., Krüger, A., ten Hompel, M.: Der Industrie 4.0 Maturity Index in der betrieblichen Anwendung. Aktuelle Herausforderungen, Fallbeispiele und Entwicklungstrends. acatech-Studienreihe (2020)

23. Seiger, R., Zerbato, F., Burattin, A., Garcia-Banuelos, L., Weber, B.: Towards IoT-driven process event log generation for conformance checking in smart factories. In: International Enterprise Distributed Object Computing Workshop, pp. 20–26 (2020)

24. Senderovich, A., Rogge-Solti, A., Gal, A., Mendling, J., Mandelbaum, A.: The ROAD from sensor data to process instances via interaction mining. In: Nurcan, S., Soffer, P., Bajec, M., Eder, J. (eds.) CAiSE 2016. LNCS, vol. 9694, pp. 257–273. Springer, Cham (2016). https://doi.org/10.1007/978-3-319-39696-5_16

25. Torkamani, S., Lohweg, V.: Survey on time series motif discovery. Wiley Interdisc. Rev. Data Min. Knowl. Discov. **7**(2), e1199 (2017)

26. Urnauer, C., Schreiber, M., Bausch, P., Metternich, J.: Anwendungen aktiver Traceability-Systeme. Zeitschrift für wirtschaftlichen Fabrikbetrieb **116**(3), 166–170 (2021)
27. Vitali, M., Pernici, B.: Interconnecting processes through IoT in a health-care scenario. In: International Smart Cities Conference. IEEE (2016)
28. Yeh, C.C.M., et al.: Matrix profile I: all pairs similarity joins for time series: a unifying view that includes motifs, discords and shapelets. In: International Conference on Data Mining. IEEE (2016)
29. Zhang, Z., Jiang, J., Liu, X., Lau, R., Wang, H., Zhang, R.: A real time hybrid pattern matching scheme for stock time series. In: Australasian Conference on Database Technologies, vol. 104, pp. 161–170 (2010)

A Framework to Support the Validation of Process Mining Inquiries

Francesca Zerbato[1,2]( ), Marco Franceschetti[1], and Barbara Weber[1]

[1] University of St. Gallen, St. Gallen, Switzerland
[2] Eindhoven University of Technology,
Eindhoven, The Netherlands
f.zerbato@tue.nl

Abstract. In exploratory process mining, analysts often start with limited knowledge of the log. As they seek to improve their understanding of the log, they develop expectations about what the results might be. Based on these expectations, they then make inquiries and translate them into queries against the log. However, during the analysis, analysts need to evaluate and compare the results of their queries to be able to validate them against their expectations. In this paper, we propose a framework to support process analysts in validating their query results and to enable them to reflect on their analytical process. The framework helps analysts to record their queries and results and allows them to characterize and compare the results obtained with different queries, thereby facilitating the validation process. We implemented the framework as a Python library that can be easily extended and integrated into existing process mining environments. We also demonstrated the usefulness of the framework through an extensive analysis of a real event log.

Keywords: Process Mining · Result Validation · Result Set Comparison · Exploratory Analysis · Meta-analysis Support

1 Introduction

Process mining is an analytical discipline that provides methods and techniques for analyzing event logs derived from the execution of work processes [1].

In exploratory process mining scenarios, although driven by high-level goals, process analysts often begin their analysis with limited knowledge of the log [18]. In this context, they often make hypotheses or inquiries to explain observations in the log that go beyond existing knowledge and form expectations about potential patterns that could help validate these inquiries [12]. Guided by these expectations, analysts then translate their inquiries into filters (or *queries*). These queries slice the event log into subsets of cases that analysts then can examine to see if they match their expectations and answer their inquiries.

By iteratively querying the log and and comparing multiple query results, analysts can deepen their understanding of the data and refine their query approach. However, this requires that analysts are able to view the results of their queries and compare them at any point during the analysis, to validate whether they meet their initial expectations.

Unfortunately, many process mining tools lack explicit support for analysts to assess and compare query results. Tools typically display one (subset of the) log at a time, forcing analysts to move back and forth between filter screens [13] and to manually record their filters operations [17]. Similarly, when examining a filtered log, analysts are presented with a partial view of the data and cannot see what has been filtered out. This can make it difficult to assess query results, especially in exploratory scenarios where there is little "baseline" knowledge of the log and analysts may struggle to keep their analysis focus [20].

In this paper, we propose a framework to support process analysts in validating the results of their queries (henceforth, *result sets*) and incrementally improve their understanding of the log. Our framework leverages properties about a result set and its elements as well as relationships among (result) sets [2] to help analysts characterize and compare query results. When running a query, analysts need to understand what the result looks like and ensure that it aligns with their expectations. We support this process by characterizing a result set through comparison with a reference log, allowing analysts to assess what has been filtered out. This helps analysts address questions such as "Did my query capture all the late cases?" and "Does my result consist exclusively of late cases, or have I missed something?" Moreover, we support the pairwise comparison of query results, helping analysts understand relationships between result sets, such as "Do these two sets of late cases overlap?" Finally, the framework maintains a record of both the results and the associated queries, which is accessible to the analyst. This is crucial for analysts to navigate and compare their results, as well as for ensuring transparency and traceability throughout the analysis.

We implemented the framework as a Python library, named LogView, which can be easily extended and integrated into existing process mining environments thanks to its modular design. Our implementation is supplemented by a tutorial demonstrating the main functionality of the framework. The usefulness of the library is also demonstrated through an extensive analysis of the Road Traffic Fine Management (RTFM) event log [7].

Our framework aims to support process analysts in reflecting on their queries and results and address common challenges in process mining analysis, such as understanding and filtering event logs [20]. Our approach can be seen as a form of *meta-analysis* that characterizes the results of process mining analysis from multiple perspectives and allows analysts to compare and trace them. In its goal, our approach differs significantly from (automated) process mining techniques that derive insights from event logs, such as comparative process mining [1], and from methods that suggest specific areas of focus within an event log [14]. Indeed, while we help analysts characterize and compare results, we let their expertise decide what to query and how to interpret the results.

The paper is as follows. Section 2 outlines preliminary concepts. Section 3 presents our framework. Section 4 describes the implementation, while Sect. 5 presents evaluation results. Section 6 discusses related work, and Sect. 7 closes the paper.

2 Background

In this section, we introduce the core concepts used in this paper in Sect. 2.1 and motivate the problem with concrete use cases in Sect. 2.2.

2.1 Preliminaries

In this paper, we focus on queries that resemble **case filters**[1], i.e., queries that allow selecting (a set of) cases in the event log that have a particular property, expressed by one or more query conditions (or predicates).

Let A be a sequence of **attributes** $A = \langle a_1, ..., a_m, a_{name}, a_{timestamp} \rangle$. An **event** e is a sequence of constant values $e = \langle k_1, ..., k_m, k_{name}, k_{timestamp} \rangle$, where k_i is the value of attribute a_i, and k_{name} and $k_{timestamp}$ are the name and the timestamp value of event e. We denote as $e.a_i$ the i-th attribute of event e.

A **case** c is a sequence of events $\langle e_1, e_2, ..., e_n \rangle$, where for each i, j where $0 < i < j \leq n$, the *timestamp* of the event e_i is smaller than the *timestamp* of e_j. We denote as $c.e_j$ the j-th event of case c. An **event log** L is a sequence of cases $\langle c_1, c_2, ..., c_l \rangle$, where i is an unique identifier for case c_i.

A **predicate** P is a universally or existentially quantified relation between an attribute of some event e and another attribute or a constant value, where the relation is in $\{=, \neq, <, \leq, >, \geq\}$, e.g., $P = \exists e : e.a_{amount} \leq 2500$.

A **query** is either a simple predicate $(Q = P)$ or a conjunction of predicates $(Q = P_1 \wedge P_2 \wedge ... \wedge P_w)$. Evaluating a query Q against a log L means evaluating all predicates in the query against all cases in L.

We denote with function $eval(L, Q) = L'$ the evaluation of Q against L, which results in L'. L' is a *view* on the event log L such that $L' \subseteq L$ and for all $c' \in L'$ for each predicate $P_i \in Q$ it holds that c' satisfies P_i. With $eval(L, Q)$, the log L is naturally partitioned into two sets: a set L' of cases that satisfy Q and a set $\overline{L'}$ of cases that do not satisfy Q (i.e., satisfy $\neg Q$). It is easy to see that $\overline{L'} = eval(L, \neg Q)$, that $L' \cap \overline{L'} = \emptyset$, and $L' \cup \overline{L'} = L$. Also, we know that L' and $\overline{L'}$ are *included* in L, i.e., $L' \subseteq L$ and $\overline{L'} \subseteq L$. As a result of applying a query, we then have three sets of cases that may be of interest to the analyst.

- L: The log before a query is applied, which we refer to as the **source log**.
- L': The set of cases that satisfy Q, i.e., the **result set** of $eval(L, Q)$.
- $\overline{L'}$: The complement of L', including cases that do not satisfy Q, which we refer to as the **complement set** of $eval(L, Q)$.

We denote as L^0 the log that is given at the beginning of the analysis process. We refer to L^0 as the **original log**. Similarly, we refer to logs derived from the original log and queried at subsequent steps using a unique index in superscript. For example, we denote $eval(L^0, Q)$ as L^1 the log derived from the evaluation of query Q against the original log, against which query R is then evaluated: $eval(L^1, R)$. Note that for $eval(L^n, Q) = L'$ the pair (L^n, Q) deterministically

---

[1] Process mining techniques also provide event filters. In this paper, we focus on cases as we consider them as the main unit of interest for the analysis.

identifies L', as the evaluation of a query against a log always returns the same set of cases. Given $(L^n, Q) = L'$, we refer to the sequence of queries and intermediate result sets obtained starting from L^0 as the **analysis context** of L'.

Relationships between a result set and the source log can tell us about the effects of a query on the log. Let us consider $Q_1 = P$ with $eval(L, Q_1) = L'$. If $L' = L$, then all the cases in L meet the criteria specified in P and Q_1 has no effect on the event log. Instead, if $L' = \emptyset$, then we know that none of the cases in L meet the criteria in P. Suppose we have another query $Q_2 = R$ with $eval(L, Q_2) = L''$. If P and R do not have any attributes in common $(A(P) \cap A(R) = \emptyset)$, we do not know whether L' and L'' have cases in common by looking at P and R alone. However, if $L' \cap L'' \neq \emptyset$, we know that P and R hold true in a common set of cases. If $L' \cap L'' = L'$, then L' is included in L'', which means that P is a refinement of R.

2.2   Motivating Scenarios

As analysts engage in an exploratory analysis session, running queries on the event log, there may be several situations in which they need to validate and compare their results. Based on our experience [18,19], we illustrate common scenarios and questions analysts might ask themselves. We use the RTFM log as a running example and imagine an analyst who wants to understand why payments have not been made using some filters to query the log.

Characterize the Results of a Query. After running their query, analysts need to understand the result and ensure it aligns with their expectations. This process allows them to validate the impact of their query on the data and decide on subsequent analysis. A common practice is to check the size of the result to determine if they are looking at frequent scenarios or to identify patterns that can confirm that filters work as desired [19]. *"Do I have all the cases in which offenders have not paid? Have I included partial payments with this query?"*

Compare the Results of Two Queries. Result comparison helps analysts understand how different filters work, especially in situations where filters act on different attributes of an event log that may be correlated.

– **Assess query refinement.** Analysts frequently refine their queries, for example adding new conditions or modifying existing ones. Evaluating the impact of query refinements is crucial to understanding how changes in the query criteria affect the result set. *"I first filtered for cases sent for credit collection, but then I added a condition for cases that do not include a payment activity. Does this filter help me identify more unpaid cases than the previous one? How does it impact the result?"*
– **Find query dependencies.** Given the multi-perspective nature of event logs and the evolving nature of inquiries, analysts often explore different dimension of the data. Yet, queries run on different attributes may produce similar results, particularly in logs with numerous correlations. Therefore, understanding the relationships between any two queries is essential for validating

the results. *"If I filter for cases where the amount paid is zero or less and filter for cases without a payment activity, will I get a similar result?"*

Summarize Analysis Results. As analysts run multiple queries, summarizing the obtained results helps them maintain an overview and identify areas for further exploration. *"What queries have I run so far to check for unpaid cases? What scenarios have I discovered? Are there unexplored parts of the log?"*

3   A Framework for Characterizing and Comparing Logs

This section outlines our framework to help process analysts validate query results and address questions such as those introduced in Sect. 2.2 through result set characterization and comparison. At a high level, the framework is organized as follows. A first component, called *Query Evaluator*, allows the analyst to execute queries on the event log using process mining capabilities such as case filters, and to generate both the result set and its complement from a given query. A component called *Registry* serves as a data repository and reference for the analysis, keeping a record of the queries run, their results, and user-defined labels. Finally, a group of *Set Comparison Components* allows characterizing and comparing query results. These components characterize a result set based on dimensions of interest while maintaining a reference log, compare any two result sets obtained during the analysis based on set relationships, and identify dependencies between queries.

3.1   Query Evaluator

The *Query Evaluator* takes as input a source log L and a query Q, and evaluates Q against L producing both the result set $eval(L, Q) = L'$ and its complement $\overline{L'}$. $\overline{L'}$ is generated to allow the analyst to compare L' with its complement without having to derive the latter manually. The component also generates unique identifiers for the logs used in the evaluation to allow accessing them later. Figure 1 shows a generic representation of the *Query Evaluator* with $Q = P$.

3.2   Registry

The *Registry* is the backbone of the framework, as it records the queries executed during the analysis, together with their evaluation by the *Query Evaluator* and

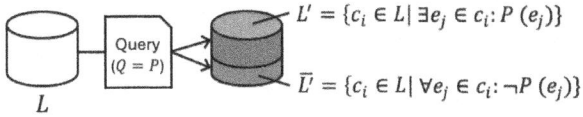

$$L' = \{c_i \in L \mid \exists e_j \in c_i : P(e_j)\}$$

$$\overline{L'} = \{c_i \in L \mid \forall e_j \in c_i : \neg P(e_j)\}$$

Fig. 1. Result set L' and its complement $\overline{L'}$ resulting from evaluating a query $Q = P$ to the source log L with the *Query Evaluator*.

the characterization of the results. It allows for traceability of results and serves as a reference for the analyst, who can view the recorded information at any time and visualize the result sets and their overlap. The functionality provided by the registry is outlined below.

- **Result Set Recording:** The registry records every evaluation done by the *Query Evaluator*, which consists of the result set, its complement, the query to produce it and the source log for that query. Additionally, when a result set is assessed with the *Characterize Result Set* component, this characterization is also recorded as a reference for further analysis.
- **Result Set Labeling:** The registry records custom labels or descriptive tags associated with result sets. This allows giving meaning to the result set or simply labeling it in a way that makes it easier to retrieve.
- **Result Set Retrieval:** The registry assists the analyst in viewing and retrieving any result set to be used as input for a new query or for the *Set Comparison Components*. A result set can be retrieved by its identifier or label. If a label refers to multiple result sets, all associated sets are retrieved.
- **Result Set Visualization:** The registry helps analysts view all the result sets generated during the analysis, the associated queries and, possibly, their overlaps. Overlaps, or lack thereof, can reveal situations where multiple queries occur simultaneously or where no query has been made, providing clues for further analysis or for combining multiple results into one.

3.3   Characterize a Result Set

The *Characterize Result Set* component characterizes the result set L' obtained from the evaluation of a query by comparing it to a reference log, which may be its complement $\overline{L'}$, the source log L, or any other result set recorded in the registry that the analyst wishes to use as a "baseline" for evaluating L'. A generic representation of the *Characterize Result Set* component is shown in Fig. 2.

The characterization is done at the level of both properties of a set and of its elements [2]. Our framework supports multiple characterization dimensions, and it can be easily extended with new dimensions, as discussed in Sect. 4.

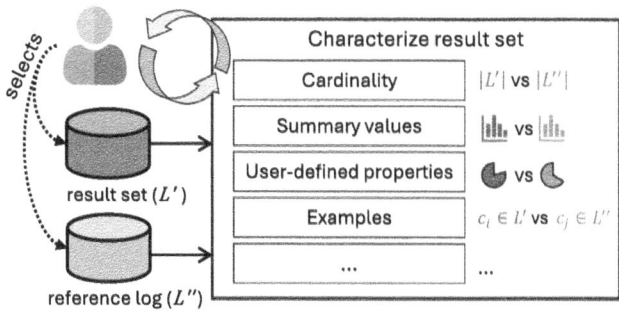

Fig. 2. *Characterize Result Set* component with result set L' and reference log L''.

Set-Level Characterization. Characterization at the set-level exploits aggregated measures, summary values, and user-defined properties to quantitatively characterize the result set. We consider the following characterization dimensions.

- **Cardinality.** Set cardinality measures the number of cases contained in a set and, being an intrinsic property, it does not require going into the specifics of the event log. Still, it allows analysts to compare the size of a result set with the reference log and estimate the impact of a query on the log.
  Example. If the reference log is L and $eval(L, Q) = L'$, then if $|L'| < |L|$ by a constant k, a large k may indicate a more focused analysis due to the significant reduction in the size of L' compared to L induced by query Q.
- **Attribute Distributions and Summary Values.** Data attributes capture key features of a log and, as such, they can help characterize a result set and identify its distinguishing features. While the membership of elements in a result set $eval(L, Q) = L'$ with $Q = P$ clearly correlates with attributes in $A(P)$, there may be other distinguishing attributes worth exploring or using to refine queries. We categorize data attributes based on whether they are (i) numeric or categorical, and (ii) domain-agnostic or -specific.
  (i) For numeric attributes, our framework includes distributions, ranges, and summary statistics that provide information about the most frequent or median values in the set, as well as the variability, outliers, and shape of the distribution. For categorical attributes, we consider the frequency distribution per category and measures such as entropy that can inform the analyst about the proportion of cases within each category and the diversity among them. (ii) For domain-agnostic attributes, our framework includes control flow and timing aspects that can be calculated for any event log, such as throughput time, case distribution over time, and activity frequency. For domain-specific attributes, in addition to standard measures such as ranges, minimum and maximum values, we allow the analyst to define derived attributes or KPIs based on their knowledge of the log. *Example.* Let us assume $Q = P$, $eval(L, Q) = L'$ and $P : \nexists e \,|\, e.a_{name} = Payment$. Elements in L' lack the activity 'Payment' but may exhibit patterns on other attributes. For instance, the range of attribute $a_{PaymentAmount}$ can be $[0, 0]$ in L', while being $[0, 3975]$ in L, indicating that elements in L' may also be correlated with a value of $PaymentAmount = 0$.

- **User-defined Properties.** Properties of interest to the user that can be expressed as predicates can also characterize a result set and based on which extent a certain property still holds after querying. Our framework supports analysts in checking to what extent a given predicate holds in the obtained result set. This is interesting when the predicate is not part of the specific query executed to obtain the result set, but could include other attributes.
  Example. Assuming $Q = P$, $eval(L, Q) = L'$ and $P : \nexists e \,|\, e.a_{name} = Payment$, let's take property $R : \exists e \,|\, e.a_{Amount} > 1000$ to identify high initial amounts. L' can be characterized based on the proportion of cases that satisfy R.

Element-Level Characterization. Characterization at the element-level relies on selected cases to qualitatively characterize sets.

- **Example Cases:** Single cases retrieved from the result set and the reference log allow the analyst to qualitatively characterize a result set at a granular level, focusing on event sequencing, timing, and data attribute updates. Our framework supports analysts in retrieving representative examples from the result set and the reference log with different sampling strategies. These strategies capture examples illustrating frequent behavior as well as "edge cases" - instances that have similarities or significant differences in certain aspects (such as control flow, data, time, or resources) that may depend on the relationship between the result set and the reference log used.

3.4   Comparing Two Result Sets

The *Compare Two Result Sets* component aims to help analysts compare any two result sets recorded in the registry (cf. [2]) and infer dependencies between the queries that generated them. Such support is needed because looking at query predicates is often not sufficient for the comparison: although two queries may differ based on their predicates, their result sets may overlap due to data-dependent query effects. For instance, consider $Q_1 = P$ and $Q_2 = R$, where $A(P) \cap A(R) = \emptyset$, and the result sets $eval(L^n, Q_1) = L'$ and $eval(L^m, Q_2) = L''$. Having $L' \cap L'' = \emptyset$ does not imply that $eval(L^0, Q_1) \cap eval(L^0, Q_2) = \emptyset$, as the previously executed queries that led to L' and L'' could have removed cases that meet both predicates P and R when they are directly evaluated with L^0.

Based on the degree of overlap between two result sets, the component helps infer dependencies between queries that may not be immediately apparent from their predicates. To this end, the component uses an *intersection matrix*, which encodes the cardinalities of the intersections between two result sets and their respective complements (cf. Fig. 3). These intersections are only meaningful if done on result sets derived from a common original log: as these derivations may involve multiple intermediate queries, the *Compare Two Result Sets* component determines a "common ancestor" source log of the result sets. On a high level, the component works as follows.

- *Check source log from registry.* First, it checks the registry to determine if the two result sets originate from the same source log. Their analysis contexts are crucial for interpreting comparison results, as they allow one to see the source log and query conditions under which the result sets were generated.
- *Find common ancestor and build intersection matrix.* If the given result sets L' and L'' are generated from the same source log, the component constructs an intersection matrix with the result sets and their complements, as shown in Fig. 3. If the result sets are not generated from the same source log, the component first identifies their closest common ancestor log with the procedure exemplified in Algorithms 1 and 2. This is essential to ensure that the

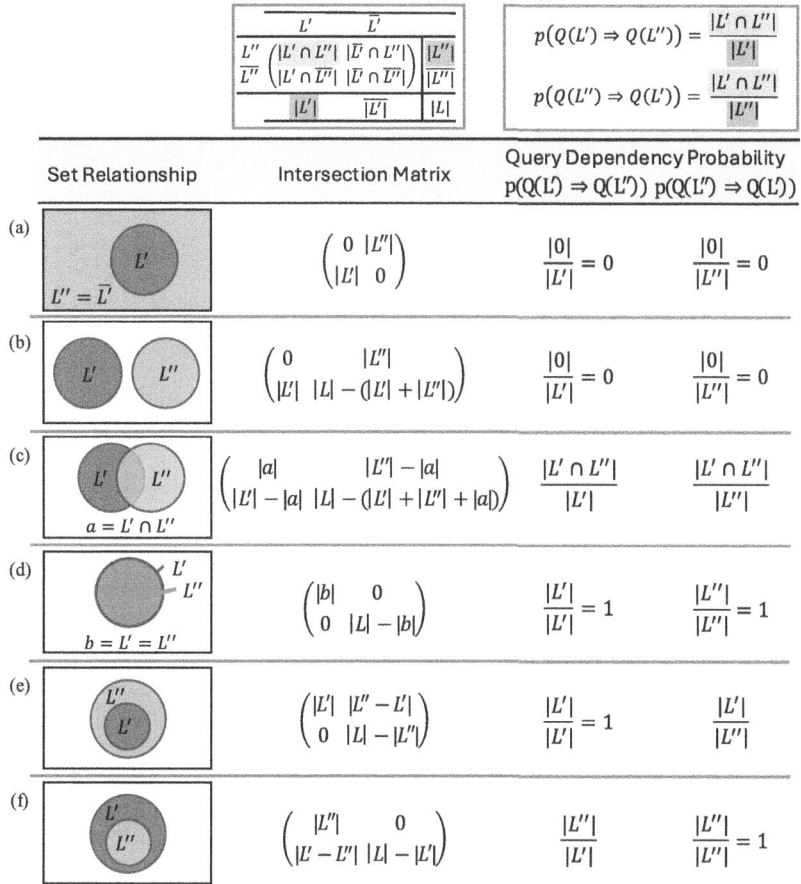

Fig. 3. Possible relationships between result sets L' and L'' and their (common ancestor) source log L, shown as a frame. For each case, we show the intersection matrix and the query dependency probability calculated based on the size of the intersection.

two result sets are compared based on the same analysis context, which consists of the closest common ancestor log and the queries used to generate the result sets from it. Indeed, relationships among result sets are subject to the query conditions that generated them from the common ancestor log. Algorithm 1 collects the recorded queries and respective source logs that led to L' and L'' from the original log L^0 in two lists with the support of Algorithm 2. Then, it iterates through the lists for both result sets simultaneously to find where they diverge. Upon identifying the point of divergence, it returns the log recorded just before this point as the common ancestor log, which in the worst case is L^0 as we assume to have only one original log for the analysis. For each result set, the queries recorded from the common ancestor log are joined by conjunction and used to determine the respective complement sets with respect to the common ancestor to build the intersection matrix.

Algorithm 1: findCommonAncestor(L', L'')

Input : Result sets L' and L'', original log L^0, Registry
Output: Common ancestor log L
$q_1 = collectRecordedQueries(L', L^0,$ Registry$)$;
$q_2 = collectRecordedQueries(L'', L^0,$ Registry$)$;
$i = -1$;
while $i < min(len(q1), len(q2))$ **do**
$\quad$ $i = i + 1$;
$\quad$ $(L_1, Q_1) = q_1[i]$;
$\quad$ $(L_2, Q_2) = q_2[i]$;
$\quad$ **if** $L_1 \neq L_2$ **then**
$\quad\quad$ | break;
$\quad$ **end**
end
$(L, Q) = q_1[i - 1]$;
return L

Algorithm 2: collectRecordedQueries

Input : Result set L', original log L^0, Registry
Output: List of tuples (L, Q) from the original log L^0 to L'
$result = [\,]$;
while $L' \neq L^0$ **do**
$\quad$ $(L, Q, L', \overline{L'}) \leftarrow$ Registry.get(L');
$\quad$ $result = [(L, Q)] + result\ L' = L$;
end
return $result$ /* As tuples are pushed at front, L^0 is the first item   */

- *Show result set relationships and query dependencies.* The intersection matrix allows establishing whether the two result sets are in any of the relationships shown in Fig. 3. Then, based on the cardinality of their intersections, the analyst can estimate the probability that the queries that generated the result sets from the ancestor log depend on each other. The probability is calculated as shown in Fig. 3 for all possible relationships between two sets.

The *Compare Two Result Sets* component serves multiple purposes. First, it enables analysts to understand the relationships between any two result sets and their analysis contexts. This knowledge is crucial when comparing result sets generated by different queries. Indeed, comparing two result sets on dimensions such as cardinality might be less informative or even confusing without knowing whether the result sets are in overlap or included in one another. Instead, if two result sets are derived from different source logs, the comparison makes sense only when considering the common ancestor log. Second, from these relationships the analyst can identify new sets of cases that may be relevant for further analysis, such as those at the intersection between previously obtained result sets. At last, the component sets the basis for analysts to assess whether apparently independent predicates exhibit any inter-dependency based on the relationships between the result sets.

4 Implementation

We implemented our framework as a Python library, named **LogView**, available on Zenodo[2]. The library is complemented by an extensive tutorial that illustrates all the implemented functionality on the RTFM event log [7].

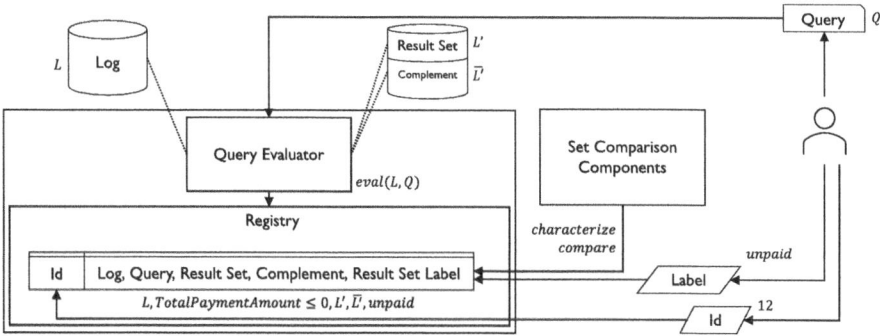

Fig. 4. Overview of the interactive framework showing the components on a high-level.

Our implementation follows a microkernel architectural style and consists of (i) a core system, formed by the Query Evaluator and the Registry, and (ii) plug-in components implementing specific characterizing and comparing methods.

The Query Evaluator implements the logic to evaluate a given query and generate the corresponding result set and its complement. We implemented query predicates using PM4Py case filters [5], which we extended to support a wider range of attribute-value and result set comparisons. The Registry implements the logic to track each query evaluation, label result sets, and view all the result sets obtained and the associated queries. The visualization of multiple result sets is instead implemented as a plug-in, as explained below.

To emphasize the purpose of the plug-ins in the framework, we have organized them into three classes, each corresponding to a Python interface. The *Result Set Characterizer* class contains plug-ins that implement set characterization along the dimensions described in Sect. 3.3. Our implementation covers all the dimensions, but focuses on numeric attributes for summary values. For example retrieval, we have implemented two plug-ins that select examples from the result set (randomly or from the most frequent variant) based on the relationships in Fig. 3. For example, if the result set L' and the reference log L'' overlap as in case (c) of Fig. 3, we pick examples from $L' \setminus L''$, $L'' \setminus L'$, and $L' \cap L''$. The *Two Result Set Comparator* class includes three plug-ins to implement the

---

[2] LogView: https://zenodo.org/doi/10.5281/zenodo.11404207.

functionality outlined in Sect. 3.4: one to find the common ancestor log, another to build the intersection matrix, and a third to determine set relationships and query dependencies. The *Multi Result Set Comparator* class integrates plug-ins for comparing and visualizing multiple result sets and their relationships, which we implemented using the UpSetPlot library [8].

The design of the framework is modular and extensible, allowing researchers and practitioners to add new functionality through plug-ins. The framework includes a custom "attach" method that allows one to add new plug-ins to the existing interfaces. We envision several potential extensions to our framework based on existing process mining research. Predicates could for instance be extended based on Declare4Py [6]. To improve the characterization of result sets, we could visualize attribute value distributions with interactive plots or log profiling approaches like DyLoPro [16]. For example retrieval, event log sampling strategies [3,4] or exemplar query approaches [10] could be integrated. However, it is important to extend the framework only with features that are suitable for interactive use that allow maintaining reasonable response times.

User Interactions. Figure 4 shows a diagram with the main components of our framework and the user. The process analyst guides the analysis workflow by deciding how and in what order to interact with the framework components. Below, we describe some of the expected interactions. The analyst starts their interaction with the framework with a query. The Query Evaluator processes the query and generates the result set along with its complement. The source log, query, result set, and complement are then stored in the Registry. The Registry is the main reference for the analysis: the analyst can view and retrieve by Id any of the logs from the Registry for further characterization and comparison, or view and reuse the query for refinement. The analyst can also label existing result sets, providing semantics to the results and communicating intent. These labels help in comparing and visualizing multiple result sets, e.g., to focus on results obtained for similar analysis goals.

5   Demonstration

We evaluated our framework through an extensive analysis on the well-known RTFM event log [7]. We chose this log for our evaluation because it is a large, real-world event log that is well-structured and relatively easy to understand without domain-specific knowledge. This allows both us and the reader to delve into aspects directly supported by our framework, beyond extracting insights from the log. Below, we present and discuss excerpts of our analysis, illustrating how our framework effectively aids in validating analysis results through characterization, comparison, and recording. Also, we show insights and patterns that can be easily spotted through result set comparisons.

The RTFM log records the process of traffic fine management by a police force in Italy. It is a large log with 561,470 events and 150,370 cases recorded over a period of 13 years. The log includes 11 different activities and 12 data attributes. Each case begins with a Create Fine event, which specifies the fine amount and other attributes. Offenders have the option to pay the fine at any time, as indicated by a Payment event and its PaymentAmount attribute, which captures the amount paid in one transaction. A TotalPaymentAmount attribute also tracks the cumulative amount paid. A Send Fine event triggers a notification letter to the offender, whose receipt is documented by an Insert Fine Notification event, followed by a possible Add Penalty event, which updates the amount to be paid. Sending the letter incurs postage costs, captured by the expense attribute. Finally, a Send for Credit Collection event sends unpaid cases to a collection agency. Offenders have the option to appeal against the fine to a prefecture or judge. If the fine is overturned, this is indicated in the dismissal attribute.

Our goal was to investigate possible payment outcomes in the process and characterize scenarios for unpaid fines. From the existing literature [7,9], we identified the following process outcomes based on payment amounts:

- *unpaid*: TotalPaymentAmount ≤ 0;
- *fully paid*: TotalPaymentAmount $\geq$ max(amount) + expense, where with max(amount) we take the last updated amount, possibly including penalties.

We began our analysis focusing on a simple query (Q_1) that returns only cases without a Payment activity. Our initial query was intentionally not precise as we were conducting an exploratory analysis. After running Q_1 on the RTFM log ($eval(RTFM, Q_1) = S$), we characterized the result set S through a property that checked for *unpaid* cases as defined above. We could clearly see that all (80655) but one case ('C12623') in S were actually unpaid based on the value of TotalPaymentAmount. From this, we learned that Q_1 was already accurate in capturing unpaid cases. Starting from this one case, we found 36 cases that have an initial amount of '0.0'. Some of them (21) have some expenses or penalty added afterwards. This small amount of cases could indicate errors in recording the data or a "clumsy" way to register only some extra amounts, such as those related to the expense. Before going deep into unpaid cases, we then focused on isolating *fully paid* cases as defined above with $eval(RTFM, Q_2) = T$, again to check whether our outcomes definition are precise. We obtained 60247 cases in T. On this set, we used the characterization by property and found that 97% of them are paid exactly, while in 1634 cases offenders pay more than they should. We then used the *Two Result Set Comparator* to relate the result sets of *unpaid* and *fully paid* cases, expecting them to be exclusive. From the intersection matrix shown (rearranged) in Fig. 5(a) we could see that 3 cases are both fully paid and unpaid, and by looking at the summary values, we confirmed that they all have an initial amount of '0.0'. Also, we could see that 9471 cases were neither fully paid nor unpaid, suggesting that they could be partially paid. Then, we took $\overline{S \cap T}$ from the matrix, and defined a new *partially paid* outcome as '$0 <$ TotalPaymentAmount $<$ max(amount) + expense'.

We then delved into unpaid and partially paid cases, merging the two result sets into one named *unpaid_or_partially*. The main result sets we obtained are depicted in the UpSet plot of Fig. 5(b), with the union set *unpaid_or_partially* highlighted in blue. The figure shows result sets ordered by cardinality and the proportion of the result sets where the query predicates hold alone or in

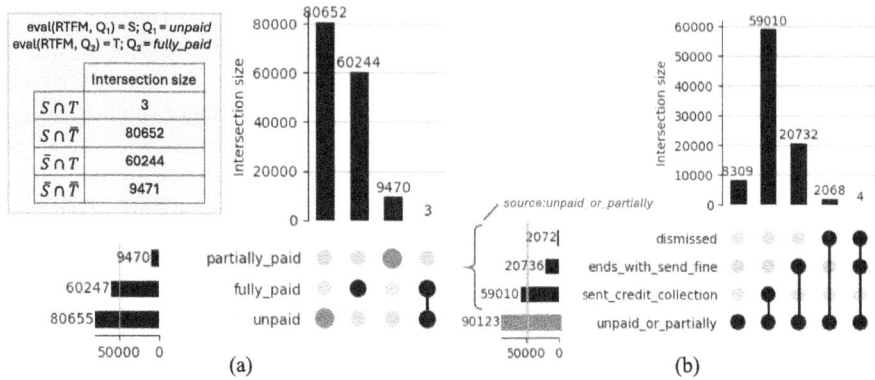

Fig. 5. (a) Intersection matrix showing *unpaid* and *fully paid* cases and UpSet plot showing the overlaps these two result sets and *partially paid* cases. (b) Main result sets obtained when exploring scenarios for *unpaid or partially paid* fines.

overlap. We created the plot using labels, which we used to tag our result sets in two ways considering: (i) the payment outcome and (ii) the source log they were obtained from. The UpSet plot in Fig. 5(b) was created using the labels '*unpaid_or_partially*' for the payment outcome and '*source:unpaid_or_partially*' for the source log. In this way, we ensure that the plot shows the source log '*unpaid_or_partially*' and all its direct subsets (in purple), i.e., we hide result sets obtained using the result sets themselves as a source to ease readability.

In Fig. 5(b) we see that 8309 cases in the *unpaid or partially* paid set are not covered by the other outcomes we checked for (cases sent for Credit Collection, ending with Send Fine, or dismissed by a prefecture or judge). This is a good example where comparing multiple result sets helps identify parts of the log that are yet unexplored. In our analysis, we went deeper into characterizing this set of 8309 cases (which we name R) checking properties on this set. We checked the normative constraint of 90 days between Create Fine and Send Fine, discovering that 47% of the cases in R do not observe it. As another property, we adjusted the payment threshold, leaving a tolerance of 10 Euros, and discovered that 9.9% of the cases in R are paid within this threshold. This 10-euro deviation in the amount paid, which is also documented in recent analyses of the RTFM log [15], informed our next queries, with which we went deeper into checking for offenders who did not pay the expense or penalty. We found 5969 cases where the penalty is not paid. In many cases, the penalty is added after the payment, suggesting that the automatic adding of the penalty, e.g., on weekends when payments are not yet registered, may need revision. Further insights from our evaluation can be found in the Jupyter notebooks available in our repository[2].

Our findings highlight the effectiveness of characterizing and comparing result sets in process mining analysis, providing deep insights into the log without yet using any process mining technique. In this analysis, we showed how user-defined properties can be used to characterize result sets across multiple process perspectives, and how the UpSet plots derived from the registry based on user labels, can help the analyst keep an overview of the result sets explored and those to still investigate. Of course, our querying approach is not meant to be used standalone but in combination with process mining algorithms and visualizations that can complement and further validate the results.

6 Related Work

Our framework aligns closely with research aimed at supporting the work of process analysts on a meta-analysis level, i.e., aiding them in understanding and validating the results obtained through iterative explorations.

In the BPM literature, we found two works that closely related to ours as they focus on supporting the work of analysts at this level. In [14], the authors present *ProcessExplorer*, a standalone interactive visual recommendation system for process discovery that automatically analyzes the log and suggests interesting subsets of cases based on process performance indicators. Stage Views store both these subsets and the recommendations, allowing analysts to navigate through recommended subsets and refine their analysis. In [17], the authors present an analytical provenance system designed to support the reproducibility and reuse of process mining results. The system provides provenance and data views as a means to interact with the event log, and a replayable history to record analysis operations and data selections and navigate them.

The LogView registry keeps track of results and related queries, much like the replayable history in [17]. However, repeated queries on the same log are not appended to the registry, unlike in the replayable history. LogView also differs significantly from other systems as it explicitly supports result set characterization and comparison, features that are missing in Data Views, and does not provide recommendations for the analysts on where to focus as done by [14].

Similar observations apply to commercial process mining tools. As far as we are aware, most existing tools lack support for recording analysis results and comparing logs without direct user intervention [13]. While existing dashboarding features in tools such as Celonis or IBM Process Mining enable the viewing of analysis artifacts or process metrics side-by-side, they allow analysts to compare process statistics and visualizations rather than results intended as event log subsets. Other tools, such as Fluxicon Disco, allow to export and persist the filtered event log, but the analyst still needs to manually compare the logs.

Since our framework includes query, characterization, and comparison functionality in addition to the registry, we see it as partially related to work in these areas. Query languages and tools that allow filtering event logs based on declarative properties or constraints are related to our predicates and characterization through user-defined properties. Examples are Declare4Py [6], which supports constraint-based filtering, or LoVizQL [13], which allows comparing and filtering Directly-Follows Graphs (DFGs) under user-defined conditions. The comparison of selected log features or subsets using visual methods is also related to our characterization and comparison approach. Examples include DyLoPro [16], which visualizes the dynamics of event logs over time, or differential perspective graphs [11], which allow comparing logs across different perspectives. While these approaches could extend our framework (as discussed in Sect. 4), they do not directly provide meta-analysis support. A similar argument can be made for comparative process mining [1], which focuses on comparing multiple processes based on metrics and analysis artifacts, while we focus on showing analysts how the event log subsets they have selected relate to each other.

7   Conclusion

In this paper, we have presented a novel framework designed to support process analysts in validating the results of their analysis and enhancing their understanding of event logs through the characterization and comparison of these results. Additionally, we have implemented the framework as a Python library and demonstrated its use in an in-depth analysis of the RTFM event log.

A notable limitation of our work is the lack of a user study with process analysts that would allow us to evaluate the practical usefulness and usability of our framework, especially to understand how they would interact with the registry. To address this limitation, we plan to conduct a user study with process analysts using our implementation in a real-world scenario.

Future Work. We see many directions for future work. One such direction is to enhance the interactivity of the registry, allowing analysts to aggregate or group result sets that share common characteristics, as well as to query for cases that belong to different result sets. Another possible direction will be to enhance the characterization of result sets with visualizations or that could enhance the support for event log understanding.

Acknowledgment. This work is part of the ProMiSE project, funded by the Swiss National Science Foundation under Grant No.: 200021_197032.

References

1. van der Aalst, W.M.: Process mining: a 360 degree overview. In: van der Aalst, W.M.P., Carmona, J. (eds.) Process Mining Handbook. LNBIP, vol. 448, pp. 3–34. Springer, Cham (2022). https://doi.org/10.1007/978-3-031-08848-3_1
2. Alsallakh, B., Micallef, L., Aigner, W., Hauser, H., Miksch, S., Rodgers, P.: Visualizing sets and set-typed data: state-of-the-art and future challenges. In: Eurographics Conference on Visualization (EuroVis), pp. 1–21. Eurographics (2014)

3. Bauer, M., Senderovich, A., Gal, A., Grunske, L., Weidlich, M.: How much event data is enough? A statistical framework for process discovery. In: Krogstie, J., Reijers, H.A. (eds.) CAiSE 2018. LNCS, vol. 10816, pp. 239–256. Springer, Cham (2018). https://doi.org/10.1007/978-3-319-91563-0_15

4. Bernard, G., Andritsos, P.: Selecting representative sample traces from large event logs. In: International Conference on Process Mining (ICPM), pp. 56–63 (2021). https://doi.org/10.1109/ICPM53251.2021.9576679

5. Berti, A., van Zelst, S., Schuster, D.: PM4Py: a process mining library for Python. Softw. Impacts **17**, 100556 (2023). https://doi.org/10.1016/j.simpa.2023.100556

6. Donadello, I., Riva, F., Maggi, F.M., Shikhizada, A.: Declare4Py: a Python library for declarative process mining. In: BPM Demos, pp. 117–121. CEUR (2022)

7. de Leoni, M., Mannhardt, F.: Road traffic fine management process. Eindhoven Unive. Technol. Dataset **284** (2015)

8. Lex, A., Gehlenborg, N., Strobelt, H., Vuillemot, R., Pfister, H.: UpSet: visualization of intersecting sets. IEEE Trans. Vis. Comput. Graph. **20**(12), 1983–1992 (2014). https://doi.org/10.1109/TVCG.2014.2346248

9. Mannhardt, F., De Leoni, M., Reijers, H.A., Van Der Aalst, W.M.: Balanced multi-perspective checking of process conformance. Computing **98**, 407–437 (2016). https://doi.org/10.1007/s00607-015-0441-1

10. Mottin, D., Lissandrini, M., Velegrakis, Y., Palpanas, T.: Exemplar queries: a new way of searching. VLDB J. **25**, 741–765 (2016). https://doi.org/10.1007/s00778-016-0429-2

11. Nguyen, H., Dumas, M., La Rosa, M., ter Hofstede, A.H.M.: Multi-perspective comparison of business process variants based on event logs. In: Trujillo, J.C., et al. (eds.) ER 2018. LNCS, vol. 11157, pp. 449–459. Springer, Cham (2018). https://doi.org/10.1007/978-3-030-00847-5_32

12. Sacha, D., Stoffel, A., Stoffel, F., Kwon, B.C., Ellis, G., Keim, D.A.: Knowledge generation model for visual analytics. IEEE Trans. Vis. Comput. Graph. **20**(12), 1604–1613 (2014). https://doi.org/10.1109/TVCG.2014.2346481

13. Salas-Urbano, M., Capitán-Agudo, C., Cabanillas, C., Resinas, M.: LoVizQL: a query language for visualizing and analyzing business processes from event logs. In: Monti, F., Rinderle-Ma, S., Ruiz Cortés, A., Zheng, Z., Mecella, M. (eds.) ICSOC 2023. LNCS, vol. 14420, pp. 13–28. Springer, Cham (2023). https://doi.org/10.1007/978-3-031-48424-7_2

14. Seeliger, A., Sánchez Guinea, A., Nolle, T., Mühlhäuser, M.: ProcessExplorer: intelligent process mining guidance. In: Hildebrandt, T., van Dongen, B.F., Röglinger, M., Mendling, J. (eds.) BPM 2019. LNCS, vol. 11675, pp. 216–231. Springer, Cham (2019). https://doi.org/10.1007/978-3-030-26619-6_15

15. Völzer, H., Zerbato, F., Sulzer, T., Weber, B.: A fresh approach to analyze process outcomes. In: International Conference on Process Mining (ICPM), pp. 97–104. IEEE (2023). https://doi.org/10.1109/ICPM60904.2023.10271968

16. Wuyts, B., Weytjens, H., vanden Broucke, S., De Weerdt, J.: DyLoPro: profiling the dynamics of event logs. In: Di Francescomarino, C., Burattin, A., Janiesch, C., Sadiq, S. (eds.) BPM 2023. LNCS, vol. 14159, pp. 146–162. Springer, Cham (2023). https://doi.org/10.1007/978-3-031-41620-0_9

17. Zerbato, F., Burattin, A., Völzer, H., Becker, P.N., Boscaini, E., Weber, B.: Supporting provenance and data awareness in exploratory process mining. In: Indulska, M., Reinhartz-Berger, I., Cetina, C., Pastor, O. (eds.) CAiSE 2023. LNCS, vol. 13901, pp. 454–470. Springer, Cha (2023). https://doi.org/10.1007/978-3-031-34560-9_27

18. Zerbato, F., Soffer, P., Weber, B.: Initial insights into exploratory process mining practices. In: Polyvyanyy, A., Wynn, M.T., Van Looy, A., Reichert, M. (eds.) BPM 2021. LNBIP, vol. 427, pp. 145–161. Springer, Cham (2021). https://doi.org/10.1007/978-3-030-85440-9_9
19. Zerbato, F., Soffer, P., Weber, B.: Process mining practices: evidence from interviews. In: Di Ciccio, C., Dijkman, R., del Río Ortega, A., Rinderle-Ma, S. (eds.) BPM 2022. LNCS, vol. 13420, pp. 268–285. Springer, Cham (2022). https://doi.org/10.1007/978-3-031-16103-2_19
20. Zimmermann, L., Zerbato, F., Weber, B.: What makes life for process mining analysts difficult? A reflection of challenges. Softw. Syst. Model. 1–29 (2023). https://doi.org/10.1007/s10270-023-01134-0

Microservices-Aware Business Process Modelling

Rene Noel[1,2](✉) ⓘ, Sergio España[1,3] ⓘ, Jose Ignacio Panach[4] ⓘ,
and Oscar Pastor[1] ⓘ

[1] Valencian Research Institute for Artificial Intelligence, Universitat Politècnica de València, Valencia, Spain
rnoel@vrain.upv.es

[2] Escuela de Ingeniería Informática, Universidad de Valparaíso, Valparaíso, Chile

[3] Information and Computing Sciences, Utrecht University, Utrecht, The Netherlands

[4] Escola Tècnica Superior d'Enginyeria, Universitat de València, València, Spain

Abstract. Microservices Architecture (MSA) is the *de facto* software architecture approach for highly scalable software systems. Organisations must design their structure and processes around business outcomes to reap MSA's benefits. Also, MSA requires the domain model for each microservice to be minimal and avoid coupling with other microservices' domain entities. However, such coupling might already occur during the design of the business process and then propagate along the development life cycle. The first opportunity to prevent coupling occurs while designing collaborations between different participants (organisational units, such as development teams) since assigning business responsibilities defines how much domain knowledge each participant must handle. This paper proposes a method to design business process models so the domain managed by each process participant matches the size and complexity required for MSA domain design, enabling the seamless use of MSA. We reviewed nine code repositories to characterise the size and complexity of MSA domain models and proposed a metamodel conceptualising the optimal microservice domain model. Then, taking as input BPMN's Choreography diagrams describing interactions among participants, we propose (i) to specify the structure of the messages interchanged by the choreography participants, (ii) a set of process modelling guidelines to avoid domain coupling by preventing coarse interactions and heavy domain-savvy process participants, and (iii) a set of transformation guidelines to systematically derive the MSA domain model from the message structures. This contribution aims to help business process designers envision the domain complexity each process participant handles and prevent coupling business domains during process design. We provide a detailed example showing the approach's feasibility and discuss the proposal's implications, benefits and limitations.

Sergio España is supported by a María Zambrano grant of the Spanish Ministry of Universities, co-funded by the Next Generation EU European Recovery Plan. This Research is supported by the Spanish State Research Agency and the Generalitat Valenciana under the project PID2021-123824OB-I00.

A. Marrella et al. (Eds.): BPM 2024 Forum, LNBIP 526, pp. 267–284, 2024.
https://doi.org/10.1007/978-3-031-70418-5_16

Keywords: Microservices · Business Process Modeling ·
Choreography Modeling · Model-Driven Development

1   Introduction

Microservice architecture (MSA) [31] has become a *de facto* standard for designing highly scalable software. It supports designing an application as a set of small and loosely coupled services, which operate independently and communicate with each other with lightweight mechanisms [20]. Field research shows that software development teams which design, develop, test, and operate software services supporting a small part of a larger business domain (e.g. microservices), with independence from other teams, deliver software more efficiently [13]. However, to achieve such decoupling in the business domain, it is necessary to start by designing a structure of teams around business outcomes with well-defined and minimal interactions among them [6,27]. Since the interactions between teams are scattered across different business processes, it is challenging to identify couplings.

BPMN choreography diagrams [24] offer a solution for having a whole perspective of multiple collaborating *participants*. In BPMN collaborations, different organisations are represented as separate pools, and their collaborations are represented as message flows among them. Choreography diagrams focus on the collaboration among participants, specifically on the information shared through messages. In this way, complex processes can be modelled in multiple collaboration diagrams, but choreography diagrams allow reason about the sequence of the collaborations and what information manages each participating organisation. The decentralised nature of choreographies reduces the gap between requirements and implementation [9], increases the degree of independence of participants handling the interactions [5], and supports microservice composition [28].

In this paper, we present a business process modelling method that is aware of the business domain dependencies among organisation units and, thus, the supporting microservice architecture. The design of the method addresses three research questions: *RQ1 - What is the reasonable range of domain complexity of a microservice?*, *RQ2 - How can business processes be designed so they facilitate the design of micro-services of a reasonable complexity?*, and *RQ3 - What is the feasibility of the proposed design approach?*. We propose 1. To model the organisation's inner units as different collaborating participants in a choreography diagram, 2. To specify the structure of fields of the messages they share, and 3. To transform the structure of the messages into participants' domain models and to warn the modeller about possible complexity issues. Following the design science methodology [30], we investigated the domain complexity of microservices and designed the method using Situational Method Engineering (SME) [17]. We assembled existing modelling methods and techniques for modelling business processes [24], specifying the structure of the messages shared by process participants [15], and for microservice design [12]. We illustrate the feasibility of the approach with a single case mechanism experiment and discuss the proposal in the light of earlier literature.

The article continues as follows. Section 2 presents the problem investigation, detailing the review of code repositories for characterizing the domain complexity in microservice implementations. Section 3 presents the proposed method, including the modelling guidelines and transformation algorithms. Section 4 presents a single mechanism experiment showing the feasibility of the approach. Section 5 discusses implications and limitations in light of previous literature. Section 6 concludes the paper.

2   Problem Investigation: Domain Complexity in Microservices

We herein define the problem following the guidelines by Wieringa in [30]. The problem's context relates to organisations with software-based services as their main value offer, needing to grow rapidly and react to environmental changes. The main stakeholders are business process analysts and software engineers. The business process analysts split the business domain into business processes, while the software engineers, particularly software architects, split the system into microservices. In this context, business process analysts typically aim to design modular processes and minimise the dependencies between the participants (e.g., teams, areas, departments). In turn, the goal of software engineers is to design loosely coupled software services to foster software scalability and delivery performance. The problem addressed by our work regards poor business process design, which hinders the growth and reaction capability of the organisation. Particularly, we aim to minimise the chances of designing (i) business processes that require intensive coordination of several participants for their implementation and (ii) business processes that delegate huge portions of domain knowledge to a few participant teams, affecting their performance due to high cognitive load. These two problems hinder the implementation of microservice architectures [20].

We consider that the root cause of the above-mentioned problem is that business process analysts have little insight into what domain information is handled by the participants. Our solution approach is to include domain information during business process modelling and provide guidance for avoiding *excessive* domain coupling between participants or overcharging participants with *too much* domain knowledge. We have investigated the reference microservice implementations to specify these quantifiers more precisely and discover what appropriate domain complexity is to be managed by business process participants.

We reviewed code repositories of software systems having an MSA. As a mean to make the repositories comparable and to ensure they have followed best design practices, we selected repositories following the domain-driven design paradigm. Domain-driven design (DDD) is a software development approach that modularises a business domain into distinct bounded contexts, often aligned with microservices, and applies specific design patterns to ensure each module is well-defined and cohesive [12]. The software industry has adopted DDD to split the business domain into *bounded contexts*; that is, highly cohesive, low-coupled parts that share the same domain language. DDD also proposes object-oriented design patterns to structure bounded contexts in *modules* that provide a microservice that handles the life cycle of a single, relevant domain entity. Inside each module, DDD proposes design patterns to characterise domain classes: *Entities* are classes relevant across the organisation for which have an identification data field, *Value Objects* represent classes with invariant attributes relevant for a particular transaction, not having an identification field, and *Aggregate Roots* is a subtype of entity which aggregates other entities and value objects. Among other patterns, DDD also proposes *Service*, which exposes the module's business logic to other modules, and *Repository*, which manages the persistence of the module's domain entities. Overall, we consider DDD the most rigorous approach to tackle domain complexity in MSA and thus find it suitable for our research.

In a convenience sampling approach, we have considered the code repositories reviewed by Rademacher [26] to design a UML profile for DDD and added two more recent ones. Our review addresses the question: *RQ1 - What is the reasonable range of domain complexity of a microservice?*. We identified the most complex microservice modules inside each code repository. For each module, we identified the packages containing the domain model classes (usually named *model* or *domain*) and inspected the code, matching the classes with the DDD patterns. Following DDD, the complexity of the domain is centred in the aggregate root classes, so we counted 1. The independent aggregate roots in the module, and 2. The number of nested aggregate roots. To ensure we correctly identified the aggregate roots in the module, we also counted the number of services by counting the service implementations in the *service* package of the modules. According to DDD, the number of Services should match the number of aggregate roots. Table 1 shows the findings, which we comment on below.

Table 1. Summary of the analysis results of domain-driven microservice architecture code repositories. (NoM: Number of Modules, MAR: Number of aggregate roots of the largest module, MNAR: Maximum number of nested aggregate roots of the largest module). Find more details in the technical report [22].

Alias	Lang	Dates active	kloc	NoM	MAR	MNAR
eShopContainers	C#	04/09/2016–27/10/2021	176.8	5	1	1
micro-company	Java	27/03/2016–10/07/2020	127.1	4	1	1
Lakeside Mutual	Java	21/02/2021–19/04/2021	157.0	3	1	1
Pit Stop	C#	24/09/2019–24/04/2023	97.4	5	1	1
mspnp	C#	02/05/2021–06/07/2022	4.1	5	4	*2
WeText	C#	27/03/2016–15/11/2017	41.9	2	1	1
FTGO	Java	10/09/2017–29/09/2018	25.5	6	3	1
sivalabs	Java	18/02/2018–16/04/0202	6.8	3	1	1
ttulka	Java	01/11/2020–06/03/2023	13.0	4	1	1

From the code review, we identified the following characteristics of a reasonable domain complexity for a microservice:

- *Aggregate roots have no nested aggregate roots.* This means that aggregate roots are composed of entities and value objects but not other aggregate roots. An exception is *msnp* project, which has two levels of aggregates, but it is just for logging the events produced by the second level aggregate.
- *Modules have a single aggregate root.* Exceptions to this are the *mspnp* and *FTGO* projects. The *mspnp* the *delivery* module handles four aggregates, but all of them aggregate value objects or enumerations and no other entities. Additionally, the domain is accessed through two closely related services: one for tracking the delivery and one for notifying changes in the delivery status. In the *FTGO* project, the *delivery* module contains three aggregates, but similarly to *mspnp*, the aggregates contain value objects or common classes and no other entities. All of them are accessed through a single service.

We acknowledge that there might be cases where increasing the microservice domain complexity can be justified by the business complexity. However, the results represent common practice and inform the design of our method. We have conceptualised the above findings in the package DDD of the method meta-model presented in Fig. 2. The DDD pattern conceptualisations are based on the UML profile for DDD patterns by Rademacher et al. [26], while the relationships and multiplicities are reasoned inductively from the code repository review findings. For simplicity, we use stereotypes to represent each pattern concept's corresponding UML class diagram elements.

3   Microservice-Aware Business Process Modelling

In this section, we address the research question *RQ2 - How can business processes be designed so they facilitate the design of micro-services of a reasonable complexity?*. An overview of our proposal is depicted in Fig. 1 using the MAP notation. Ellipses denote the method's intentions, while arrows denote the strategies to achieve such intentions. The dashed arrows highlight the contributions of our proposal. The method starts with the *business process modelling* intention achieved *by choreography*, which implies the use of BPMN's choreography diagrams. Choreography diagrams depict the interactions between process participants, and can be designed from scratch or derived from the message interchanges between the participants of BPMN's collaboration diagrams.

The *message structure specification* intention achieved by *choreography analysis* supports the definition of the detailed structure of fields for the messages interchanged in each choreography task. The method contributes with a set of guidelines to specify such structures, keeping in mind the complexity of the domain model. After specifying the message structure, the method user can refine the *business process modelling* by *domain-driven modularisation*. The method contributes with guidelines for modularising the participants in the BPMN model in case message structure specifications reveal that one or more participants handle a large portion of the domain model.

Having specified the message structures, the method user can continue with *microservice domain modelling* by *message structure transformation*. In this case, the method contributes with a model-to-model transformation algorithm that takes the choreography diagram and the message structures as input to generate UML class diagrams, one for each participant, representing their respective domain models. The generated classes are stereotyped using the pattern language of DDD. Though the derivation of class operations is not considered, it could be achieved following a procedure similar to the proposed in [11].

3.1   Method Metamodel

The metamodel presented in Fig. 2 supports the proposed method, which integrates three methods and techniques. On the one hand, the BPMN package contains the main elements of the collaboration and choreography diagrams from the BPMN 2.0 specification [24]. It is noteworthy that BPMN.ChoreographyTask is associated with the BPMN.MessageFlow of a BPMN.Collaboration thus supporting the *choreography task analysis* step of our proposal; however, although BPMN.MessageFlow has an associated BPMN.Message, BPMN does not define how to specify messages.

On the other hand, the MS package presents the metamodel of the Message Structure technique, initially introduced in [11]. We connectedMS.MessageStructure with BPMN.Message to support the *message structure specification* step of our proposal. We are aware that BPMN supports defining a message through XSD; the proposed metamodel aims to highlight the concepts behind the MS technique regardless of technological support. As can be seen, message structures can specify MS.DataFields, but also more complex structures such as MS.Aggregations of fields or other structures, as well as MS.Iterations to support multiple instances of a structure, e.g., the items of an order. A special type of field is MS.ReferenceField, which allows referencing existing structures defined in the same message or other messages, e.g., an order item can reference a product created on a different message. Importantly, MS.ReferenceFields could shed light on coupling domain concepts between different participants, so its use must be carefully assessed.

Finally, the package DDD presents a proposed conceptualisation of the pattern language of the domain-driven design [12] approach for designing microservices. According to DDD, organisation units that share a common business vocabulary define a DDD.BoundedContext. Inside a bounded context, there can be many microservices, which are grouped in DDD.Modules. Each module contains a portion of the business domain that is managed by the microservice, which is organised into DDD.AggregateRoots that are entities that group and manage the content and the life cycle of other DDD.Entities and DDD.ValueObjects in a way that ensures data integrity. Aggregate roots define a boundary, so its constituents are not directly accessible to external clients. DDD.Entities are classes the business is interested in tracking throughout its life cycle so they have an identifier attribute, while DDD.ValueObjects group invariant data that are purposeful for the service logic but not for other services. DDD considers other classes for exposing and persisting the domain (`DDD.Repository`, `DDD.RepositoryImplementation`, `DDD.Service`, and `DDD.ServiceImplementation`, among others), that are directly associated with a `DDD.AggregateRoot`.

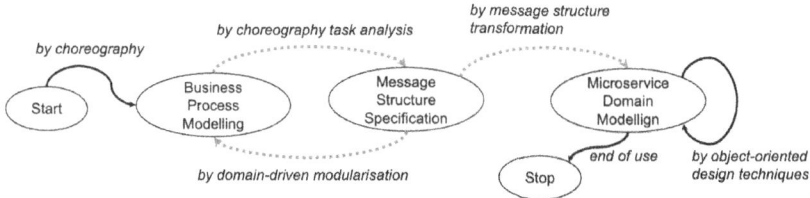

Fig. 1. Method requirements map.

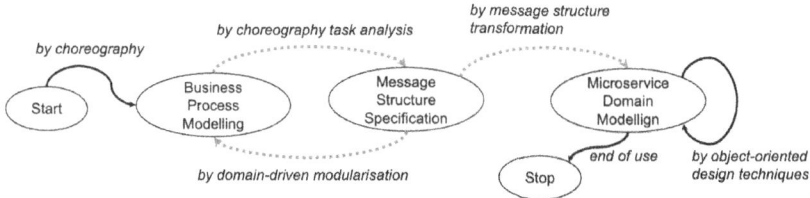

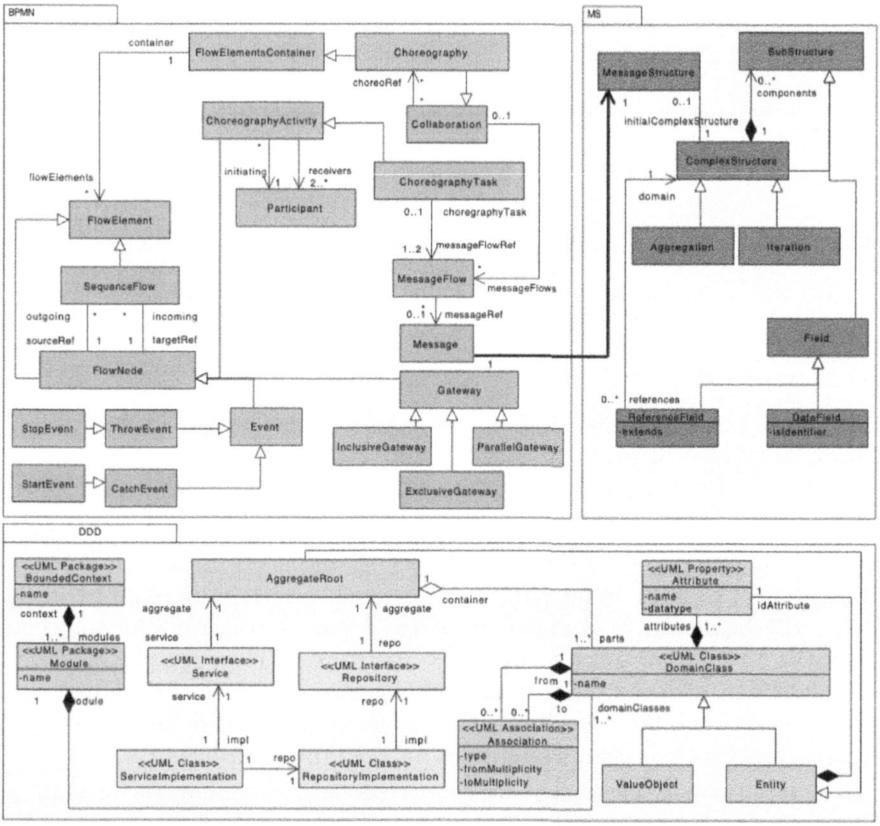

Fig. 2. Method metamodel integrating metamodel fragments of BPMN Choreography, Message Structures, and a proposed conceptualisation of microservice patterns. The full BPMN metamodel can be found in [24]; Message Structures metamodel is available in [15].

3.2   Guidelines Specification

As depicted in Fig. 1, the proposed method integrates business process modelling, message structure specification, and microservice domain modelling through three strategies: *choreography task analysis*, *domain-driven modularisation*, and *message structure transformation*. To specify how to achieve the intentions using the proposed strategies, we use the SMEs' intention achievement guidelines (IAG). An IAG specification provides support for describing the guidelines to go from one intention to another using a specific strategy [17].

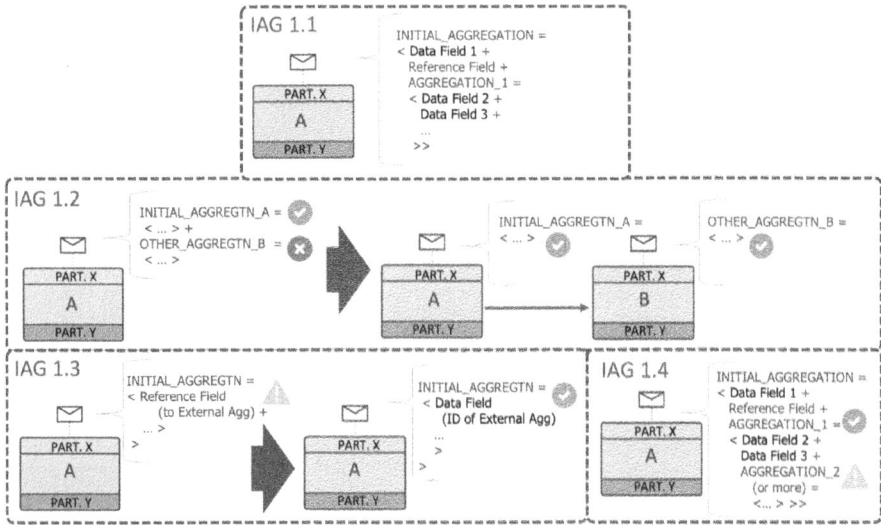

Fig. 3. Intention achievement guidelines for message structure specification through choreography task analysis.

Guidelines for Choreography Task Analysis: The motivation for these guidelines is to exploit the business knowledge gathered during the business process design to elicit **what information each participant should know when collaborating in a choreography**. Below, we describe the guidelines IAG 1.1 to IAG 1.4 for specifying message structures in choreography diagrams. The guidelines reference the metamodel elements depicted in Fig. 2. In Fig. 3, we provide an example of how to specify the message structure for a message in a choreography task and representations for the three guidelines.

> ***IAG 1.1:*** For each BPMN.ChoreographyTask in the choreography diagram, elicit the structure of the initiating message (e.g. through stakeholder interviews, system archaeology, or any other requirements elicitation technique [25, chapter 3.3]) and then specify it using MS.MessageStructures (see [15] for detailed instructions).
>
> ***IAG 1.2:*** Specify message structures considering that each MS.Message Structure can contain a single MS.Aggregation within the first level of the

message. A message with more than one aggregation could denote that a choreography task is performing more than one business interaction, which should be separated into different tasks and, thus, different messages.

IAG 1.3: Specify message structures considering that a message **should** have MS.ReferenceFields only for referencing MS.Aggregations that are part of the domain of the target participant. A message referencing MS.Aggregations which are unknown for the target participant would couple the target participant with the domain of other participant. We recommended providing the identifier field of the aggregation and leaving the receiver participant to decide to get the rest of the information from the other participants.

IAG 1.4: A message **should** not contain more than two levels of nested *aggregations* or *reference fields*. A message with multiple levels of nested aggregations denotes (i) the domain managed by the target participant is complex and (ii) the initiating participant must know more about the structure of the receiving participant's domain than is advisable, according to DDD.

Guidelines for Domain-Driven Modularisation: These guidelines aim to help analysts redesign business process models so the domain managed by each process participant matches the desired size and complexity for the microservices domain design. The guidelines inform the business process modelling activity with the microservice domain modelling practices elicited in the problem investigation, taking the message structure specifications as input. The guidelines are motivated by the fact that **the information received by a participant across its interactions reveals how much of the business domain the participant handles**. A participant managing a large portion of the domain threatens the modularisation of the business processes, hampering the low coupling and cohesion of the microservices design [13]. The guidelines are based on our conceptualisation of microservices domain modelling (see Sect. 2). The guidelines are illustrated in Fig. 4.

IAG 2.1: Different organisational units (OUs) (e.g., areas, departments, development teams) **must** be modelled as separate participants in the fashion proposed by earlier research that uses BPMN within microservice developments [28]. This means OUs must be modelled as pools in the BPMN collaboration diagram, not as pool lanes.

IAG 2.2: If even considering the previous guideline, a participant receives *many* messages, it **should** be separated into different participants. A participant receiving many messages means it concentrates a great part of the business domain, hindering the implementation of MSA, as it is assumed that the participant has a common memory (e.g. database) for persisting the received messages. As seen in Table 1, the maximum number of modules in the reference implementation is six (one per domain entity). Splitting a participant would re-configuring the organisational structure and strategy. A method to do so is proposed in [23].

IAG 2.3: In case two participants share messages having nested aggregations or reference fields (thus, it is impossible to follow IAG 1.3 and IAG 1.4), the collaboration **should** be separated into two or more collaborations. The collaboration could be coupling more than one business transaction and thus yielding complex aggregations within the same microservice. As seen in Table 1, a module's typical number of nested aggregate roots is one.

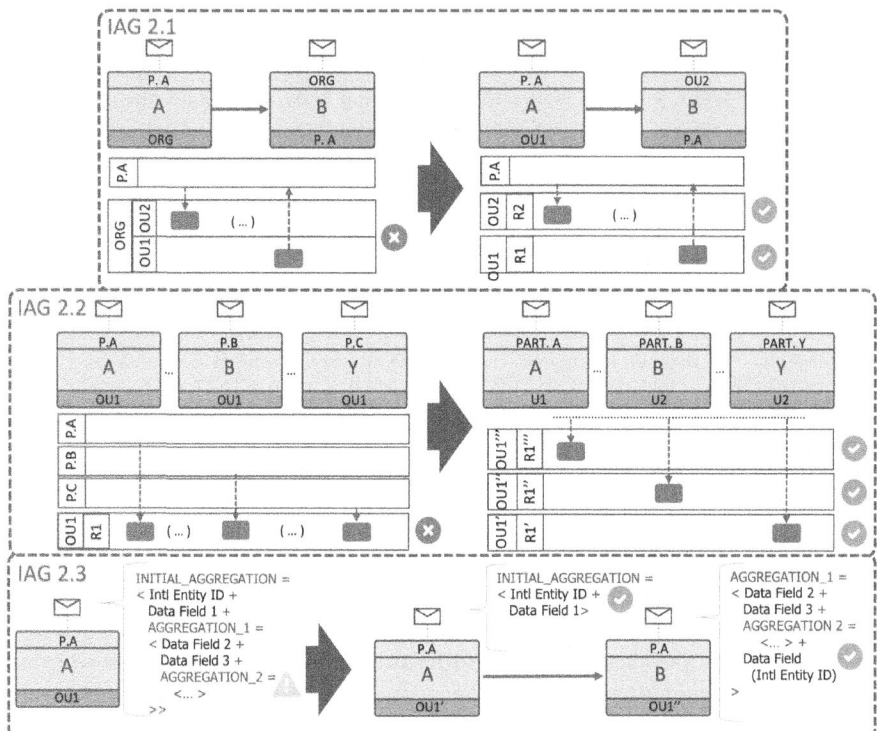

Fig. 4. Intention achievement guidelines for business process modelling through domain-driven modularisation.

Guidelines for Message Structure Transformation: these guidelines aim to help analysts glimpse the complexity of the domain that each participant in the business process model is handling. Each domain model contains a set of classes derived from the message structures, which are stereotyped according to the microservice domain design patterns elicited during the problem investigation in Sect. 2. The guidelines take the set of messages received by each participant as input. The guidelines are based on the systematic derivation of class diagrams from message structures, proposed in [16], and have been adapted to the context of DDD and MSA. We have specified the Message Structure Transformation guidelines in two flavours. In the technical report [22], we offer a textual specification similar to the earlier guidelines. Herein, we specify them with three algorithms. Algorithm 1 describes the overall approach for generating the domain models for the participant's microservices. Algorithm 2 details the transformation of message structures into domain classes using the DDD patterns. Finally, Algorithm 3 assesses the complexity of the generated domain models based on the findings from the code reviews presented in Sect. 2. Within the algorithms, the comments provide the definition for the guidelines (e.g., #IAG 3.3.1 Create an entity from a message structure, which can be traced to their imperative description in the technical report. Figure 5 illustrates the transformation.

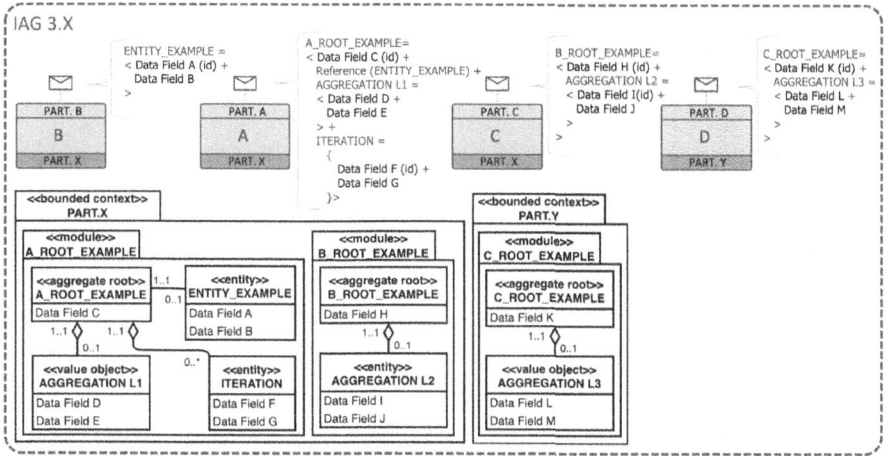

Fig. 5. Example of the application of intention achievement guidelines for microservice domain modelling through message structure transformation.

Algorithm 1. Intention achieving guidelines for message structure transformation.

Require: Inputs: Models *BPMN* with the choreography diagram and *MS* with the companion message structures, and an empty domain model *DDD*.

Require: newBoundedContext(*d*) creates and returns a new bounded context in the domain model *d*.

Require: receivedBy(*m*, *p*) returns true if message *m* is received by participant *p*, or false otherwise.

Require: processComplexStructure(*cs,bco*) is defined in **Algorithm 2**. Creates a domain class from the complex structure *cs* in the bounded *bco*. Returns the domain class created from the complex structure.

Require: getIsolatedEntities(*bco*) returns all the domain classes in the bounded context *bco* without participating on aggregation or association relationships.

Require: prototypeAggregateRoot(*e*) creates and returns a new aggregate root with the name and attributes of the entity *e* passed as parameter.

Require: newModule(*bco*, *n*)creates a new Module with the name *n* in the bounded context *bco*.

Require: moveDomainAggregation(*c*, *m*) moves the domain class *c* and all its directly and indirectly related classes to module *m*.

Require: assessDomainModel(*DDD*) is defined in **Algorithm 3**. Assesses the complexity of the domain model *DDD* by checking if aggregate roots have other aggregate roots as components.

```
1  for all par ∈ BPMN.Participant:
2    #IAG 3.1 Create a bounded context for each participant
3    bco = newBoundedContext(DDD)
4    #IAG 3.2 Process each message structure received by the participant
5    for all mst in MS.MessageStructure where receivedBy(mst, par):
6      #Recursively generate domain classes from initial complex structure
7      initialDomainClass = processComplexStructure(mst.initial_complex_structure, bco)
8    #IAG 3.6 Transform isolated entities into aggregate roots
9    for all ent in getIsolatedEntities(bco):
10     ar = prototypeAggregateRoot(ent)
11     ent.delete()
12   #IAG 3.7 Create modules from aggregate roots.
13   for all ar_top in bco.AggregateRoot where ar_top.container = NULL :
14     mod = newModule(bco, ar_top.name)
15     moveDomainAggregation(ar_top, mod)
16 #Asssess whether aggregate roots have other aggregate roots as components
17 assessDomainModel(DDD)
18 return DDD
```

Algorithm 2. processComplexStructure(*com_st*, *bc*) - Recursively process complex structure *com_st* creating a domain class *dc* in the bounded context *bc*. Returns a domain class *dc*.

Require: Inputs: *com_st* a complex structure of a message structure, *bc:* The bounded context in which the domain classes created from the complex structure will be created.
Require: newEntity(*b, n*) creates and returns a new entity in the bounded context *b*, named *n*.
Require: newValueObject(*b,n*) creates and returns a new value object in the bounded context *b*, named *n*.
Require: newAggregateRoot(*b,n*) creates and returns a new aggregate root in the bounded context *b*, named *n*.
Require: newAttribute(*e, n, t*) creates a new attribute in entity *e*, with name *n* and data type *t*.
Require: typeToDomain(*d*) is a user-defined function mapping the message structure field domains used during MS specification into attribute data types (e.g. number → integer, text → string, date → datetime, money → double)
Require: newDomainLink(*s, t, a, n, m, c*) creates an association between from class *s* to class *t*, with *fromMultiplicity n* and *toMultiplicity m*. The *type* of association *a* could be association, aggregation, or composition.
Require: warning(*m*) shows a the warning message *m* to the user.

```
1   for all sub_st in com_st.direct_items :
2       #Check for identifier data fields and select them
3       ide = select sub_st where specialisation_type = data_field and is_identifier = True
4       #Check for reference fields and select them
5       ref = select sub_st where specialisation_type = reference_field
6       #Check for any first-level aggregations and select them
7       agg = select sub_st where specialisation_type = aggregation or complex_substructure
8       domainClass = null
9       #IAG 3.3 Create domain classes from message structures
10      if ide ≠ ∅ and ref = ∅ and agg = ∅ then
11          #IAG 3.3.1 Create an entity from a message structure with an identifier and only
              data fields.
12          domainClass = newEntity(bco, com_st.name)
13      if ide = ∅ and ref = ∅ and agg = ∅ then
14          #IAG 3.3.2 Create a value object from a message structure with only data fields.
15          domainClass = newValueObject(bco, com_st.name)
16      if ide = ≠ ∅ and (ref ≠ ∅ or agg ≠ ∅) then
17          #IAG 3.3.3 Create an aggregate root object from a message structure with an
              identifier field. Inner structures are processed recursively below.
18          domainClass = newAggregateRoot(bco, com_st.name)
19          #IAG 3.3.4 warn about possible modelling problems
20      if domainClass = null then
21          warning('Please check if an identifier data field is needed for' + com_st.name)
22      #Add data fields in the structure as attributes to the created domain class
23      for all fie in sub_st where fie.specialisation_type = data_field :
24          typ = typeToDomain(fie.domain)
25          newAttribute(domain_class, name = fie.name, type = typ)
26      #IAG 3.4, IAG 3.5 Process inner structures of aggregate roots recursively
27      if ide ≠ ∅ and (ref ≠ ∅ aor agg ≠ ∅) then
28          #IAG 3.4.1 Aggregations
29          for all inner_agg in com_st.direct_items where specialisation_type = aggregation
              :
30              #Recursively process the inner aggregations and creates 1 to 1 composition
                links
31              nextDomainclass = processComplexStructure(inner_agg, bc)
32              domainLink = newDomainLink(domainClass, nextDomainClass, com, 1, 1, bc)
33          #IAG 3.4.2 Iterations
34          for all inner_it in com_st.direct_items where specialisation_type = iteration :
35              #Recursively process inner iterations. Create a 1 to many composition
36              nextDomainclass = processComplexStructure(inner_it, bc)
37              domainLink = newDomainLink(domainClass, nextDomainClass, com, 1, m, bc)
38          #IAG 3.4.3 Reference Fields not extending aggregations
39          for all inner_rfie in com_st.direct_items where specialisation_type =
              reference_field :
40              referencedSt = inner_rfie.domain
41              #Reference field not extending the structure: create association
42              if inner_rfie.extends = False then
43                  nextDomainclass = processComplexStructure(referencedSt, bc)
44                  domainLink = newDomainLink(domainClass, nextDomainClass, asso, 1, 1, bc)
45              else
46                  #IAG 3.4.4 Reference field extending a structure: add fields
47                  for all dfs in com_st.direct_items where specialisation_type = data_field :
48                      typ = typeToDomain(dfs.domain)
49                      newAttribute(referencedSt, name = dfs.name, type = typ)
```

Algorithm 3. assessDomainModel(*DDD*) - Checks the complexity of the domain models in *DDD*. Returns a list of warnings *w* *dc*.

Require: Inputs: Model *DDD* with the microservices domain model produced by the transformation in Algorithm 1.
Require: warning(*m*) shows a the warning message *m* to the user.

```
1  for all bco ∈ DDD.BoundedContext :
2    for all mod ∈ bco.Module :
3      for all ar in mod.DomainClasses where specialisation_type = AggregateRoot :
4        #Check whether aggregate roots contain other aggregate roots
5        sub_ar = select ar.parts where specialisation_type = AggregateRoot
6        if sub_ar ≠ ∅ then
7          warning("Aggregate_root" + ar.name + "contains_other_aggregate_roots._Check_if
  _they_can_be_separated_into_different_operations.")
```

4   Preliminary Treatment Validation

To address *RQ3 - What is the feasibility of the proposed design approach?*, we present a single case mechanism experiment.

As part of the increasing popularity of the sharing economy, we have experienced a rise in private car rental initiatives, also known as peer-to-peer car sharing. For this fictional case, we have drawn inspiration from carsharing business models, where the company offers a platform that manages a virtual fleet made up of vehicles from participating car owners, who charge a fee to rent out their cars when they do not plan to use them. Participant renters can rent available cars at affordable prices. In our case, we consider the existence of salespersons facilitating the rentals. For brevity and understandability, rather than aiming to define a case of realistic size and complexity, we intend to define the minimal case that illustrates the method and its guidelines well.

Figure 6 presents the fragment of the business process that is relevant to this single-case mechanism experiment. A customer who has been in contact with a salesperson closes the deal to rent a specific car on a given date. The customer is offered the chance to rate the attention of the salesperson; the Quality Department will use this information to define key performance indicators on service quality. The salesperson contacts the car owner to inform him/her about the rental request. Additionally, a delivery clerk (freelancers who work part-time for the company) also receives the message, so they know where to pick up the car from, where to deliver it, and what additional services they need to ensure (e.g. a maxi-cosi seat adapter for a baby, in-depth cleaning before delivery). Upon delivery at the specified location, the customer must confirm the delivery.

Following guideline IAG 1.1, we specify one initiating message with a representative name (e.g. Sale closure). Each initiating message is further specified with a message structure. For brevity, we only present the message for the first choreography; the rest can be found in the technical report [22]. As shown in Table 2, there is only a single MS.InitialAggregation (RENTAL), complying with IAG 1.2. Since there are no reference fields to aggregations outside the domain of the Sales Department (BPMN.TargetParticipant), the message structure also complies with guideline IAG 1.3. No nested structures are specified, complying with IAG 1.4.

Regarding guidelines for business process modelling through domain-driven modularisation, the BPMN.TargetParticipants of the BPMN.Choreo graphyTasks in Fig. 6.A represent three different organisational units: Sales Department, Delivery, and Quality Department, according to IAG 2.1. All the participants receive no more than two messages, so IAG 2.2 is also fulfilled. The message structures have been designed with no nested aggregations, as seen in Table 2, to meet IAG 2.3.

Figure 6.B shows the result of microservice domain modelling by message structure transformation. Next, we explain the application of the guidelines IAG 3.1 to 3.7 for the BPMN.ChoreographyTask Close Car Rental in Fig. 6.A, and for the structure of it message Rental specified in Table 2. Following IAG 3.1, the MSA.BoundedContexts Sales Department, Delivery, and Quality are created from the target BPMN.Participants in Fig. 6.A. The target participant Car Owner is excluded for not being part of the organisation. From IAG 3.2, the BPMN.Messsage Rental is associated with the MSA.BoundedContext Sales Department while Rating is associated with Quality Department. The messages Delivery Request and Delivery Confirmation are associated with the bounded context Delivery.

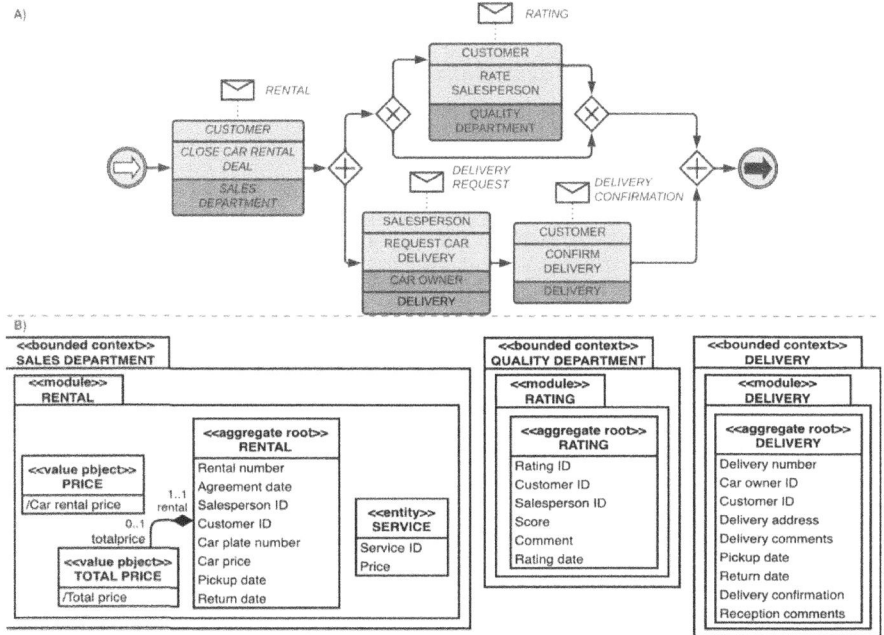

Fig. 6. A. Choreography diagram for the single-case mechanism experiment; B. Domain model resulting from applying the guidelines for Microservice Domain Modelling by message structure transformation.

Table 2. Message structure of the choreography task Close Car Rental Deal.

Field	Op	Id	Domain	Example
RENTAL =				
< Rental number +	g	id	number	202300345
Agreement date +	i		date	24-05-2023
Salesperson ID +	i		text	prat002
Customer ID +	i		text	8857657-Z
Car plate number +	i		text	465679-FGT
Car price +	i		money	25.00 €/day
Pick-up date +	i		date	15-08-2023
Return date +	i		date	31-08-2023
CAR PRICE =				
< Car rental price > +	d		money	400.00 €
SERVICE LINES =				
{ SERVICE LINE =				
< Service ID +	i		text	mx
Service price > } +	i		money	2.00 €/day
TOTAL PRICE =				
< Total price > >	d		money	432.00 €

We use the message structure in Fig. 2 for the remaining guidelines. For the set of guidelines IAG 3.3, addressing the MS.InitialAggregation, IAG 3.3.3 can be applied to RENTAL since it has theMS.DataField Rental Number marked as identifier and nested MS.Aggregations, producing the MSA.AggregateRoot RENTAL. For the set of guidelines IAG 3.4 addressing the internal elements of the message structure, IAG 3.4.2 applies to the MS.Iteration SERVICES, producing the MSA.Entity SERVICE. In this case, the multiplicity is 0..* from the created class to the MSA.aggregateRoot class redRENTAL. Guidelines IAG 3.4.3 and IAG 3.4.4 do not apply to the example. Guideline IAG 3.5 cannot be applied since there is no nested MS.Aggregations, MS.ReferenceFields or MS.Iterations inside the second level elements previously studied, in compliance with IAG 2.3. Guideline IAG 3.6 cannot be applied to the message structure in Table 2; however, as depicted in Fig. 6.B, the MSA.AggregateRoot DELIVERY and RATING were created as MSA.Entity since they do not have nested MS.Aggregations MS.Iterations or MS.ReferenceFields. Finally, the IAG 3.7 guides us to move each MSA.AggregateRoots and related classes inside a newMSA.Module. In the example, the MS.AggregateRoot RENTAL is moved inside the MSA.Module RENTAL, similarly, DELIVERY and RATING are moved into their respective MSA.Modules.

5   Discussion

Results in the Light of Earlier Literature. From the point of view of deriving domain models from business process models or requirements specifications, our work can be framed as a model transformation approach. This is quite conventional in the area of model-driven engineering, where transformation guidelines or rules are common. Some methods provide guidance for creating a UML class diagram, either taking a use a use case model as the sole input [8] or extending the requirements models with sequence diagrams [18], activity graphs [19], or information flow specifications [14,29]. Requirements expressed as user stories have also been used to derive Class Diagrams either automatically [7,21] or manually [3]. Other methods use BPMN Collaborations as a starting point [4]. This paper is inspired by [16], where the authors define guidelines to derive UML class diagrams from Communication Analysis specifications. Not only has this approach been experimentally validated [10], but it has inspired other authors in their own empirical research [1,2,4]. We have opted for modelling business processes as BPMN Choreography since they are similar to Communication Analysis process models but have wider adoption in industry and academia. Furthermore, while all approaches mentioned above are aimed at developing centralised, monolithic information systems, we have adapted the guidelines to MSA and DDD.

Limitations and Future Work. There are some limitations in the method we propose herein that require further research. Firstly, we do not cover MSA services; so far, these need to be defined by the domain analyst after deriving the domain models in order to complete the specification of the MSA. However, services required for managing the lifecycle of domain classes and complex transactions can be derived from the approach proposed in [11]. We are also aware that some complex business domains require more than two levels of aggregation nesting; for instance, an order that is structured in destinations, each of which has one or several order lines (see [16]). This would require more than two levels of entities in the corresponding MSA domain model, something so far not allowed in our guidelines (see IAG 2.2) since it contradicts the good industrial practices in DDD. We plan to empirically investigate cases of such complexity and nuance our guidelines. An obvious trade-off of our proposal is that it requires domain analysts to learn and apply modelling languages they might not be currently familiar with. Also, we expect that the quality of the output domain models is affected by the quality of the input models, but an empirical sensitivity analysis is needed to confirm this and measure the size of the effect.

6   Conclusions

This study introduces a method to integrate business process modelling with microservices architecture (MSA) through BPMN choreography diagrams and the Message Structure technique. We provide a systematic approach to derive UML class diagrams for MSA domain models by addressing domain complexity and minimising coupling. The guidelines and algorithms designed facilitate the design of business processes that align with MSA principles, ensuring modularity and scalability. The single case experiment demonstrates the feasibility of our approach, highlighting its potential to enhance the efficiency of MSA development teams. Future work will involve empirical validation to refine our guidelines and assess their effectiveness in various business contexts. This method offers a promising direction for organisations seeking to optimise their software architecture and business process integration.

References

1. Al-Fedaghi, S.: Communication-oriented business model based on flows. Int. J. Bus. Inf. Syst. **15**(3), 325–337 (2014)
2. Berkhout, M., Leewis, S., Smit, K.: Translating business process models to class diagrams. In: BLED 2020, p. 21 (2020)
3. Bragilovski, M., Dalpiaz, F., Sturm, A.: Guided derivation of conceptual models from user stories: a controlled experiment. In: Gervasi, V., Vogelsang, A. (eds.) REFSQ 2022. LNCS, vol. 13216, pp. 131–147. Springer, Cham (2022). https://doi.org/10.1007/978-3-030-98464-9_11
4. Brdjanin, D., Banjac, G., Banjac, D., Maric, S.: An experiment in model-driven conceptual database design. Softw. Syst. Model. **18**, 1859–1883 (2019)
5. Butzin, B., Golatowski, F., Timmermann, D.: Microservices approach for the internet of things. In: ETFA 2016, pp. 1–6 (2016)
6. Conway, M.E.: How do committees invent. Datamation **14**(4), 28–31 (1968)
7. Dahhane, W., Zeaaraoui, A., Ettifouri, E.H., Bouchentouf, T.: An automated object-based approach to transforming requirements to class diagrams. In: WCCS 2014, pp. 158–163 (2014)
8. Díaz, I., Sánchez, J., Matteo, A.: Conceptual modeling based on transformation linguistic patterns. In: Delcambre, L., Kop, C., Mayr, H.C., Mylopoulos, J., Pastor, O. (eds.) ER 2005. LNCS, vol. 3716, pp. 192–208. Springer, Heidelberg (2005). https://doi.org/10.1007/11568322_13
9. Dragoni, N., et al.: Microservices: yesterday, today, and tomorrow. In: Mazzara, M., Meyer, B. (eds.) Present and Ulterior Software Engineering, pp. 195–216. Springer, Cham (2017). https://doi.org/10.1007/978-3-319-67425-4_12
10. España, S., Ruiz, M., González, A.: Systematic derivation of conceptual models from requirements models: a controlled experiment. In: RCIS 2012, pp. 1–12 (2012)
11. España, S.: Methodological integration of communication analysis into a model-driven software development framework. Ph.D. thesis, Universitat Politècnica de València (2011)
12. Evans, E., Evans, E.J.: Domain-Driven Design: Tackling Complexity in the Heart of Software. Addison-Wesley Professional (2004)
13. Forsgren, N., Humbpotifle, J., Kim, G.: Accelerate: the Science of Lean Software and DevOps Building and Scaling High Performing Technology Organizations. IT Revolution Press (2018)
14. Fortuna, M.H., Werner, C.M., Borges, M.R.: Info cases: integrating use cases and domain models. In: RE 2008, pp. 81–84 (2008)
15. González, A., Ruiz, M., España, S., Pastor, Ó.: Message structures: a modelling technique for information systems analysis and design. In: WER 2011 (2011)

16. González, A., España, S., Ruiz, M., Pastor, Ó.: Systematic derivation of class diagrams from communication-oriented business process models. In: Halpin, T., et al. (eds.) BPMDS/EMMSAD -2011. LNBIP, vol. 81, pp. 246–260. Springer, Heidelberg (2011). https://doi.org/10.1007/978-3-642-21759-3_18
17. Henderson-Sellers, B., Ralyté, J., Ågerfalk, P., Rossi, M.: Situational Method Engineering. Springer, Heidelberg (2014). https://doi.org/10.1007/978-3-642-41467-1
18. Insfrán, E., Pastor, O., Wieringa, R.: Requirements engineering-based conceptual modelling. Requirements Eng. **7**, 61–72 (2002)
19. Kösters, G., Six, H.W., Winter, M.: Coupling use cases and class models as a means for validation and verification of requirements specifications. Requirements Eng. **6**, 3–17 (2001)
20. Lewis, J., Fowler, M.: Microservices: a definition of this new architectural term (2014). https://martinfowler.com/articles/microservices.html. Accessed 20 June 2023
21. Lucassen, G., Robeer, M., Dalpiaz, F., Van Der Werf, J.M.E., Brinkkemper, S.: Extracting conceptual models from user stories with visual narrator. Requirements Eng. **22**, 339–358 (2017)
22. Noel, R., España, S., Pastor, O., Panach, J.I.: From choreography diagrams to microservice architecture domain models: technical report (2024). https://doi.org/10.5281/zenodo.11624682
23. Noel, R., Panach, J.I., Ruiz, M., Pastor, O.: Stra2Bis: a model-driven method for aligning business strategy and business processes. In: Ralyté, J., Chakravarthy, S., Mohania, M., Jeusfeld, M.A., Karlapalem, K. (eds.) ER 2022. LNCS, vol. 13607, pp. 255–270. Springer, Cham (2022). https://doi.org/10.1007/978-3-031-17995-2_18
24. OMG: Business Process Model and Notation (BPMN) version 2.0.2. Technical report, Object Management Group (2013)
25. Pohl, K., Rupp, C.: Requirements Engineering Fundamentals. Rocky Nook (2016)
26. Rademacher, F., Sachweh, S., Zündorf, A.: Towards a UML profile for domain-driven design of microservice architectures. In: Cerone, A., Roveri, M. (eds.) SEFM 2017. LNCS, vol. 10729, pp. 230–245. Springer, Cham (2018). https://doi.org/10.1007/978-3-319-74781-1_17
27. Thoughtworks: Inverse Conway maneuver (2016). https://www.thoughtworks.com/es-es/radar/techniques/inverse-conway-maneuver. Accessed 09 Nov 2021
28. Valderas, P., Torres, V., Pelechano, V.: A microservice composition approach based on the choreography of BPMN fragments. Inf. Softw. Technol. **127** (2020)
29. de la Vara, J.L., Sánchez, J.: System modeling from extended task descriptions. In: SEKE 2010, pp. 425–429 (2010)
30. Wieringa, R.J.: Design Science Methodology for Information Systems and Software Engineering. Springer, Heidelberg (2014). https://doi.org/10.1007/978-3-662-43839-8
31. Zimmermann, O.: Microservices tenets. Comput. Sci.-Res. Dev. **32**(3), 301–310 (2017)

Management

Human Friendly Automation: A Literature Review on the Role of the Human Factor in AI-Driven Business Process Automation

Manfred Baer[(⊠)] and Ralf Plattfaut

University of Duisburg-Essen, Duisburg, Germany
manfred.baer@stud.uni-due.de, ralf.plattfaut@icb.uni-due.de

Abstract. Business Process Management (BPM) enables organizations to boost efficiency, improve quality, cut costs, accelerate process times, and maintain competitiveness. With the advent of Artificial Intelligence (AI) and Generative AI (GenAI), the focus has moved from rules-based process automation for clearly defined processes to more nuanced cognitive automation of complex processes. These innovations clearly have huge influence on the role of the human in organizations. Job profiles, routines and job satisfaction are affected. Interestingly enough, the current role of the human or employee in the BPM, AI and automation literature seems to be mainly a role of a production factor. Our literature review highlights the importance of considering the human factor in AI-driven business process automation and identifies a research gap on the role of the human factor. To structure the current state of literature we introduce a holistic human-friendly automation framework. We see a demand for future research to address gaps in understanding the human factor in automation, particularly in terms of ethical considerations, demographic shift, and the changing relationship between humans and machines. Specifically, investigating the roles of management, HR, and employees in AI-driven automation initiatives may provide valuable insights for organizations seeking to balance process optimization with employee engagement and satisfaction.

Keywords: Human factor · BPM · AI · literature review

1 Introduction – The Role of the Human Factor in BPM

Business Process Management (BPM) is essential for organizations to increase efficiency and quality, to reduce cost, to speed up process time, and to become more competitive. One means to achieve these goals is process automation. With the new and immense possibilities of Artificial Intelligence (AI) and Generative AI (GenAI), the recent focus shifted from rules-based process automation (using, e.g., Robotics Process Automation, RPA) for clearly defined and structured processes to more and more cognitive automation of complex processes [32]. Nowadays, the discussion in research and practice is broader, spreading from classical BPM and systems that assist employees over human-machine collaboration (augmentation) to autonomous systems which need limited supervision

A. Marrella et al. (Eds.): BPM 2024 Forum, LNBIP 526, pp. 287–301, 2024.
https://doi.org/10.1007/978-3-031-70418-5_17

and control by humans (e.g. [7]). Organizations evaluate increasingly complex processes for AI-powered process automation to further increase efficiency. Especially the role of autonomous or agentic information systems [3] in organizations is part of a vivid discussion on the background of GenAI. Here, GenAI opens up completely new areas of system support also in complex front office processes.

The statement that everything what can be automated will be automated can be seen with new eyes after the introduction of ChatGPT at the end of 2022. New areas of use are the support of highly skilled jobs like journalists (e.g. [26]), creative jobs, medical experts and doctors [24], case workers and accountants or employees of the professional services firms [23].

In the light of these new possibilities in BPM with AI/GenAI, organizations must deal with ethical standards [21], trust and transparency of algorithms. Task augmentation and autonomous systems have a massive impact on the role of employees and humans interacting with their new teammates [26]. Managers in both private and public sector need to find answers to questions relating to the acceptance of new processes [24], the potential skill erosion [22, 25], or the changed role and self-identity of employees as parts of the traditional processes are no longer executed by humans. This also pertains to the overall attractiveness of the organization as an employer. Especially the last point of being an attractive company for applicants and new talents will be more and more critical in industrialized markets that are severely affected by the demographic changes and a shrinking workforce. In these markets, companies will need to see both sides of the process automation medal: Increased use of AI and automation to stay competitive and a focus on employees needs to remain attractive as employer. In addition, underestimating the human factor can lead to employees not accepting or blocking process optimizations, which can lead to significant negative financial effects [24].

Interestingly enough, the current role of the human or employee in the BPM, AI and automation literature seems to be mainly a role of a production factor. The focus is how the human factor can be optimized to increase process efficiency and success-fully implement process changes to reduce cost and headcount [11]. However, the need to design benevolent processes has been identified [20]. Therefore, there might be a gap in the IS literature about the classical role of the human factor in BPM given the recent AI/GenAI developments and the current and future challenges for organizations to remain attractive as employer. To address this gap, we aim to answer the following research question (RQ):

RQ: What is the current state of research in literature in terms of the human factor in process automation through AI?

This paper is structured as follows: Sect. 2 describes the methodology and our app-roach for the literature review. Section 3 explains the results of the literature research and introduces a holistic framework for the role of the human and the employee in the overall setup of BPM in an organization. Section 4 discusses the results of the literature research and suggests a future research agenda. In the last Sect. 5, we end with a short conclusion and outlook.

2 Methodology

To address our research question and to assess the state of the art of research we conducted a systematic review of the IS literature. We followed the guidelines of vom Brocke et al. (2009) [6]. While we acknowledge that full coverage of the literature is not possible, we still tried to ensure a broad coverage of the relevant literature.

As a first step we considered which major journals and conferences are essential and needed to be included in our search. This review sits in the fields of IS and BPM and should accordingly include its top-tier sources. As such, we opted to conduct our first database search in the Electronic Library of the Association for Information Systems (AISeL), which contains premier journals in this field (e.g., Management Information Systems Quarterly, Journal of the Association of Information Systems, Scandinavian Journal of Information Systems, etc.) as well as the proceedings of relevant conferences (e.g., International Conference on Information Systems (ICIS), European Conference on Information Systems (ECIS), Americas Conference on Information Systems (AMCIS), Pacific-Asia Conference on Information Systems (PACIS) =. In addition, we searched the proceedings of the International Conference on Business Process Management (BPMC) on SpringerLink and the Business Process Management Journal (BPMJ) on the database of the Emerald Group Publishing.

As a second step we defined the keywords for the search. We built our search string using three parts related to our research question. To cover the human factor, we not only included human friendly into our search but also used common synonyms to ensure that we capture all relevant paper in this space: human friendly, user friendly, people oriented, customer centric, human centered, user centric, human oriented, customer friendly, user-focused, client-centric or humanized. Theoretically, other synonyms that are rarely used for human friendly would have been possible, but we have deliberately excluded them, which must be seen as a limitation. To ensure the fit to the theme of process automation, we selected this as a corresponding keyword. We chose process automation other than Business Process Automation to ensure a broader search result. The last part of the search string is related to AI and GenAI. Again, we selected the broader term AI.

We finally used the following search string: (`"human friendly"` OR `"User friendly"` OR `"People oriented"` OR `"Customer centric"` OR `"Human centered"` OR `"User centric"` OR `"Human oriented"` OR `"Customer friendly"` OR `"User focused"` OR `"Client centric"` OR `"Humanized"`) AND `"process automation"` AND `"AI"`.

The literature search was performed on December 18th 2023. The search in the 3 databases (AISeL, Springer, Emerald) resulted in a total of 292 publications based on the requirements of the search query defined. Due to the focus on BPM and automation in the context of AI and due to the fact that we did not find relevant articles in our search before 2018 after initial title screening, we excluded all publications before 2018. This is of course a clear limitation of our literature research. Therefore 249 papers remained. As a further step, we conducted a qualitative analysis and studied the titles and the abstracts of the 249 remaining articles. We used no automation tools for the screening. Papers not dealing with the key topics like automation, AI or without any relation to people, humans or employees were excluded. We documented our results with a short justification or summary in a complete list. After this phase we ended up with 63 papers where we saw a relevance for our RQ in general. The complete list with assessment relating the exclusion and inclusion criteria can be found on figshare (https://doi.org/10.6084/m9.figshare.25358449.v2) (Fig. 1).

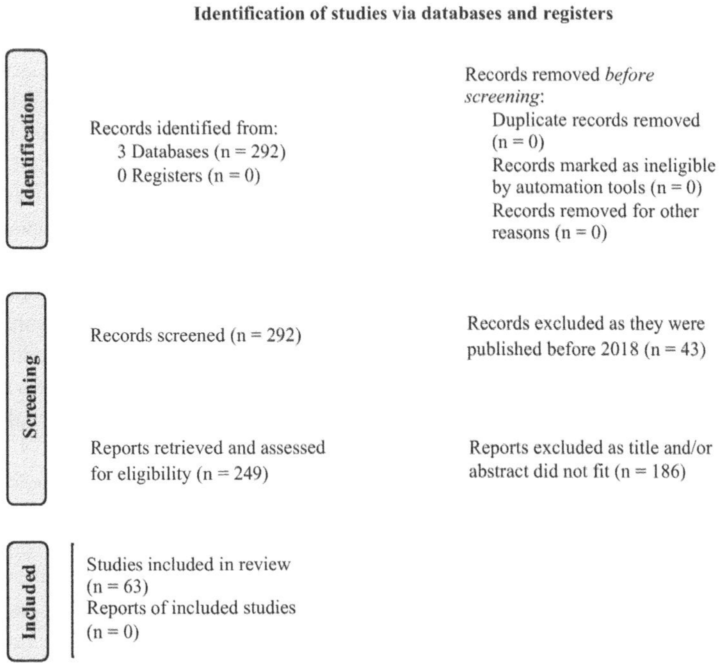

Identification of studies via databases and registers

Identification	Records identified from: 3 Databases (n = 292) 0 Registers (n = 0)	Records removed *before screening*: Duplicate records removed (n = 0) Records marked as ineligible by automation tools (n = 0) Records removed for other reasons (n = 0)
Screening	Records screened (n = 292)	Records excluded as they were published before 2018 (n = 43)
	Reports retrieved and assessed for eligibility (n = 249)	Reports excluded as title and/or abstract did not fit (n = 186)
Included	Studies included in review (n = 63) Reports of included studies (n = 0)	

Fig. 1. Flow chart (based on PRISMA 2020 flow diagram)

To analyze the literature, we relied on a bottom-up approach to identify themes in the literature. This approach was inspired by Wolfswinkel et al. (2013) [36]. Six themes emerged from the data. Based on a deeper analysis of these themes, we were able to develop a framework of human-friendly automation. This framework allows to both structure the current state of research and to outline major areas for future research in the IS and BPM domain.

3   Results

3.1   Bottom-Up Analysis of the Identified Literature

During our bottom-up analysis of the 63 articles, six themes emerged from the data. These themes outline perspectives when studying the human factor in process automation through AI. Table 1 gives a short overview of the themes and their occurrence in literature.

Firstly, papers take a pure instrumental perspective of process automation and focus on reducing human work to achieve monetary goals in BPM. In this context, automation is seen as the concept of letting non-human actors, i.e. a virtual workforce, perform regulated and repetitive work [30]. Here, a human focus is a mere means to an end to the ultimate goal of cost reduction and efficiency. The claims of human-friendliness or user centricity are included to ensure quick and seamless adoption of automated processes and to finally increase job performance (e.g. [8]).

Secondly, papers also take the perspective of task augmentation. Augmentation means that humans are not being replaced by machines, they are interacting and communicating with machines in form of collaborative work (e.g. [17]). As such, the capabilities of humans are extended through AI-driven automation. The next step of the evolution would be autonomous systems which can independently execute actions (e.g. [26]).

Thirdly, literature also takes the clear angle of demographic change. Here, it is acknowledged that organizations continue to rely on humans. However, due to a shortage of skilled labor, process automation through AI is used to free up capacity (e.g. [23]). While these first three themes appear comparably often in the scholarly literature analyzed, the next three themes are less often covered by publications.

Fourthly, articles also discuss on a higher level how BPM should be organized and implemented with human-friendliness in mind. As such, articles are about broadening the scope of BPM from an instrumental focus on cost, quality, time, or flexibility to a general benevolence [20] also fulfilling humanistic goals. Human-friendliness in this context is seen as the concept of putting the human or employee into the center of the automation activity, which means a focus on people and not on the machine or the solution.

Fifthly, some articles inspect the human element in specific. We use the broad term human in our research. However, the human element can pertain to individual employees, customers, specific (human) skills, or the specific role of process owners (e.g. [24]).

Lastly, some articles study the effects of human-friendliness in process automation through AI. They discuss, e.g., whether human-friendly process automation programs are more successful. These papers also analyze the effects of automation and AI on specific tasks and the satisfaction of the employees with their work.

It also includes approaches how tasks shall be designed to have a higher level of satisfaction, creativity and wellbeing. This all shall contribute to a more attractive workplace (e.g. [19]).

Table 1. Results of the qualitative bottom-up analysis.

Theme	Appearance in papers studied
1) Human-friendliness as a means to reduce costs	30
2) Human-friendliness in task augmentation	51
3) Human-friendliness to cope with demographic changes	25
4) Organizing BPM for human-friendliness	16
5) Discussing the human in human-friendliness	6
6) Effects of human-friendliness in process automation	8
Papers could cover more than one theme (with three papers covering all six themes)	

3.2 Towards a Conceptual Framework of Human-Friendly Automation

New technologies like AI have a significant impact on how work is designed, performed and managed on individual, organizational and societal level with transformational impact on the labor market [35]. Although all market participants and humans are affected by these major changes in the end, we could not find an overarching concept or description which would make the interactions and dependencies on the different level transparent.

Therefore, based on the insights gained through the bottom-up analysis of the papers, we were able to derive a conceptual framework of human friendly automation. Following the approach of Jabareen [15], our conceptual framework demonstrates our understanding and interpretation of the different concepts we have found in the literature search. To build the framework we followed an iterative approach as a grounded theory technique, identifying the key concepts in the relevant articles first and then evaluating the dependencies between the concepts as a second step. The framework describes our current understanding and interpretation of the literature analyzed. It will therefore be dynamic and may be revised according to new insights, comments and literature in the future [15]. The core elements of the framework have been identified and described by the first author and then being discussed and further developed with the second author.

The framework can be used both to describe the current state of knowledge on the human factor in AI-driven process automation and to outline potentially fruitful areas of future research (see Fig. 2).

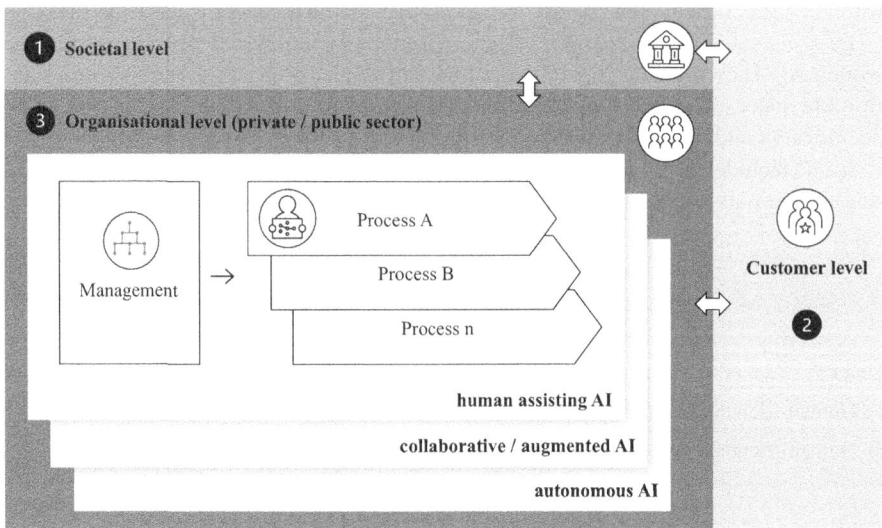

Fig. 2. Human Friendly Automation (HFA)-Framework

The framework shows the core elements of human friendly automation on three different levels: the societal level (1), the customer (consumer and citizens) level (2) and the organizational level in the public and the private sector (3). On the organizational level, we are in the center of process automation differing between the grades of AI driven automation: AI assisting humans, augmented human machine collaboration and autonomous systems. The corresponding processes within an organization are defined and controlled by the responsible management which includes the human resources (HR) functions. In the following paragraphs we describe the relevance of each element within the framework and summarize the associated literature with specific examples.

Societal Level

On societal level, we need to consider influences of AI-driven automation in two directions. On the one hand, the external environment, e.g., laws, ethical principles and demographic trends, defines the frame for organizations, customers and in the end all humans. On the other hand, organizational behavior is influenced by the behavior of other organizations and individuals, e.g., adoption of or resistance towards AI-driven automation. All these elements have therefore a significant influence on the use of AI and the pressure on the automation agenda of organizations.

Regarding the external environment, we observe a demographic change in industrial countries leading to a war for talents and a high competition among organizations in the private and the public sector. Employees have the choice between a variety of vacant positions. They are more and more looking for meaningful work and attractive working conditions. While the baby boomer generation tried to separate their work and home life with a focus on face-to-face communication, the digital natives integrate work and private life [13]. The technology adoption curve of the digital natives for innovative technologies is much more exponential [13]. Obviously, this has an impact on how fast organizations can implement AI and automation in their process landscape.

Moreover, there is a lot of debate on the ethics of using AI which will be translated into legislature in the near future. Both legal and ethical frameworks are thus creating guidelines for dealing with this technology. As AI has the potential to replace or reduce human control, responsibilities, skills and self-determination, ethical guidelines can help to ensure trust and transparency. The ethical dimensions of AI are seen as ambiguous and the discussion on how to best address this topic are seen as unstructured in current IS literature, clear definitions are missing with a huge potential for future research [21].

A further aspect of AI-driven automation on the societal level is the behavior of other organizations and individuals. Behavior of other organizations might lead to institutional isomorphism [9]. As an example, the effect of de-skilling or skill-erosion. As a phenomenon the erosion of certain skills is a normal factor in human history and directly related to all kind of industrial progress. However, the general possibilities of AI-driven automation have the potential to massively disrupt the role of knowledge workers and make parts of their role and identity irrelevant [25]. This can be seen as a major future challenge for highly skilled societies.

Customer Level

The customer plays a central role for all organizations. Still, the classical view of BPM focuses on the organization itself with the main target of cost reduction. With the evolution of process automation from systems assisting employees to more augmented and autonomous systems with direct interaction with end clients, the focus will further shift towards the customer level.

A great example of increased customer focus is the benevolent business processes approach [20]. Benevolent means in general the well-being of the other party. It enlarges the view from the dominating pure cost optimization of processes and includes the customer and his specific requirements (user-centricity). Benevolence indeed goes beyond the usual user-centricity (design-thinking, customer journeys etc.) and is designed with the clear target to put the customer´s wellbeing above the short-term interests of the organization [20].

The customer is also seen as a success factor for process digitalization projects. Customer knowledge allows a more customer centric design and an early integration allows a process adaptation to the customer's needs [2].

From a customer perspective, another level of AI ethics becomes important, too. Here it is less about the ethical use of AI in general but about the ethics employed within the algorithms. Customers interacting with an AI supported process in general expect a clear definition of responsibilities of the actions and results of the systems, a bias free decision process, protection of data privacy and transparency of the decision process [27]. Therefore, clear ethical guidelines for AI use in BPM are essential.

Organizational Level

Organizations in the private and the public sector are investing massively into AI driven BPM and automation to stay competitive, compensate demographic changes and skill shortages and to overall transform their workplaces. Large corporations like Siemens are initiating programs to infuse AI broadly into the organization [33]. Siemens is targeting to democratize AI. This means AI is no longer seen as a topic for specialists and data scientists but for a broad range of people within the organization. To make this transformation happen, management, information technology (IT), human resources (HR) and the employees on process level need to work together. In the following paragraphs we will analyze the current state of research in terms of the role of the management and HR and afterwards in terms of role of the employees in AI driven processes.

Management Level

At the management level, the strategy of the organization is defined and implemented. In terms of AI driven automation and BPM, the responsible managers need to define the overall direction of AI use and the level of automation and human-machine collaboration or human replacement. The details and speed may vary significantly by sector (public vs. private) and industry (e.g., healthcare vs. manufacturing) and is also very much dependent on the market and the societal environment (demographic changes, ethical guidelines, legal requirements etc.). Although the management level is implicitly part of the BPM and AI driven automation literature, we found only a few papers addressing the management explicitly. This is interesting because the CEO's view and attitude has a major impact on the company's automation and digitalization agenda, the mindset and culture. In this context, Schaeper et al. (2023) [28] analyze and describe the impact of the CEOs level of digital strategy on innovation within the organization. In a nutshell they describe that a CEO should keep a balanced level of digital investments and not overdo it. Furthermore, it is acknowledged that BPM and digital transformation shall be driven with top-management support and an appropriate organizational structure [14]. The right level of digital transformation and new digital technologies free up additional time for innovation among the employees due to streamlined processes and increased efficiency. A practical example is a project at EY in the area of a contract reading tool. The tool reduces the required time to analyze a contract by 60%–80% and allows EY employees to focus on higher value tasks [23]. It is also described that leaders need to increasingly focus on giving strategic directions, empowering employees and taking care of their needs and acting as mentors and coaches [5].

The organizational changes based on digital transformation and AI driven automation have also an impact on the human resources functions. Gierlich-Jonas et al. therefore suggests a transformation of the HR-function from pure resource managers to growth catalysts [12]. They postulate a new mindset of HR characterized by innovation, an attitude as a service provider, flexible structures and an enablement of the employees. This need for change in HR is underpinned by works focusing on skills erosion of highly skilled knowledge workers by cognitive automation [25]. As a consequence, short term process optimization can lead to skill erosion leading to decreasing services quality, additional risks (e.g., in healthcare, aviation) and the question how an employee shall be responsible for the outcome without the key knowledge of the (automated) decision process. Another aspect is that AI can now also automate routine tasks formerly performed by junior team members e.g. in the legal space [4], The question will be how younger team members are able to develop their expertise and can be supported by HR. In conclusion, it can be stated that the (early) involvement of HR in cognitive automation (i.e., AI-driven automation) projects is needed to retain talent by adopting job descriptions, salary schemes and create individual growth opportunities [16].

Process Level
The role of the employee in BPM is significantly changing. The classical role has been that the employee is in full control of the process. In the recent years more and more repetitive process steps have been taken over by machines/robots, like Robotics Process Automation (RPA) [16]. Recent approaches are more focusing on complex processes where we see a strong collaboration between humans and machines up to fully automated tasks.

The scientific BPM literature focuses mainly on the process level. A key question is, e.g., which processes should be performed by humans and which by machines [32]. The answer to this question is of course dependent on the potential of AI and has changed in the recent years. The clarification of the collaboration model between humans and machines is also essential for the whole organization. Chedrawi and Haddad [7] describe the development of AI and the human-machine collaboration from assisted over augmented to autonomous intelligence and the effects on an organization. They believe that "quasi-humans" have the potential to deeply impact organizational practices. Based on new technologies like RPA, automation, machine learning, chatbots etc. organizations can become "AI fueled organizations" with limited or no involvement of humans. Van der Aalst (2018) [32] sums it up when asking: "Automate or not to automate, that is the question" under the light of "hybrid intelligence". The process complexity is without any doubt one of the core influencers of the automation potential [34]. Vilmalkumar et al. [34] introduce a task complexity typology organizing tasks into five categories from simple tasks over decision tasks to fuzzy tasks and describe their automation potential, opacity and fairness concerns.

But in the end, the technology is only one part of the digital transformations. Employees and their interaction with one another stay a differentiating factor [5]. Therefore, in addition to the technical perspective, the human side of automation must also be considered. The cultural and the human dimension are seen as success factors in digitalization and BPM projects [2] stating the need of enough detailed knowledge about the domain and employee agility. The employee perspective in BPM and automation is also an element in the 10 principles of good BPM ([1, 6]). An integration of BPM stakeholders like employees into all stages of BPM is critical as employee resistance is a clear threat. Under the principle of joint understanding, it is described that employees shall share the same understanding how and why BPM creates value. Scholars as well try to answer research questions like: *"Which factors determine the adoption of intelligent process automation by employees?"* [18]. They identified an employee's attitude as a main driver for the adoption of intelligent process automation and also trust and transparency in form of explainable AI. Seiffer et al. (2021) [29] come to a similar conclusion when they review how employees respond to software robots. They state that most studies on software robots focus on the implementation and the economic outcome and do not pay much attention on the employee perspective. In this paper, they also describe the early involvement of employees in the pre-implementation phase as an important context factor to share their professional and process knowledge. They clearly suggest more research in the direction of the interplay between employees, software robots and the context factors.

To summarize the human element on the process level: We can hold on to the fact that the human factor in BPM and automation is primarily seen as a production factor which needs to be optimized for a successful outcome of the process changes.

4   Discussion: Towards a Research Agenda on Human Friendly Automation

Our literature review showed that the human factor in AI driven automation is still predominantly seen as an optimization factor. In recent papers we recognized a change towards a higher focus on employees and their role in the automation programs. How to involve employees earlier, keep them motivated, how to ensure a higher acceptance of technical change or how to deal with skill erosion. Nevertheless, we saw a number of papers addressing a gap in research in terms of the increasingly important role of the human factor in automation and the changes in the nature of work (e.g., [4]). Based on our own analysis of IS literature we clearly see several areas for future research under the focus of human friendliness in AI driven automation. A summary of specific research questions can be found in Table 2.

Starting at the **societal** level: A number of papers addressed the ethical dimension of AI (e.g., [21]). They confirmed that key questions are still open, e.g., what kind of tasks shall not be delegated to AI-driven systems or how data-bias can be prevented.

Another important question would be around the area of demographic change. The pressure on organizations due to skill shortages can increase the speed and intensity of automation. This pressure can also lead to more focus on a more human-friendly workplace design? We did not find a paper addressing this topic in our analysis. Due to the significant effects of the demographic changes on the workforces this is a question with high practical relevance for organizations.

Table 2. Overview: Towards a research agenda for Human Friendly Automation.

level	research area	exemplary research questions
societal level	ethical guidelines	*What kind of tasks must not be delegated to machines and shall remain under human control only?*
		How can data-bias be prevented in AI-driven process automation?
	culture	*How does increased automation influence the communication and social interaction between humans?*
	demographic change	*What effects has demographic change on the speed and intensity of AI-driven automation?*
		How do demographic change and skill shortages impact the workplace design?
customer level	user-centricity	*How does a double (human friendly) user centricity positively affect the customer satisfaction?*
		In how far do customers care for the human-friendliness of automated processes?
organizational level	digital transformation	*What opportunities and risks do management, HR, IT and employees see in AI-driven (human friendly) process automation?*
	management involvement	*What role does the top management play in the success of human-friendly automation projects?*
	human resources	*What role does HR play to ensure a successful human friendly process automation?*
		How does HR deal with the phenomenon of skill erosion as an effect of AI driven automation?
	human-machine collaboration	*How can humans and bots work together?*
		What impact have bots on the overall human responsibility for process outcomes?

(continued)

Table 2. (*continued*)

level	research area	exemplary research questions
		What design principles would lead to a higher acceptance of software bots in automation?
	employee involvement	*What impact would have an end-to-end involvement of the employee in AI-driven process automation projects on success and acceptance?*
	trust in AI	*How can trust be measured constantly in a human friendly AI driven process automation?*
		What role does trust play in the acceptance of AI-driven automation?
	guidelines	*How does the ten principles of good BPM need to be adopted in the light of human friendly automation?*
		Which success factors exist for the implementation of human-friendly automation?

Moving to the **customer** level, we also see room for further research. The key question on customer level is how to create human (customer) friendly processes and applications by using AI. The benevolence concept already focusses on the customer and describes design guidelines for benevolent business processes [20]. The paper suggests future research on how these guidelines impact retention, advocacy or trust. Another interesting aspect would be how a double user centricity (customer and employee) in the design of business processes would positively impact the customer satisfaction: *How does a double (human friendly) user centricity positively affect the customer satisfaction?*

On the **organizational** level future research could focus, e.g., on management, HR, IT and employee aspects. To better understand the current view on human friendly automation a suitable question would be: *What is a holistic view from management, HR, IT and employees on AI driven human friendly process automation and what needs for action can be identified?*

Another aspect would be the role of the C-level to drive human friendly automation: *What role does top management play in the success of human-friendly automation projects?*

The important and changing role of HR in digital workplace transformation has already been addressed [12]. Further questions in this area are e.g.: *What role does HR play to ensure a successful human friendly process automation? How does HR deal with the phenomena of skill erosion as an effect of AI driven automation?*

In the light of human-machine collaboration and the so-called new colleague "bot" questions are: *How can humans and bots work together? What impact has colleague bot on the overall human responsibility for process outcomes? What design principles would lead to a higher acceptance of software bots in automation?* These topics are already addressed in several papers, but the employee perspective seems to be under-researched (e.g. [29]).

Another aspect would be the ongoing involvement of the employee in the transformation of processes and workplaces. The identification of the right task and process owners is already seen as a crucial factor to achieve organizational acceptance and to make cognitive automation human-centered [10]. A concept for a structured and ongoing involvement of the employees seems to be still missing. Therefore, a research question could be: *What impact would have an end2end involvement of the employee in cognitive process automation projects on success and acceptance?*

The enormous potential of GenAI for process automation also leads to uncertainties and lack of trust on employee's side. They are in fear of being replaced by bots, losing jobs and responsibilities or ending up with jobs that are not favorable any more. In this setup trust into the organization and the management plays is an important factor. The importance of trust for the technology acceptance has already been researched [18]. In this context a further research question could address the importance of trust in HFA: *How can trust be measured constantly in a human friendly AI driven process automation?*

A last aspect we would like to add is the question how existing implementation guidelines and principles need to be enriched by the human-factor. As an example, we could ask: *How does the ten principles of good BPM need to be adopted in the light of human friendly automation?*

5 Concluding Remarks

In our paper we studied scholarly literature on the role of the human factor in AI-driven process automation. Our literature review underscores the crucial role of the human factor in business process automation. Traditionally, researchers have emphasized optimizing the human factor for enhanced process efficiency. However, emerging literature reveals a rising curiosity about the shifting roles of employees and the consequences of automation on their work experiences. For a better and holistic understanding of the human role in automation and the dependencies within organizations we introduced an HFA framework.

As a literature review, our paper is beset with some limitations. We have concentrated on selected outlets and databases. It is possible that a broader search could lead to additional insights. Therefore, the selection of our search-keywords and the focus on papers starting with 2018 are limitations as well. Moreover, the creation of codes and categories that culminated in the presented framework is a creative research process. While we took care to reduce any potential researcher bias, it is possible that other researchers would arrive at different codes, categories, or an adapted framework. As the framework describes our current understanding and interpretation of the literature analyzed, it may be revised according to new insights, comments and literature in the future.

To fill existing gaps in comprehending the human factor in automation, future research ought to delve deeper into ethical considerations, demographics, and the evolving relationship between humans and machines. More specifically, examining the roles of management, HR, and employees in AI-driven automation endeavors can offer invaluable insights for organizations aiming to strike a balance between process optimization and employee engagement and satisfaction.

References

1. Badakhshan, P., Scholta, H., Schmiedel, T., vom Brocke, J.: A measurement instrument for the "ten principles of good BPM". Bus. Process Manag. J. **29**(6), 1762–1790 (2023). © Emerald Publishing Limited 1463–7154
2. Baier, M.-S., Lockl, J., Röglinger, M., Weidlich, R.: Success factors of process digitalization projects – insights from an exploratory study. Bus. Process. Manag. J. **28**(2), 325–347 (2022)
3. Baird, A., Maruping, L.M.: The next generation of research on IS use: a theoretical framework of delegation to and from agentic IS artifacts. MIS Q. **45**(1), 315–341 (2021)
4. Benbya, H., Pachidi, S., Jarvenpaa, S.: Special issue editorial: artificial intelligence in organizations: implications for information systems research. J. Assoc. Inf. Syst. **22**(2) (2021). https://doi.org/10.17705/1jais.00662
5. Bitzer, M., Hinsen, S., Jöhnk, J., Urbach, N.: Everything is IT, but IT is not everything–what incumbents do to manage their digital transformation towards continuous change. In: ICIS 2021 Proceedings, vol. 6 (2021)
6. vom Brocke, J., et al.: Reconstructing the giant: on the importance of rigour in documenting the literature search process. In: ECIS 2009 Proceedings, vol. 372 (2009)
7. Chedrawi, C., Haddad, G.: The rise of quasi-humans in AI fueled organizations, an ultimate socio-materiality approach to the lens of Michel Serres. Pac. Asia J. Assoc. Inf. Syst. **14**(2), 5–24 (2022)
8. Duan, S., Wibowo, S., Deng, H.: Affordances of digital technology for enhancing job performance in digital work. In: PACIS 2021 Proceedings, vol. 216 (2021)
9. Di Maggio, P.J., Powell, W.W.: The iron cage revisited: institutional isomorphism and collective rationality in organizational fields. Source: Am. Sociol. Rev. **48**(2), 147–160 (1983)
10. Engel, C., Elshan, E., Ebel, P.: Moving beyond rule-based automation: a method for assessing cognitive automation use cases. In: ICIS 2021 Proceedings, vol. 19 (2021)
11. François, P.A., Borghoff, V., Plattfaut, R., Janiesch, C.: Why companies use RPA: a critical reflection of goals. In: Di Ciccio, C., Dijkman, R., del Río Ortega, A., Rinderle-Ma, S. (eds.) BPM 2022. LNCS, vol. 13420, pp. 399–417. Springer, Cham (2022). https://doi.org/10.1007/978-3-031-16103-2_26
12. Gierlich-Jonas, M., Zimmer, M.P.: Digital workplace transformation triggers a shift in the HR function: from resource manager to growth catalyst. In: ECIS 2023 Research Papers, vol. 310 (2023)
13. Hadidi, R., Power, D.: Technology adoption and disruption – organizational implications for the future of work. J. Midwest Assoc. Inf. Syst. **2020**(1) (JMWAIS) (2020). Article 1
14. Helbin, T., Van Looy, A.: Process innovation in the EU public sector through the lens of BPM ambidexterity. In: Marrella, A., Weber, B. (eds.) BPM 2021. LNBIP, vol. 436, pp. 141–152. Springer, Cham (2022). https://doi.org/10.1007/978-3-030-94343-1_11
15. Jabareen, Y.: Building a conceptual framework: philosophy, definitions and procedure. Int. J. Qual. Methods **8**(4), 49–62 (2009)
16. Lacity, M., Willcocks, L.: Becoming strategic with intelligent automation. MIS Q. Exec. **20**(2) (2021). Article 7
17. Laut, P., Dumbach, P., Eskofier, B.M.: Integration of artificial intelligence in the organizational adoption – a configurational perspective. In: ICIS 2021 Proceedings (2021)

18. Mayr, A., Stahmann, P., Nebel, M., Janiesch, C.: Unified theory of acceptance and use of technology (UTAUT) for intelligent process automation. In: Rising like a Phoenix: Emerging from the Pandemic and Reshaping Human Endeavors with Digital Technologies ICIS, vol. 6 (2023)

19. Mendling, J., Decker, G., Hull, R., Reijers, H.A., Weber, I.: How do machine learning, robotic process automation, and blockchains affect the human factor in business process management? Commun. Assoc. Inf. Syst. 43 (2018). Article 19

20. Rosemann, M., Ostern, N., Voss, M., Bandara, W.: Benevolent business processes - design guidelines beyond transactional value. In: Di Francescomarino, C., Burattin, A., Janiesch, C., Sadiq, S. (eds.) BPM 2023. LNCS, vol. 14159, pp. 447–464. Springer, Cham (2023). https://doi.org/10.1007/978-3-031-41620-0_26

21. Mirbabaie, M., Brendel, A.B., Hofeditz, L.: Ethics and AI in information systems research. Commun. Assoc. for Inf. Syst. 50 (2022)

22. Mirispelakotuwa, I., Syed, R., Wynn, M.T.: Is RPA causing process knowledge loss? Insights from RPA experts. In: Köpke, J., et al. (eds.) BPM 2023. LNBIP, vol. 491, pp. 73–88. Springer, Cham (2023). https://doi.org/10.1007/978-3-031-43433-4_5

23. O'Leary, D.E.: Using AI to read contracts. In: ICIS 2022 Proceedings, vol. 7 (2022)

24. Reis, L., Maier, C., Mattke, J., Creutzenberg, M., Weitzel, T.: Addressing User resistance would have prevented a healthcare AI project failure. MIS Q. Exec. 19(4) (2020). Article 8

25. Rinta-Kahila, T., Penttinen, E., Salovaara, A., Soliman, W., Ruissalo, J.: The vicious circles of skill erosion: a case study of cognitive automation. J. Assoc. Inf. Syst. 24(5), 1378–1412 (2023). https://doi.org/10.17705/1jais.00829

26. Rix, J., Hess, T.: Hello, Mate! Insights from the field on leveraging machine teammates in organizations. In: PACIS 2022 Proceedings, vol. 344 (2022)

27. Rothenberger, L., Fabian, B., Arunov, E.: Relevance of ethical guidelines for artificial intelligence – a survey and evaluation. In: Proceedings of the 27th European Conference on Information Systems (ECIS), Stockholm, Uppsala, Sweden, 8–14 June 2019 (2019). ISBN 978-1-7336325-0-8. Research-in-Progress Papers

28. Schaeper, T., Maibaum, F., Schulz, C., Foege, J.N.: Decoding the mindset: a neural network approach for analyzing CEO's digital strategy and its innovation implications. In: Rising like a Phoenix: Emerging from the Pandemic and Reshaping Human Endeavors with Digital Technologies ICIS 2023, vol. 10 (2023)

29. Seiffer, A., Gnewuch, U., Maedche, A.: Understanding employee responses to software robots: a systematic literature review. In: ICIS 2021 Proceedings, vol. 5 (2021)

30. Söderström, F., Johansson, B., Toll, D.: Automation as migration? – Identifying factors influencing adoption of RPA in local government (2021). In: ECIS 2021 Research-in-Progress Papers, vol. 38 (2021)

31. van der Aalst, W.M.P.: Hybrid Intelligence: to automate or not to automate, that is the question. Int. J. Inf. Syst. Proj. Manag. 9(2), 5–20 (2021)

32. van der Aalst, W.M.P., et al.: Views on the past, present, and future of business and information systems engineering. Bus. Inf. Syst. Eng. 60(6), 443–477 (2018)

33. van Giffen, B., Ludwig, H.: How Siemens democratized artificial intelligence. MIS Q. Exec. 22(1) (2023). Article 3

34. Vimalkumar, M., Gupta, A., Sharma, D., Dwivedi, Y.: Understanding the effect that task complexity has on automation potential and opacity: implications for algorithmic fairness. AIS Trans. Hum.-Comput. Interact. 13(1), 104–129 (2021)

35. Wibowo, S., Deng, H., Duan, S.: Understanding digital work and its use in organizations from a literature review. Pac. Asia J. Assoc. Inf. Syst. 14(3), 29–51 (2022)

36. Wolfswinkel, J., Furtmueller, E., Wilderom, C.: Using grounded theory as a method for rigorously reviewing literature. Eur. J. Inf. Syst. 22, 45–55 (2013)

Design Principles for Bots Supporting Case-Based Reasoning

Mikhail Monashev[1]([✉]) [iD] and Jan Mendling[2,3,4] [iD]

[1] Masaryk University, Lipová 41a, 602 00 Brno, Czech Republic
`mikhail.monashev@mail.muni.cz`
[2] Humboldt-Universität zu Berlin, Unter den Linden 6, 10099 Berlin, Germany
`jan.mendling@hu-berlin.de`
[3] Weizenbaum-Institut e. V., Hardenbergstraße 32, 10623 Berlin, Germany
[4] Wirtschaftsuniversität Wien, Welthandelsplatz 1, 1020 Vienna, Austria

Abstract. Robotic process automation (RPA) initiatives bring process improvements by substituting human actors with algorithms also referred to as "software robots" or "bots" in tasks that require interactions with information systems. RPA initiatives are only considered economically feasible when they focus on improving frequently repetitive sequences of activities in routine processes. We suggest that nonroutine processes that lack such patterns can also prove to be suitable RPA candidates focus of automation efforts is shifted from automating repetitive sequences of activities that are rarely found in nonroutine processes to automating meta-routines human actors engage in when performing nonroutine processes that usually exhibit such repetitiveness. In this paper, we turn our thinking towards case-based reasoning (CBR), a specific type of meta-routine of creative problem solving, which is deemed important in a wide range of nonroutine processes. Having formulated a design problem as applying RPA to support CBR, we take a first step towards finding its solution by formulating, evaluating, and revising a set of design principles capable of providing such support. These design principles represent a main theoretical contribution of our work, extending a body of design object knowledge on IS support of nonroutine processes. Our design principles are intended to guide the development of RPA systems promising effective maintenance of organizational memory, and time savings in searching for similar cases to inform solutions to present problems.

Keywords: Nonroutine Processes · Design Science · Design Principles · Case-Based Reasoning · Robotic Process Automation

1 Introduction

Organizations engage in robotic process automation (RPA) when they employ algorithms to substitute human actors in tasks that require interaction with information systems (IS) [1]. These algorithms, often referred to as "software robots" or "bots," operate faster than humans and make no human errors allowing organizations to improve their processes time- and quality-wise [2]. Since bot development is an effortful and costly enterprise,

© The Author(s), under exclusive license to Springer Nature Switzerland AG 2024
A. Marrella et al. (Eds.): BPM 2024 Forum, LNBIP 526, pp. 302–318, 2024.
https://doi.org/10.1007/978-3-031-70418-5_18

RPA initiatives become economically feasible only when they focus on automating processes that repeat frequently, follow predefined structure, and have unambiguous rules for making decisions [3]. These criteria are met by routine processes characterized by a restricted set of input and output varieties [4]. So far, most (if not all) RPA initiatives have been focusing on routine processes [5].

At the same time, utilizing bots to automate nonroutine processes such as patient treatment [6] or product development [7] is not considered economically feasible [2]. Nonroutine processes have an unrestricted set of input varieties [4], which leads to a lack of patterns in the sequences of their activities [6] that would reproduce frequently enough. However, we argue that such patterns are present in meta-routines human actors engage in when executing nonroutine processes. One such meta-routine is creative problem solving, to which we turn our thinking in this paper due to its importance for the successful execution of a wide range of nonroutine processes [7].

From the perspective of creative problem solving, figuring out an approach to transform an unpredictable set of inputs into a desired output in nonroutine processes can be considered as finding a solution to an ill-structured problem [8]. People solve such problems by applying various types of reasoning that can be mimicked by artificial intelligence (AI), including rule-based reasoning, model-based reasoning, and case-based reasoning (CBR) [9]. Among these types of reasoning, we decided to focus our attention on CBR. The decision was based on two considerations. First, CBR is alike analogical thinking, which is important to our discussion since it represents "one of the hallmarks of creative problem solving" [10, p. 113]. Second, CBR plays a central role in designing, planning, and diagnosing [11] that represent a vast majority of nonroutine processes. When applying CBR, people retrieve, reuse, revise, and retain [9] relevant pieces of knowledge about past solutions to solve present problems. These activities repeat from one instance of a nonroutine process to another and across processes, which makes them suitable RPA candidates [3].

These observations led us to engage in a design science research (DSR) project aimed at solving an exaptation problem [12] of using RPA (i.e., existing means) to assist human actors in CBR (i.e., new ends). This paper describes actions we took to formulate, evaluate, and revise design principles for a class of RPA systems aimed at providing such assistance. These design principles represent a main theoretical contribution of our work, extending a body of design object knowledge [13] on IS support of nonroutine processes [14–16]. We also deem the design principles proposed in this paper valuable for practice, as they guide the development of RPA systems that promise improvements in maintaining organizational memory [17] and decreased search time for exploring past cases [18].

2 Theoretical Background

2.1 Information Systems Support of Nonroutine Processes

Nonroutine processes, also referred to as "knowledge-intensive," are characterized as knowledge-driven, collaboration-oriented, unpredictable, emergent, goal-oriented, event-driven, rule-driven, and non-repeatable [6]. These characteristics make it challenging to improve nonroutine processes using conventional business process management

(BPM) methods and tools [19]. As a response, conventional methods are adapted, and new tools are developed to address the needs of nonroutine processes. Since knowledge creation plays a central role in the successful execution of nonroutine processes [7], research efforts aimed at supporting the execution of nonroutine processes tend to focus on improving the creativity of human actors through knowledge provision. To illustrate this tendency, we briefly present three studies that, in our view, significantly contributed to achieving this aim.

Müller-Wienbergen et al. [15] proposed a design theory on building IS that support convergent and divergent thinking by providing access to potentially relevant knowledge and sources of inspiration. These systems utilize knowledge from both internal and external sources while enabling ongoing extension of the former. They represent knowledge items from different perspectives by including them in multiple tag trees, enable convergent thinking by allowing humans to apply dynamic filters, and boost divergent thinking by suggesting unexpected links between knowledge items. Sarnikar and Deokar [16] approached the problem of knowledge provision in nonroutine processes from a different angle by suggesting design procedures for building process-based knowledge management systems. Unlike systems described by Müller-Wienbergen et al. [15] that focus on connecting knowledge items to knowledge items, systems described by Sarnikar and Deokar [16] specify knowledge requirements of tasks within nonroutine processes thus linking knowledge items with these tasks. Moreover, these systems link each knowledge item to its source and depict how knowledge is transferred within organizations.

Knowledge items also became a central concept in a study by Löhr et al. [14], who were concerned with adapting the process mining approach to the needs of nonroutine processes and proposed design principles for devising this adaption. For process mining to provide actionable insights into nonroutine processes, the authors suggested extending event logs with relevant unstructured data that would capture process goals and contextual factors connected to their execution. Moreover, they suggested considering and implementing prescriptive actions during process execution based on analyzing previously enacted process instances similar to the instance at hand. These suggestions were based on CBR, which we describe in more detail below.

2.2   Case-Based Reasoning

CBR is a reasoning methodology [20], which utilizes relevant fragments of knowledge about past solutions to solve present problems [11]. This makes CBR similar to analogical thinking, which allows people to learn about a new situation by relating it to a more familiar one from a different domain [10]. However, CBR is not the same as analogical thinking, since unlike the latter, it draws relevant knowledge from the domain to which a present problem belongs [9]. One difference leads to another as, unlike in analogical thinking, general domain knowledge plays a supportive role in CBR, while specific knowledge about previously enacted process instances comes to the forefront. The latter type of knowledge is often objectified in the form of case records or cases that usually contain details about (i) an initial situation in terms of the problem and the context in which it needed solving, (ii) a solution and approach to its derivation, and (iii) a resulting outcome in terms of success or failure [11].

As illustrated in Fig. 1, cases transform along the flow of a general CBR process. A process is triggered by receiving a notification about a new problem that needs to be solved. An initial description of a problem defines a new case, which is used as a reference for retrieving the most similar cases from the organizational memory. Then, a process actor creatively combines pieces of knowledge from similar cases to generate a solution for a present problem, thus transforming the status of the case from "new" to "solved." Upon further revision, the status of the case is changed to "successful" or "failed." Successful cases suggest solutions that can either be directly applied or adapted to address future problems, while failed cases help to anticipate potential issues in developing a solution before they occur [11]. At the final step, knowledge generated during the execution of the case is retained in the organizational memory, thus extending a pool of similar cases that can be reused in solving future problems.

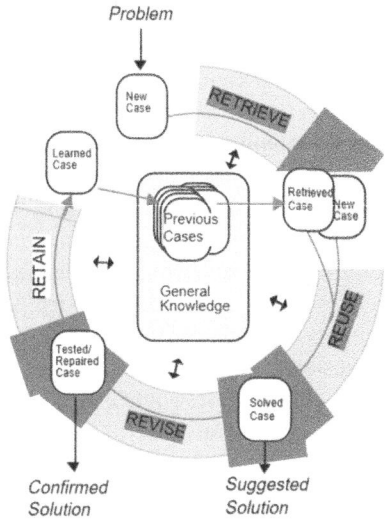

Fig. 1. The case-based reasoning cycle [9, p. 8].

Due to the limited capabilities of people in remembering relevant cases, and the constrained creative capacity of automation, multiple human-automation integration systems were built to improve CBR in nonroutine processes [11]. Many such systems, referred to as "case-based reasoners," were designed in the 1980s-1990s. For example, the GREBE system used "detailed knowledge of the facts and reasoning of specific past [legal] cases, together with legal rules and common-sense knowledge, to determine and justify the legal consequences of new cases" [21, p. 797]. Another example is the PERSUADER system, which applied CBR to suggest compromise terms of a labour contract, which would satisfy parties with conflicting interests [22]. Review of these and other examples of case-based reasoners reported by Aamodt and Plaza [9] reveals the tendency of their designers to automate retrieve, reuse, and retain activities of the CBR cycle, while letting humans revise automatically generated outputs.

2.3 Robotic Process Automation

Case-based reasoners can be built using any technology [20], and RPA does not make an exception. Various types of bots can be applied to automate activities forming the CBR cycle. Bots differ in their ability to handle structured and unstructured data as well as in the amount of human control required to ensure their proper functioning [2].

When it comes to data handling capabilities, bots are classified into basic and cognitive ones. On the one hand, basic bots strictly follow predefined rules and outperform humans when it comes to modifying existing data structures. For instance, basic bots in Wikipedia perform multiple such modifications fixing files, creating hyperlinks, tagging articles, updating statistics, and archiving content [23]. Nevertheless, basic bots can only operate with structured data inputs, which doesn't allow companies to use them to automate perceptual and judgment-based tasks that rely on human cognition. On the other hand, cognitive bots strengthened by artificial intelligence (AI) can work with unstructured data inputs thus compensating for limitations of basic bots. Examples of tasks performed by cognitive bots include extracting and structuring information from various media including audio, text, or images, as well as making various sorts of predictions and recommendations [24].

When it comes to the extent of human control required for their operation, bots are classified into unattended and attended ones. A bot employed in the retirement scheme switching process, which takes information from customers' requests and updates relevant databases [25] exemplifies an unattended bot, which works independently from a human supervisor. Bots capturing customer feedback [5] and running regular cleaning of cash files [23] represent other examples of unattended bots. At the same time, a human triggers and controls the outputs of a bot employed in the credit alternation approval process, which searches for customer details across multiple IS [25]. Bots that provide suggestions to users [23] or ask users to review requests [5] represent other examples of attended bots relying on human supervision.

3 Methodology

Complex DSR projects, such as the one we have engaged in to solve an exaptation problem of using RPA to support CBR, normally consist of multiple self-contained segments referred to as "design echelons" [26]. In this paper, we describe activities we conducted within the requirements and objectives definition echelon, which focuses on producing design objectives validated against multiple criteria including their applicability to the design problem and feasibility. Design objectives or design principles represent prescriptive statements that describe a class of systems aimed at solving a specific problem [27] by linking generalized design requirements to generalized design features [26]. To produce a set of validated design principles for RPA systems supporting CBR, we, first, derived them from the body of justificatory knowledge [27], then evaluated them through expert interviews [26], and, finally, revised design principles based on the insights gained from interviews.

We started by defining four design requirements for a class of systems supporting CBR. To do so, we used the CBR cycle [9] as a kernel theory, which would explain why these systems are expected to work [12]. Then, drawing from justificatory knowledge [27], we defined four design principles aimed at fulfilling our design requirements by guiding the development of RPA systems. We proceeded by evaluating an initial set of design principles through semi-structured interviews with five practitioners experienced in designing bots and conducting RPA initiatives (hereafter referred to as experts E1, E2, E3, E4, and E5). The evaluation was guided by four criteria in the light reusability evaluation framework proposed by Iivari et al. [28], as operationalized in Table 1. Interviews were conducted in the period from the 26th of January to the 20th of February, 2024; they lasted between 47 and 59 min.

Table 1. Operationalization of reusability evaluation criteria based on [28].

Evaluation criterium	Operationalization
Accessibility	Q1. Do you find the provided description easy to understand? Are there any fragments in the table that caused confusion?
Novelty	Q2. Do you find the idea behind the design principle new to yourself?
Actability	Q3. Do you think that the design principle can realistically be implemented in practice?
Appropriate guidance	Q4. Do you think that the design principle provides sufficient guidance for designing bots able to support problem-solving tasks? Q5. Do you think that the design principle would provide bot engineers sufficient design freedom when designing the bots for problem-solving support?

Interviews were transcribed and analyzed using qualitative coding, which aimed at achieving two objectives. Firstly, we characterized evaluation criteria in degree (from low to moderate to high) by combining structural and magnitude coding procedures [29]. Secondly, we obtained additional insights from our data by employing grounded theory method procedures of open and selective coding. This resulted in the emergence of three categories that informed the revision of an initial set of design principles: (i) boundary conditions, (ii) limitations, and (iii) suggested improvements. Finally, we revised the initially formulated design principles by reflecting upon the insights gained from interviews.

4 Formulation of Design Principles

Considering that in our DSR project, we aim to prescribe guidelines for designing RPA systems aimed at supporting CBR in nonroutine processes, we started by formulating design requirements for these systems. Reflecting on the literature reviewed in Sect. 2, we concluded that supporting CBR, essentially, means enabling activities that form the CBR cycle [9]. This led us to formulate four design requirements for a general class of IS supporting CBR as follows:

Design Requirement DR1 – Enable retention of cases: Make bots retain information about implemented cases in the organizational memory.

Design Requirement DR2 – Enable retrieval of similar cases: Make bots retrieve cases similar to the case at hand.

Design Requirement DR3 – Enable reuse of relevant cases: Allow human actors to establish the relevance of similar cases to then reuse them in solving the case at hand.

Design Requirement DR4 – Enable revision of cases: Allow humans to reflect on the success or failure of implemented solutions.

We proceeded by defining design principles, which would allow designers to fulfill the above requirements in a specific class of IS, that is RPA systems. Figure 2 summarizes the above discussion by linking design principles with design requirements and outlining justificatory knowledge used in formulating the former.

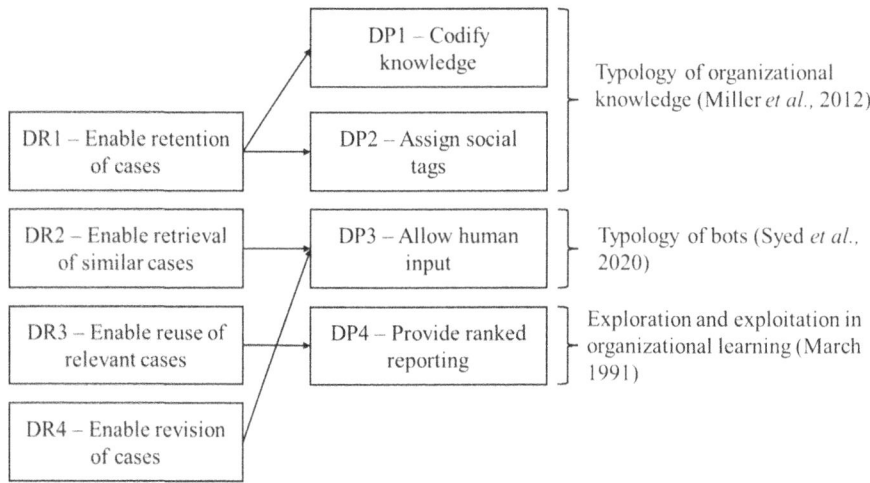

Fig. 2. Design requirements and initial design principles towards RPA systems supporting CBR.

When thinking about ways of enabling the retention of cases, and thus fulfilling design requirement DR1, we considered that knowledge retained in organizational memory is classified into three types: (i) declarative, (ii) procedural, and (iii) transactive [30]. Declarative knowledge, that is know-what, includes descriptions, stories, and propositions. When suggesting an extension of event logs of nonroutine processes with unstructured data, Löhr et al. [14] were referring to this type of knowledge. By applying declarative knowledge, people interpret situations and infer appropriate responses [30]. Historically, two types of bots were used to retain declarative knowledge: (i) unattended bots responsible for content archiving [23], and (ii) attended bots allowing users to manually enter content in the UI [5]. Correspondingly, we formulate the following design principle:

Design Principle DP1 – Codify knowledge: Record declarative knowledge produced by bots and human actors during process execution in relevant databases. To do so, specify rules that need to be followed by bots when entering data into relevant databases.

Procedural knowledge, that is know-how, is "the capacity for individual skillful action" demonstrated by individuals as they execute processes [30, p. 5]. Procedural knowledge contains a significant tacit component and thus cannot be codified. When the procedural knowledge of a single individual does not suffice to proceed with the process execution, he/she utilizes transactive knowledge, that is know-who, to access expertise possessed by others. When talking about knowledge sources, Sarnikar and Deokar [16] also considered human actors as those who retain tacit procedural knowledge about process executions. Applying social software functionalities such as social tagging is commonly considered to be an effective means of capturing and communicating transactive knowledge [31]. Tagging in general is a standard task performed by bots [23]. Therefore, we formulate the following design principle:

DP2 – Assign social tags: Make it possible for human actors to track relevant cases back to the users involved in their execution. To do so, ensure bots retain information about users executing nonroutine process instances, which will allow them to identify these users in the future.

In the CBR cycle, a new problem triggers the retrieval of one or more previously experienced cases that might be relevant in the present context [9]. Cases are retrieved from organizational memory. Therefore, retrieval is enabled by getting access to the organizational memory. RPA use cases aimed at data retrieval were reported, for instance, in the finance domain, where attended bots [2] queried customer details from a database based on the input provided by the user to the specific field in the UI [25]. Correspondingly, we formulate the following design principle:

DP3 – Allow human input: Make it possible for human actors to guide bots in querying for similar cases in databases. To do so, ensure that the user interface (UI) contains fields, where the user can enter input data that are later used by bots to execute search queries; if similar cases are identified, ensure that bots display them on the UI.

Upon retrieval, relevant cases are reused to solve present ones. In situations, when no similar cases were executed in the past, individuals generate new ideas by intuiting, experimenting, or acquiring knowledge from external sources [32]. Neither intuiting nor experimenting is automatable [11], while knowledge acquisition is not related to CBR, and therefore, is out of the scope of our paper. When one similar case is retrieved, it needs to be displayed on the UI. Alternatively, when multiple similar cases are retrieved, more effective reuse would be achieved by decreasing search efforts involved in the exploration of similar cases [18]. These observations lead us to formulate the design principle as follows:

DP4 – Provide ranked reporting: Display reports outlining similar cases on the UI of an RPA system. If multiple similar cases are identified, ensure that bots represent them in the order of decreasing similarity.

The last step in the CBR cycle is revision. It implies evaluation of the case solution generated by reuse and retaining it as is if the solution was successful or repairing the case solution using domain-specific knowledge [9]. The revision itself does not have to be enabled by the system, however, the system needs to enable users to enter the results of their revision by classifying cases either as success or failure. Design principle DP3, which enables human input, is deemed suitable to enable revision in the CBR cycle.

5   Evaluation of Design Principles

Once an initial set of design principles was formulated, we conducted semi-structured interviews with five practitioners experienced in designing bots and conducting RPA initiatives. The goal of these interviews was twofold. Firstly, we aimed to evaluate the reusability of the initially formulated design principles in terms of accessibility, novelty, actability, and appropriate guidance [28]. Secondly, we sought insights that would allow us to increase the reusability of design principles.

All experts neither found any of the descriptions of initial design principles hard to understand nor reported being confused regarding specific parts of provided descriptions, thus confirming their high accessibility. A consensus was also reached regarding the novelty of design principles. All experts characterized each design principle taken separately as not novel, confirming that we were adopting existing RPA solutions. Simultaneously, they highlighted the novelty of the idea of combining these design principles to support CBR in nonroutine processes using RPA, confirming the applicability of our design principles to solving an exaptation problem [12].

Experts differed in their opinions regarding the actability of initial design principles. All principles were noted as being of moderate or higher in terms of actability. However, the actability of some design principles was remarked to be achieved within specific boundary conditions. Experts agreed that the actability of design principles would be higher with structured data while operating with unstructured data would require strengthening bots with AI. This consideration goes in line with the theoretical distinction of bots into basic and cognitive [2] as outlined in Sect. 2.3.

When discussing design principle **DP1 – Codify knowledge**, experts pointed out different approaches to data retention. According to expert E2, in some cases, data retention can be automated. This is true, for instance, for data on user actions that he/she executed in respective IS. In this regard, expert E4 mentioned an RPA tool, which documents all the actions a user performs on the UI. The problem arises when it comes to the manual entry of case-related data into the system. Expert E2, for instance, mentioned that data on meetings and calls as well as their results is normally not captured automatically. Expert E1 also mentioned the need to ensure that people fill in the required fields, by, for instance, making this procedure mandatory.

Design principle **DP2 – Assign social tags** was regarded by all experts as easily implementable with RPA. Experts E2 and E5 cautioned us about GDPR considerations related to storing sensitive personal data. Moreover, expert E2 mentioned that principle DP2 does not explicitly specify what is happening if the person involved in the execution of the past case has left the organization.

Design principle **DP3 – Allow human input** was discussed with respect to sources and forms of data input. It was noted that in some cases, data input can be achieved by unattended bot and thus will not require user intervention, decreasing human workload. Expert E4 recalled use cases from the accounting domain, where the preparation of the input data was mostly done "outside RPA" in databases and Excel files, and later automatically passed to the bot. In line with that, expert E2 remarked that "the context where the use is in can be passed to the bot directly without going through the UI." Expert E4 also mentioned that in addition to providing pre-defined fields, structured data input can be enabled by providing a drop-down list with pre-defined options. Moreover, expert E5 remarked that providing input in a free format of plain text would require the implementation of cognitive bots.

When discussing design principle **DP4 – Provide ranked reporting**, experts noted variability of technical feasibility depending on the degree to which data are structured. In the case of structured data, the implementation of the design principle DP4 has been deemed technically feasible. Moreover, Expert E3 noted that "the technical feasibility would decrease in proportion to the increased complexity of the similarity ranking algorithms." In the case of unstructured data, experts agreed that implementing principle DP4 with basic bots alone won't be technically feasible, whereas cognitive bots have been deemed suitable for such implementation. Specifically, expert E1 noted the importance of employing machine learning algorithms for automatically preparing reports with unstructured data. Experts E2 and E5 recalled use cases of duplicate checking and invoice matching in finance. Additionally, expert E4 cautioned us that regardless of data type, bots need to be granted access rights to retrieve these data from relevant databases.

As for the appropriate guidance, in its current specifications, some design principles gave rise to alternative interpretations, and imagining different scenarios, as described above, signalizing that design principles were formulated broadly to provide sufficient design freedom [28]. At the same time, experts mostly agreed that each design principle offered sufficient guidance for implementation. The only exception was the design principle **DP4 – Provide ranked reporting**, regarding which experts E2, E3, and E4 remarked that mechanisms and criteria of ranking could have been defined more precisely to allow better guidance.

6   Revision of Design Principles

We used insights gained from expert interviews to revise the initial set of design principles. Table 2 provides an overview of changes that we made in each design principle while justifying these changes.

Table 2. Changes made during the revision of design principles.

ID	Design principle	Changes made and their justification
DP1.1	Codify knowledge automatically	Reassembles former DP1 – Codify knowledge
DP1.2	Allow manual knowledge codification	A new sub-principle, prescribing a possibility of human actors to codify knowledge manually (E1, E2) – **New mechanism:** Users can create a log file and add there notes as a process instance emerges [33, 34] – New mechanism: Bots nudge users to maintain the log file by sending e-mail notifications (E1, [23])
DP2	Assign social tags	– **New constraint:** Tracking relevant cases back to the users involved in their execution is only possible when these users are still available for knowledge sharing (E2) – **New constraint:** Bots retain only non-sensitive contact details of these users to comply with GDPR (E2, E5)
DP3	Allow human input	– **New mechanism:** Consider an option of allowing users to select standardized input parameters from (drop-down) lists (E4) – **New mechanism:** Consider an option of providing a free format data input (E1, E5) – **New mechanism:** Allow users to assign weights to specified input parameters for the purposes of similarity comparison (Revision of design principle DP4)
DP4	Provide ranked reporting	– **New mechanism:** Bots need to be granted sufficient access right to retrieve data from relevant databases (E4) – **New mechanism:** Similarity is estimated against input parameters, considering weights assigned to these parameters by users (E2, E3, E4) – **New mechanism:** Retrieval of unstructured data is ensured by AI plug-ins (Revision of design principle DP3)

Upon reflecting on insights gained in interviews, we have decided to break down the design principle **DP1 – Codify knowledge** into two sub-principles. The first sub-principle **DP1.1 – Codify knowledge automatically** reassembles aims and mechanisms considered in the initial formulation of design principle DP1 that were validated during expert interviews. The sub-principle DP1.1 is thus formulated as follows:

Design Principle DP1.1 – Codify knowledge automatically (former DP1 – Codify knowledge): Record declarative knowledge produced by bots and human actors during process execution in relevant databases. To do so, specify rules that need to be followed by bots when entering data into relevant databases.

The second sub-principle **DP1.2 – Codify knowledge manually** has been formulated to react to the comments of experts E1 and E2 regarding the inability to capture all the knowledge relevant to the execution of nonroutine processes by bots alone. Drawing from literature, we considered allowing users to take notes on the process execution as it unfolds by introducing features for creating a log file, a practice proposed for journaling design science [34]. The log file entries, among other things, might include (i) information on events relevant to the execution, such as meetings and calls, as proposed by expert E2, (ii) a summary of event outputs, and (iii) personal reflections on the process by a human actor similar to analytical memos used in qualitative coding [29]. Next, our data suggested that the class of systems needs to have mechanisms in place to ensure that manual data entries are provided. Possible mechanisms increasing the chances of knowledge being codified manually include making this process mandatory, as suggested by expert E1, or sending notifications nudging users to enter the data [23]. Considering all the above, the sub-principle DP1.2 is formulated as follows:

Design Principle DP1.2 – Allow manual data codification (newly added): Make it possible for human actors to codify knowledge relevant to the case execution. To do so, ensure that users can create a log file and maintain it by recording relevant events and reflections as the process instance emerges. Moreover, nudge users to maintain the log file, for instance, by sending e-mail notifications.

Design principle **DP2 – Assign social tags** was revised to reflect the GDPR considerations as well as the availability of users involved in past process executions. It was specified that the retention of user-related data is limited to non-sensitive contact details, such as work e-mails. Moreover, it was emphasized that if the user is no longer available, for example, if he/she left the organization, his/her contact details are not reported. In its revised version, design principle DP2 is formulated as follows:

DP2 – Assign social tags (revised): Make it possible for human actors to track relevant cases back to the users involved in their execution if these users are still available for knowledge sharing. To do so, ensure bots retain non-sensitive contact details of users executing nonroutine process instances, which will allow them to identify these users in the future.

Revision of the design principle **DP3 – Allow human input** concerned specifying additional mechanisms for entering data in different formats. We followed a suggestion by expert E2 and considered an automatic pre-fill of an input form with data from relevant systems. Following a suggestion by expert E4, we considered (drop-down) lists as an additional mechanism for providing data in a standardized format. Then, we also considered the option of providing free format input, for example, as plain text, thus incorporating suggestions of experts E1 and E5. Finally, following up revision of design principle **DP4 – Provide ranking reporting**, we considered an option to specify weights of input parameters to rank retrieved cases by similarity. In its revised form, design principle DP3 is formulated as follows:

> DP3 – Allow human input: Make it possible for human actors to guide bots in querying for similar cases in databases. To do so, ensure that UI contains mechanisms for capturing input data in different formats, including fields and dropdown lists for capturing standardized data, and fields for entering free format data. Moreover, ensure these fields are automatically pre-filled with data from relevant systems, and allow users to assign weights to specified input parameters for case similarity comparison.

When revising the design principle **DP4 – Provide ranked reporting**, our focus was twofold. Firstly, we specified additional mechanisms for retrieving structured data, that is granting bots with sufficient access rights. Secondly, we provided additional details on mechanisms and criteria for ranking in situations when two or more similar cases are identified. As a general guideline, regardless of data type, cases need to be ranked by similarity to the input parameters and considering the weight of these parameters if specified by users. Finally, we specified that as in the case of the design principle **DP3 – Allow human input**, mechanisms enabling matching cases by similarity within unstructured data models will require the application of cognitive bots. In its revised form, design principle DP4 is formulated as follows:

> DP4 – Provide ranked reporting (revised): Display reports outlining similar cases on the UI of an RPA system. To do so, ensure that bots are granted sufficient access rights allowing them to retrieve data from relevant databases. If multiple similar cases are identified, ensure that bots represent them in the order of decreasing similarity to the input parameters considering the weights users assign to these parameters. Moreover, ensure that AI is plugged in to enable retrieval of unstructured data.

7   Discussion

In this paper, we set out to produce a set of validated design principles for RPA systems supporting CBR in nonroutine processes. We achieved this goal through the initial formulation of design principles based on several sources of justificatory knowledge and their subsequent revision based on the insights gained from five expert interviews. Our research resulted in a series of theoretical contributions and practical implications.

A main theoretical contribution of our work is a set of validated design principles for RPA systems supporting CBR in nonroutine processes, by which we extend a body of theoretical design object knowledge [13] on IS support of nonroutine processes. A class of systems described by our design principles leverages and extends an internal body of knowledge thus going in line with the implementation principles suggested by Müller-Wienbergen et al. [15]. However, unlike in the design theory by Müller-Wienbergen et al. [15], search for relevant knowledge is automated using RPA. We also adapt the idea by Sarnikar and Deokar [16] when proposing the design principle **DP2 – Assign social tags** to capture the source of knowledge, however, our adaption is based on different justificatory knowledge as it leverages the concept of transactive memory enabling knowledge transfer. Moreover, we align with Löhr et al. [14], when using the CBR cycle as a kernel theory for our design requirements, however, we go beyond the suggestions of these authors by proposing detailed mechanisms of enabling activities forming this cycle.

In addition to advancing theoretical design knowledge, our design principles have the potential to bring value to practice. Four design principles taken together are intended to guide practitioners in developing RPA systems supporting CBR in nonroutine processes, promising several benefits. Once implemented in development, principles **DP1.1 – Codify knowledge automatically**, **DP1.2 – Allow manual knowledge codification**, and **DP2 – Assign social tags** would allow organizations to effectively develop a body of declarative and transactive knowledge about nonroutine processes in a semi-automatic fashion while nudging human actors to engage in manual knowledge codification. This is especially valuable considering that manual knowledge codification is often regarded as a hard task and therefore often neglected resulting in fragmented process knowledge [17]. Implementing principles **DP3 – Allow human input** and **DP4 – Provide ranked reporting** would allow organizations to correctly recognize problems as novel (if no similar cases are found) or potentially reproduced (if some similar cases are found) thus avoiding bureaucracy and chaos traps that decrease process performance [35]. Moreover, implementation of these principles would decrease the search time required for exploring similar cases [18].

As this paper describes only the requirements and objectives definition echelon of an ongoing DSR project, future research efforts will focus on building upon results achieved so far. The requirements and objectives definition echelon is followed by the design and development echelon, which focuses on producing a validated design of the artifact [26]. In line with the idea of design patterns [36], we aim to build and evaluate at least three prototypes of envisioned RPA systems to support CBR in three different nonroutine processes. In addition to developing the project chronologically by conducting activities within further design echelons, our project can also benefit from considering alternative ends and means. So far, we have solely focused on CBR, while considering how RPA can be used to support other types of reasoning [9] might lead to the development of a design theory on building RPA systems that support creative problem solving. At the same time, one of the potential means that was mentioned multiple times in expert interviews and is currently causing a growing interest in the BPM community [37] is generative AI. Considering how generative AI can support multiple types of reasoning might result in the development of a design theory on building a general class of IS that supports creative problem solving.

8 Conclusion

RPA initiatives bring significant benefits to routine processes improving them along time- and quality dimensions, while attempts to use bots to automate nonroutine processes are not made. In this paper, we suggested that nonroutine processes can prove to be suitable candidates for economically feasible automation if the focus is shifted from finding patterns in their sequences of activities to meta-routine human actors engage in when executing these processes. We then showcased our suggestion to be generative by developing design principles for RPA systems supporting the most important type of reasoning, CBR, in the most important of such meta-routines, creative problem solving. Evaluation of our design principles by experts in designing bots and conducting RPA initiatives showed them to be technically feasible and potentially useful for practice. Continuation of work we started as well as further efforts to support creative problem solving and other meta-routines important for performing nonroutine processes is encouraged to bring RPA benefits to nonroutine processes.

Acknowledgements. The research of Mikhail Monashev was supported by Masaryk University internal grant No. MUNI/A/1400/2023, entitled "An Integration of Business Process Management and Knowledge Management." The research of Jan Mendling was supported by the Einstein Foundation Berlin under grant EPP-2019-524, by the German Federal Ministry of Education and Research under grant 16DII133, and by Deutsche Forschungsgemeinschaft under grants 496119880 (VisualMine) and 531115272 (ProImpact).

References

1. van der Aalst, W.M.P., Bichler, M., Heinzl, A.: Robotic process automation. Bus. Inf. Syst. Eng. **60**, 269–272 (2018). https://doi.org/10.1007/s12599-018-0542-4
2. Syed, R., Suriadi, S., Adams, M., et al.: Robotic process automation: contemporary themes and challenges. Comput. Ind. **115**, 103162 (2020). https://doi.org/10.1016/j.compind.2019.103162
3. Viehhauser, J., Doerr, M.: Digging for gold in RPA projects – a quantifiable method to identify and prioritize suitable RPA process candidates. In: La Rosa, M., Sadiq, S., Teniente, E. (eds.) CAiSE 2021. LNCS, vol. 12751, pp. 313–327. Springer, Cham (2021). https://doi.org/10.1007/978-3-030-79382-1_19
4. Lillrank, P.: The quality of standard, routine and nonroutine processes. Organ. Stud. **24**, 215–233 (2003). https://doi.org/10.1177/0170840603024002344
5. Storey, M.-A., Zagalsky, A.: Disrupting developer productivity one bot at a time. In: Proceedings of the 2016 24th ACM SIGSOFT International Symposium on Foundations of Software Engineering, pp. 928–931. Association for Computing Machinery, New York (2016)
6. Di Ciccio, C., Marrella, A., Russo, A.: Knowledge-intensive processes: characteristics, requirements and analysis of contemporary approaches. J. Data Semant. **4**, 29–57 (2015). https://doi.org/10.1007/s13740-014-0038-4
7. Aureli, S., Giampaoli, D., Ciambotti, M., Bontis, N.: Key factors that improve knowledge-intensive business processes which lead to competitive advantage. Bus. Process. Manag. J. (2019). https://doi.org/10.1108/BPMJ-06-2017-0168
8. Simon, H.A.: The structure of ill structured problems. Artif. Intell. **4**, 181–201 (1973). https://doi.org/10.1016/0004-3702(73)90011-8
9. Aamodt, A., Plaza, E.: Case-based reasoning: foundational issues, methodological variations, and system approaches. AI Commun. **7**, 39–59 (1994). https://doi.org/10.3233/AIC-1994-7104
10. Thomas, J.C., Lee, A., Danis, C.: Enhancing creative design via software tools. Commun. ACM **45**, 112–115 (2002). https://doi.org/10.1145/570907.570944

11. Kolodner, J.L.: An introduction to case-based reasoning. Artif. Intell. Rev. **6**, 3–34 (1992). https://doi.org/10.1007/BF00155578
12. Gregor, S., Hevner, A.R.: Positioning and presenting design science research for maximum impact. MIS Q. (2013). https://dl.acm.org/doi/https://doi.org/10.25300/MISQ/2013/37.2.01. Accessed 5 Nov 2023
13. Seckler, C., Mauer, R., Vom Brocke, J.: Design science in entrepreneurship: conceptual foundations and guiding principles. J. Bus. Ventur. Des. **1**, 100004 (2021). https://doi.org/10.1016/j.jbvd.2022.100004
14. Löhr, B., Brennig, K., Bartelheimer, C., et al.: Process mining of knowledge-intensive processes: an action design research study in manufacturing. In: Di Ciccio, C., Dijkman, R., del Río Ortega, A., Rinderle-Ma, S. (eds.) BPM 2022. LNCS, vol. 13420, pp. 251–267. Springer, Cham (2022). https://doi.org/10.1007/978-3-031-16103-2_18
15. Müller-Wienbergen, F., Müller, O., Seidel, S., Becker, J.: Leaving the beaten tracks in creative work – a design theory for systems that support convergent and divergent thinking. J. Assoc. Inf. Syst. **12** (2011). https://doi.org/10.17705/1jais.00280
16. Sarnikar, S., Deokar, A.V.: A design approach for process-based knowledge management systems. J. Knowl. Manag. **21**, 693–717 (2017). https://doi.org/10.1108/JKM-09-2016-0376
17. Cabrera, A., Cabrera, E.F.: Knowledge-sharing dilemmas. Organ. Stud. **23**, 687–710 (2002). https://doi.org/10.1177/0170840602235001
18. March, J.G.: Exploration and exploitation in organizational learning. Organ. Sci. **2**, 71–87 (1991)
19. França, J.B.S., Baião, F.A., Santoro, F.M.: Towards characterizing knowledge intensive processes. In: Proceedings of the 2012 IEEE 16th International Conference on Computer Supported Cooperative Work in Design (CSCWD), pp. 497–504 (2012)
20. Watson, I.: Case-based reasoning is a methodology not a technology. Knowl.-Based Syst. **12**, 303–308 (1999). https://doi.org/10.1016/S0950-7051(99)00020-9
21. Branting, L.K.: Building explanations from rules and structured cases. Int. J. Man-Mach. Stud. **34**, 797–837 (1991). https://doi.org/10.1016/0020-7373(91)90012-V
22. Sycara, K.: Resolving Goal Conflicts via Negotiation (1988)
23. Zheng, L., Albano, C.M., Vora, N.M., et al.: The roles bots play in Wikipedia. Proc. ACM Hum.-Comput. Interact. **3**, 1–20 (2019). https://doi.org/10.1145/3359317
24. Schatsky, D., Muraskin, C.: Robotic process automation. Deloitte Insights (2021). https://www2.deloitte.com/content/www/xe/en/insights/focus/signals-for-strategists/cognitive-ent erprise-robotic-process-automation.html. Accessed 29 Oct 2023
25. Techatassanasoontorn, A.A., Waizenegger, L., Doolin, B.: When Harry, the human, met Sally, the software robot: metaphorical sensemaking and sensegiving around an emergent digital technology. J. Inf. Technol. (2023). 02683962231157426. https://doi.org/10.1177/026839622 31157426
26. Tuunanen, T., Winter, R., Vom Brocke, J.: Dealing with complexity in design science research: a methodology using design echelons. MIS Q. **48**, 427–458 (2024). https://doi.org/10.25300/MISQ/2023/16700
27. Gregor, S., Kruse, L.C., Seidel, S.: Research perspectives: the anatomy of a design principle. J. Assoc. Inf. Syst. **21** (2020). https://doi.org/10.17705/1jais.00649
28. Iivari, J., Rotvit Perlt Hansen, M., Haj-Bolouri, A.: A proposal for minimum reusability evaluation of design principles. Eur. J. Inf. Syst. **30**, 286–303 (2021). https://doi.org/10.1080/0960085X.2020.1793697
29. Saldaña, J.: The Coding Manual for Qualitative Researchers. Sage Publications Ltd., Thousand Oaks (2009)
30. Miller, K.D., Pentland, B.T., Choi, S.: Dynamics of performing and remembering organizational routines. J. Manag. Stud. **49**, 1536–1558 (2012). https://doi.org/10.1111/j.1467-6486.2012.01062.x

31. Erol, S., Granitzer, M., Happ, S., et al.: Combining BPM and social software: contradiction or chance? J. Softw. Maint. Evol. Res. Pract. **22**, 449–476 (2010). https://doi.org/10.1002/smr.460

32. Zietsma, C., Winn, M., Branzei, O., Vertinsky, I.: The war of the woods: facilitators and impediments of organizational learning processes. Br. J. Manag. **13**, S61–S74 (2002). https://doi.org/10.1111/1467-8551.13.s2.6

33. Urquhart, C.: Grounded Theory for Qualitative Research: A Practical Guide, 2nd edn. SAGE Publications Ltd., Los Angeles (2022)

34. Vom Brocke, J., Gau, M., Mädche, A.: Journaling the design science research process. Transparency about the making of design knowledge. In: Chandra Kruse, L., Seidel, S., Hausvik, G.I. (eds.) DESRIST 2021. LNISA, vol. 12807, pp. 131–136. Springer, Cham (2021). https://doi.org/10.1007/978-3-030-82405-1_15

35. Brown, S.L., Eisenhardt, K.M.: Competing on the Edge: Strategy as Structured Chaos. Harvard Business School Press, Boston (1998)

36. Wurm, B., Goel, K., Bandara, W., Rosemann, M.: Design patterns for business process individualization. In: Hildebrandt, T., van Dongen, B.F., Röglinger, M., Mendling, J. (eds.) BPM 2019. LNISA, vol. 11675, pp. 370–385. Springer, Cham (2019). https://doi.org/10.1007/978-3-030-26619-6_24

37. Vidgof, M., Wurm, B., Mendling, J.: The Impact of Process Complexity on Process Performance: A Study Using Event Log Data (2023)

A Nascent Taxonomy of Machine Learning in Intelligent Robotic Process Automation

Lukas Laakmann[iD], Seyyid A. Ciftci[iD], and Christian Janiesch[✉][iD]

TU Dortmund University, Dortmund, Germany
{lukas.laakmann,seyyid.ciftci,
christian.janiesch}@tu-dortmund.de

Abstract. Robotic process automation (RPA) is a lightweight approach to automating business processes using software robots that emulate user actions at the graphical user interface level. While RPA has gained popularity for its cost-effective and timely automation of rule-based, well-structured tasks, its symbolic nature has inherent limitations when approaching more complex tasks currently performed by human agents. Machine learning concepts enabling intelligent RPA provide an opportunity to broaden the range of automatable tasks. In this paper, we conduct a literature review to explore the connections between RPA and machine learning and organize the joint concept intelligent RPA into a taxonomy. Our taxonomy comprises the two meta-characteristics RPA-ML integration and RPA-ML interaction. Together, they comprise eight dimensions: architecture and ecosystem, capabilities, data basis, intelligence level, and technical depth of integration as well as deployment environment, lifecycle phase, and user-robot relation.

Keywords: Robotic process automation · machine learning · literature review · taxonomy

1 Introduction

In recent years, the field of process automation has undergone a remarkable transformation, with robotic process automation (RPA) revolutionizing how businesses across various industries manage and streamline their operations in a timely fashion. RPA, with its ability to automate repetitive, rule-based tasks, has emerged as a crucial tool for enhancing efficiency, reducing costs, and improving overall productivity [40]. Despite its widespread market acceptance, RPA has limitations. Traditional RPA robots are confined to executing repetitive rule-based processes as configured by recordings or flowcharts. They lack the flexibility to adapt to changes in context or process unstructured data [36].

Concurrently, machine learning (ML) has witnessed unprecedented advancements, offering powerful capabilities for example in data analysis, pattern recognition, and text generation [11,19]. The integration of RPA and ML presents a

A. Marrella et al. (Eds.): BPM 2024 Forum, LNBIP 526, pp. 319–336, 2024.
https://doi.org/10.1007/978-3-031-70418-5_19

logical next step, promising more comprehensive and intelligent levels of automation in business processes [17], for example to enable sharing data among multiple parties within neutral data trust models.

As both, RPA and ML technologies, continue to evolve and possibly converge, a need arises for a systematic taxonomy that can effectively categorize and delineate the multifaceted landscape of RPA implementations that harness ML concepts. The absence of a standardized frame for discussion hinders not only the clarity of communication within and across the RPA and ML communities but also the effective evaluation, comparison, and selection of RPA solutions by organizations seeking to intelligentize their automation journey. By providing a structured and standardized vocabulary, our taxonomy will enable practitioners, researchers, and organizations to communicate more precisely about their RPA-ML initiatives, fostering a deeper understanding of the field's potential and limitations.

With this article, we seek to address this gap in the literature by advocating for the development of a nascent taxonomy for RPA with ML integration. It will help to systematize the new and substantive developments in RPA and ML and - foremost - it synthesizes knowledge from different lines of research. It may even assist in identifying missing or neglected themes. Consequently, we pose our research question as follows:

RQ: *"What are the dimensions and characteristics of a taxonomy of RPA and ML integration and use?"*

To answer this question, we conducted a hermeneutic literature review [5] and developed a taxonomy based on a sample of 45 publications following a iterative taxonomy development process [29]. The resulting nascent taxonomy is useful, comprehensive, extensible, and explaining for IS researchers as well as for practitioners.

Our paper is organized as follows: In Sect. 2, we introduce common concepts of RPA and ML separately. Section 3 contains an outline of our research methodology. We present the taxonomy in Sect. 4 before we close with a brief discussion and summary.

2 Background and Related Literature

2.1 Robotic Process Automation

We characterize RPA as an umbrella term for automation technology operating on the user interface (UI) level, evolving from the traditional screen scraper method that recorded and replicated precise user mouse interactions. It uses software robots to mimic human interaction with legacy software so that there are no adjustments necessary to existing software [17]. RPA addresses the limitation of UI shifts by processing user interactions at a logical UI element level, rather than based on coordinates [31]. RPA also follows in the tradition of older lightweight automation methods like macros or scripting, which consolidate individual user interactions.

RPA is particularly suitable for mundane tasks where humans extract structured data from one or more applications, process it rule-based, and re-enter it into another system, known as "swivel-chair processes" [24]. The goal is to relieve users of such tasks, allowing them to focus on higher-value activities [25].

In contrast to other traditional automation techniques, RPA operates "outside-in", that is it builds upon the existing information system architecture without altering it. This allows RPA to connect various applications without the for need for an explicit application programming interface (API) but the UI [40]. Despite this technical difference, RPA also changes the organizational and social aspects of automation by allowing user-side RPA robot development via low-code flowcharts outside a dedicated project in the IT department [36].

In a nutshell, symbolic RPA suits rule-based, non-interactive tasks and is not ideal for dynamic environments needing flexibility [27]. It is limited by structured data requirements and may overlook exceptional cases due to "happy path" recording. In the context of intelligent RPA, especially technical limitations of common symbolic RPA shall be addressed through ML integration.

2.2 Machine Learning

Various nuances of definitions for the terms artificial intelligence (AI) and ML have been identified. Common to all these approaches is the definition of AI as the ability of machines to solve cognitively human-like tasks [4]. For instance, Haenlein and Kaplan [15] define AI more specifically as the ability of a system to accurately interpret external data, learn from it, and apply the learned knowledge through flexible adaptation to specific tasks and objectives.

ML is one method that machines can potentially employ to achieve AI. While early approaches like expert systems required rules and axioms to be explicitly defined and used logic to infer knowledge, ML autonomously generates knowledge from training data through experience, eliminating the need for explicit programming [19]. Its current state of development can be characterized as *narrow artificial superintelligence* that can exhibit super-human capabilities for specific tasks that it was selectively trained for such as playing board games or analyzing medical scans.

In the field of ML, distinctions can be made based on algorithmic learning types, tasks, and model architectures. Depending on the data provided to the learning model, we distinguish three learning types [19]: supervised learning, unsupervised learning, and reinforcement learning. Furthermore, ML algorithms can be differentiated based on their structure as shallow or deep learning [19].

2.3 Related Work on Intelligent RPA

The literature provides numerous categorizations of how RPA and ML can be combined [1,28,31,38,41]. However, these categorizations often focus on specific aspects and do not provide a comprehensive differentiation. To gain a complete understanding of the complexity of the possibilities for enhancing RPA with ML, it is necessary to interconnect and relate these various dimensions.

On the other hand, not all dimensions found in the literature offer relevant insights into the conceptual integration with ML. For example, the distinction between whether UI recorders operate at the browser, system, or application level is merely descriptive of RPA, regardless of whether it is enriched with ML [1]. Similarly, the literature does not always differentiate between the specific ML algorithms used, as it is more important in the context of RPA to focus on their capabilities and general functionality. Technical implementation details are typically not emphasized in the literature because the use in the RPA context is a transfer application.

Our undertaking is further complicated by the widespread use of the term AI in industry, often applied to all innovative approaches that are marketed as AI regardless of the actual intelligence involved. Competing definitions must also be reconciled. For example, while Beerbaum [3] attributes the same meaning to *smart process automation*, cognitive RPA, and intelligent RPA in the sense of combining RPA with AI, Lacity and Willcocks [25] explain that intelligent RPA is the convergence of RPA and cognitive automation (CA), where CA broadly signifies automation through AI.

Further, Lacity and Willcocks [25] argue that a strict one-dimensional separation between forms like symbolic RPA and CA is not the right perspective. Instead, we observe a continuum, where RPA is gradually enriched by cognitive elements. We address this finding by elaborating different levels of ML integration and the complexity of integration approaches through a multidimensional characterization.

Given the nascent state of the research field, it is expected that terminologies will further evolve but become more standardized over time. Our nascent taxonomy can help to shape this journey.

3   Research Design and Methodological Approach

We conducted an integrative literature review to explore concepts for structuring the connection between RPA and ML employing a hermeneutic approach [5]. We searched multiple times for relevant literature in the databases *AIS eLibrary*, *IEEE Xplore*, *ACM Digital Library*, *SpringerLink*, and *ScienceDirect* with an evolving search string eventually resembling: *("AI" OR "ML" OR "Artificial Intelligence" OR "Machine Learning") AND ("RPA" OR "Robotic Process Automation")*.

This search string takes into account the insight gained in the preliminary research that there is often no clear differentiation between the terms AI and ML. Therefore, limiting the search to the term ML might omit relevant publications. Furthermore, the search results were narrowed down according to the capabilities of each database and filtered. Due to the lack of scientific literature, as mentioned in overview articles such as those by Syed et al. [36] or Chugh et al. [7], there was a deliberate choice not to further restrict to high-quality journals or similar sources. Publications in languages other than English or German were not found during the search and, therefore, did not need to be explicitly excluded.

Subsequently, we conducted an initial review of the search results based on titles and abstracts. We excluded publications that were not research articles from computer science or information systems, did not explicitly link RPA and ML in the title or the abstract, adopted a highly individual perspective, focused on other AI concepts than ML or concentrated only on implications, but lacked a conceptual perspective on connecting RPA and ML.

This process resulted in a reduction to 51 potential articles. The remaining publications were analyzed following the methodology outlined by Boell and Cecez-Kecmanovic [5], and they were classified based on their central ideas and methodologies. After removing duplicates, a backward and forward search was conducted. As suggested by Boell and Cecez-Kecmanovic, literature searches may never truly conclude, but at some point saturation is reached. In this case, that point of saturation occurred with 45 publications that we reviewed in depth. Therefore, this literature review can only be considered representative and not exhaustive.

To gain a structure of the discovered concepts, we applied the taxonomy development method proposed by Nickerson et al. [29] that shall avoid ad hoc classification by iterating and defined ending conditions. As meta-characteristic, we choose conceptual properties of *structure* and *process* to specify how ML augments RPA: *RPA-ML integration* to represent structural topics and *RPA-ML interaction* to represent the augmentation in use.

In the present case, we strived to develop a taxonomy based on literature sources. However, the method by Nickerson et al. [29] is designed for synthesis based on individual exemplars. After a fundamental analysis of the selected publications for the review, it was observed that many of the publications ($n = 26$) not only discuss individual instances but also provide a framework or conceptual differentiations for connecting RPA with ML. Therefore, we organized the approach as follows: first we collected dimensions and characteristics based on these works in the conceptual-deductive sub-circle, and then we analyzed the instances in application-oriented publications using the empirically-inductive sub-circle. Another 10 publications describe individual conceptions or approaches on how RPA and ML can be combined.

For more information concerning the research process, please see the available dataset comprising the detailed literature search settings including exclusion criteria, a list of the reviewed literature and our concept matrix [23].

4   A Nascent Taxonomy of ML-RPA Integration and Use

4.1   Overview

In total, we identified 8 dimensions with 24 characteristics. Table 1 presents the results in an aggregate form. The table is organized as follows: the first column contains our meta-characteristics, while the second column provides the associated dimensions. We documented the frequency of mentions of these dimensions in the literature in the third column. The fourth column contains our characteristics. The literature under consideration was initially categorized into the three types practical reports, conceptions, and frameworks. In the four columns on the right, we recorded the frequency of each characteristic in the respective publication type and as a total.

We will delve into our specific findings for each dimension in the subsequent sections, providing a detailed exploration of their individual characteristics and a comprehensive analysis of their occurrences. According to Nickerson et al. [29], a taxonomy with 8 dimensions is sufficiently differentiated yet concise. It thoroughly explains the differences in comparison to individual literature bases and can also be extended to include additional characteristics, such as other process environments.

Table 1. Taxonomy of machine learning in intelligent robotic process automation. *Legend: MC meta-characteristics, M mentions, # total, P practitioner reports, C conceptions, F frameworks*

MC	Dimensions	M	Characteristics	#	P	C	F
				45	9	10	26
RPA-ML integration	Architecture and ecosystem	2	External integration	13	5	7	1
			Integration platform	11	2	2	7
			Out-of-the-box (OOTB)	3	1	0	2
	Capabilities of ML	4	Computer vision	21	6	1	14
			Data analytics	13	3	5	5
			Natural language processing	22	3	4	15
	Data basis for ML	1	Structured	2	1	0	1
			Unstructured	25	8	4	13
			UI logs	6	0	5	1
	Intelligence level	8	Symbolic	13	1	1	11
			Intelligent	33	8	3	22
			Hyperautomation	22	0	8	14
	Technical depth of integration	1	High code	11	5	5	1
			Low code	10	4	2	4
RPA-ML interaction	Deployment area	1	Analytics	7	1	2	4
			Back office	17	8	4	5
			Front office	7	0	3	4
	Lifecycle phase	1	Process selection	8	0	2	6
			Robot development	10	1	5	4
			Robot execution	18	8	2	8
			Robot improvement	5	1	1	3
	User-robot relation	5	Attended	12	5	4	3
			Unattended	10	4	2	4
			Hybrid	4	0	2	2

4.2   Architecture and Ecosystem

First, we can make a distinction in how RPA software is architecturally and conceptually linked with ML capabilities. The question arises of whether ML components are distributed as separate parts of RPA software by the RPA provider, if RPA software opens up to become open an platform and ecosystem for third-party providers, or even adopts open-source models [10]. Alternatively, they can offer interfaces for individual external integration with ML. While frameworks particularly emphasize the platform approach ($n = 7$), practical reports still predominantly favor external linkage ($n = 5$).

External Integration. Users can develop and train ML models themselves. Connecting to these models can be done through standardized APIs, but it requires programming effort and specialized knowledge for ML model development.

Integration Platform. In this approach, RPA software transforms into an execution platform or engine [25]. This approach involves not only ML, but also additional modules for the direct integration of applications and services, eliminating the need for UI interaction or API-based programming [18]. The RPA software opens up to third-party specialized AI providers and open-source ML libraries, aligning with the rise of artificial-intelligence-as-a-service (AIaaS). AIaaS provides AI capabilities, including pre-trained models, through cloud-based training and configuration environments [10]. External ML capabilities can be integrated as offered modules into workflows. This approach allows for quick adoption of new ML trends and flexible module licensing.

Out-of-the-Box (OOTB). This category encompasses integrations where ML capabilities are either built directly into the RPA software or can be added later via a robot store controlled by the software provider [33]. Crucially, the availability of ML capabilities depends entirely on the provider's offering.

4.3   Capabilities of ML

Four of the examined frameworks differentiate ML integration based on the specific ML capabilities utilized. However, most sources remain at a relatively high-level perspective without delving into technical details. According to Nickerson et al. [29], characteristics within a taxonomy dimension should ideally be mutually exclusive. Nevertheless, it has proven challenging to determine characteristics in this dimension in such a way that an approach can be unambiguously assigned to only one capability while not unbalancing our taxonomy by introducing excessively many characteristics. Given the context of *multi-skilling* [25], it is worth questioning whether intelligent robots might possess multiple ML capabilities and could thus be characterized by the combination of these capabilities. Further, some publications explicitly mention domains but do not specify the underlying algorithmic types. It is a dimension particularly affected by contemporary changes in AI capabilities due technological advancements.

Computer Vision (CV). Here, we subsume all skills related to the capture and analysis of images in the broadest sense, such as optical character recognition (OCR) or intelligent character recognition (ICR), or generally, CV. In the case of OCR and ICR, the goal is to create machine-readable documents with a text layer from document images. While OCR is limited to printed documents, ICR can also recognize handwriting, for example. According to Kanakov and Prokhorov [20], approximately 90 % of all use cases of AI-augmented RPA use these techniques for document processing. In fact, 17 out of the 45 publications mention the possibility of using OCR. Among the 8 practical reports, OCR is used in 5. Outside the context of documents, ML models are used to recognize faces, classify images based on their content, pre-process text recognition, and also identify UI elements [20, 34, 41]. In the frameworks, CV, along with OCR, is the second most commonly mentioned capability.

Data Analytics. In this case, ML does not serve to unlock data that would otherwise be inaccessible for automation, but rather facilitates the processing of data that is already usable. A common theme is the use of ML-based classification. This encompasses all abilities that classify data into various categories based on a previously trained ML model, serving, for example, in the preprocessing of emails or other documents such as splitting and prioritizing inquiries [25]. In this characteristic, we also capture models that, for example, aim to recognize and extract patterns from log files based on unsupervised learning.

Natural Language Processing (NLP). The use of ML has two perspectives in the domain of natural language. Traditionally, ML models in the sub-discipline of NLP are designed to process natural language texts and automatically extract their meaning [38]. NLP can be used for example not only in the execution phase to categorize user inquiries based on user sentiment, but it can be used also in other phases such as process selection for identifying automation candidates [26]. Other concepts involve configuring entire robots through natural language interaction [6]. In 22 out of 45 publications, the use of NLP in conjunction with RPA is considered, and in 3 out of 9 practical reports, NLP is included. Currently, due to the recent emergence of generative AI [11], the second perspective, namely natural language generation, is of significance. The integration of conversational agents in RPA primarily aims at enabling users, both end customers and employees, to interact with software robots in natural language, thereby making it as accessible as possible [33].

The use of natural language generation is closely intertwined with natural language understanding, where, for example, a three-part architecture initially uses NLP to determine the user's intent, then implements the user's request with RPA, and finally generates an appropriate response to the user, either rule-based or through a large language model (see *understand - act - respond*, [32]). Through the integration with RPA, a conversational agent can transform from a simple question-answer machine into a comprehensive assistant by triggering actions or transactions in existing application systems. By bridging the gap between the input and output of the conversational agent, the respective limitations are addressed as RPA can only process structured data into structured results, while conversational agents handle unstructured inputs leading to their natural language responses [8]. In our literature review, we found that connections to conversational agents are currently almost exclusively found in conceptual publications.

4.4   Data Basis for ML

Since ML is strongly dependent on the underlying data, it is important to note which data is used in the context of RPA for both learning and applying a model.

Structured. RPA can retrieve data for the ML model, for instance, from legacy systems through the UI, making it usable for ML. The results of the ML model application can then be further processed based on rules. An example is insurance fraud detection, where RPA first extracts data from an Excel sheet, then the data records are classified by a pre-trained model, and depending on the results, various process variants are executed by RPA [30]. This way, ML can not just overcome the limitations of RPA, but RPA can also address the issue of data availability for ML.

Unstructured. Documents, video and audio files, as well as emails, need to be processed in 95% of all companies; 80% of the generated data is said to be semi- or unstructured [2]. However, RPA cannot process them in a rule-based manner without ML capabilities. ML makes the data usable for automation [41]. In the practical reports, the unlocking of unstructured data clearly dominates the other characteristics of this dimension.

User Interface Logs. To automate RPA itself, the system must learn autonomously from user interactions. In this case, ML models are not applied to external structured or unstructured data but rather to internal log files that would otherwise not be processed, or to representations of the robots themselves [33]. This characteristic is mainly observed in advanced conceptual approaches, while no practical report mentions learning from UI logs.

4.5   Intelligence Level

The conceptual classification based on intelligence levels is the most common approach in the literature, appearing in various forms ($n = 8$). Other terms used for this classification are *automation types* or *stages* [7]. These classifications differentiate how intelligent the automation artifact is and how intelligent it behaves. These types are hierarchical, with higher stages expanding the capabilities of lower stages. Further proposed three-level classifications, such as the automation waves according to PricewaterhouseCoopers International, can be recognized as strongly related to this dimension [39]. Table 1 indicates that practical reports are almost exclusively associated with the stage *intelligent*, while frameworks and conceptions also focus on *hyperautomation*.

Symbolic. This stage includes traditional rule-based RPA, as described in Sect. 2. Systems in this category rely on structured data and can only perform tasks for which they were deterministically configured [7]. The corresponding automation wave is the *algorithm wave*, where simple arithmetic and data analysis tasks are automated [39]. Basic Automation operates relatively mechanically, replacing the "arms and legs" of human workers [41]. Robots in this category are referred to as *doing bots*, indicating that they perform routine tasks and are otherwise dependent on humans.

In analogy to symbolic AI, Herm et al. [17] refer to this stage as *symbolic RPA* because it faces similar constraints as knowledge-based and expert systems in the AI domain. Symbolic RPA robots can only handle tasks for which they have been explicitly programmed or configured, but complex and cognitive tasks are challenging to formalize in this way.

Nevertheless, even in symbolic stage tools without externally noticeable intelligent features, ML techniques are being used to adapt, for example, to changing UI surfaces [41]. To reflect this internal usage in our taxonomy, we have decided to include this characteristic.

Intelligent. In this second stage, robots can perform more complex processes by incorporating specific cognitive abilities. This includes processing unstructured or semi-structured data. Robots in this stage are referred to as *thinking bots*. They can, for example, process natural language because they have been trained for this specific task [7]. These robots are specialists in their respective domains and problems and are limited to those areas. They cannot autonomously transfer their skills to other tasks, learn new things, or self-improve. This corresponds to the concept of narrow AI [10]. Nevertheless, CA and intelligent automation (IA) no longer operate strictly deterministically but with well-defined probabilistic components. The execution logic remains rule-based to remain predictable and meet manually specified requirements [41]. The corresponding automation wave is the *augmentation wave*, predicted by PricewaterhouseCoopers International for the late 2020s [39]. The term *intelligent robotics* also falls into this category [36].

Hyperautomation. In advanced automation, robots learn adaptively from data and experience, and they manage and improve autonomously. These robots are referred to as *learning bots* [7] or autonomous agents [28]. *Hyperautomation* is also part of this type. The Greek prefix "hyper" implies that automation occurs at a higher level. It is about automating the process of automating individual processes through a higher-level perspective, thus improving it [17]. Hyperautomation aims to automatically generate suitable automation artifacts such as scripts, robots, or workflows [35]. The corresponding automation wave here is the *autonomy wave* predicted for the mid-2030s, where problem-solving is automated [39]. Currently, there is no agreed upon definition and scope of hyperautomation as it originated from recent business practice. Its detailed specification and appearance will emerge in the coming years.

4.6   Technical Depth of Integration

Closely tied to the topic of architecture is the question of the technical depth into which an RPA developer must delve to integrate ML in RPA software. Consequently, this dimension depends to some respect on the architecture on the provider side, but takes the complementary perspective of the users. Agostinelli et al. [1] distinguish between "strong coding", graphical user interface (GUI), and low-code tools. Due to the minimal difference between GUI (viz. no code) and low code, we only differentiate between high code (viz. classical formal language programming) and low code. While low-code paradigms clearly dominate in frameworks ($n = 4$ vs. $n = 1$), in the examined practical reports, no paradigm has clearly prevailed ($n = 4$ vs. $n = 5$).

High Code. Integrating ML capabilities to RPA robots requires programming in high-level, domain-specific programming or scripting languages. The skills of the integration experts would need to be those of a programmer rather than of a user.

Low Code. No-code or low-code environments do not require knowledge of programming languages as robots are generated through UI-based modeling. The skills of the person responsible for the integration is that of a power user and may be available in business departments rather than the IT department.

4.7 Deployment Area

Integrated RPA and ML solutions are opening up to process tasks from the back office up to the front office as technology advances [8]. Rather than automating transactional tasks intelligently, a strong focus on intelligent automation of decisions can be observed. With these three characteristics, we try to subsume these developments.

Analytics. AI-enhanced RPA robots can make decisions more flexibly than rule-based robots and are faster and more consistent than humans. The decision-making is informed based on a data foundation, but it is not without its challenges. These processes are no longer repetitive and straightforward tasks but have become so complex that they would typically be handled by experts [22]. Consequently, rules-based approaches do not suffice.

Back Office. Even though structured data is no longer mandatory, simple, structured, and matured processes in back-office contexts remain highly suitable for RPA [41]. In fact, [20] state that 90% of intelligent RPA applications would still perform back-office tasks, which can now also be based on unstructured documents through OCR and intelligent document processing as well. This is supported by 8 out of 9 practical reports that can be attributed to this characteristic.

Front Office. For example in combination with conversational agent technology, RPA can be used not only in the back office but also to interact directly with customers. Automation enables faster customer service and more availability [8].

4.8 Lifecycle Phases

ML techniques are employed at various stages when creating, using, and improving RPA robots. These phases are distinct from the project management phases of an RPA project, as outlined by [16]. The explicit distinction among these phases is evident in the framework proposed by Agostinelli et al. [1].

Process Selection. Before creating an RPA robot, it is crucial to identify which routines to automate. ML techniques can automatically detect these tasks. Typically, process selection relies on manual methods like interviews, observations, or document analysis, which can overlook unconscious routines and involve significant effort [9]. Alternatively, process mining or task mining can be used, involving the use of UI logs to automate process discovery and data-driven analysis [1]. This is referred to as *inter-routine self-learning*, where various routines with automation potential are autonomously identified. However, the subsequent handling of these routines traditionally requires implementation by human experts.

Robot Construction. In standard RPA software, robots are created in low-code environments by developers who build models based on recorded UI interactions [33]. Automating this process would require RPA software to autonomously derive rules from user observations to instantiate a robot [12]. The challenge lies in operational user interaction data, which may contain non-linear, interchangeable routines and disruptive noise. Automatic segmentation is difficult, especially as UI logs typically lack a label (or case ID) for the underlying business case. This type of learning is termed *intra-routine self-learning* [1]. For instance, [37] suggest using shallow ML techniques to segment similar operation sequences into potential routines. To make a robot ready for operation, the actual model (i.e., flowchart) must be constructed from analyzed user interaction data. User-driven flowchart construction poses risks, particularly in complex processes, as it may lead to faulty robots interacting with production systems and causing harm due to human errors or the neglect of exceptional cases, mainly due to a lack of testing environments [1].

Robot Execution. In the examined publications describing real-world applications, ML is used in 8 out of 9 cases during the execution phase of the robots. This application does not improve the automation process itself but enhances the robot's task execution. Pre-trained models are employed to process unstructured data such as documents, or conversational agents are used to interact with customers using natural language. Moreover, in extensive RPA application scenarios, complex sequences of interdependent routines are automated. Concepts of automatic planning can be used to automatically determine intelligent execution strategies [1,41].

Robot Improvement. In RPA software, even without apparent intelligent features, CV techniques are often integrated to tackle the challenge of evolving UI interfaces. They serve to ensure the continued identification of UI elements, particularly through virtual interfaces like Citrix [38,41]. More advanced concepts propose the use of experiential learning from mistakes during this phase, aiming to generalize individually created but similar robots. This addresses the lightweight nature of RPA, which tends to lead to localized and bespoke solutions. To reintroduce standardization across organizational processes, robots with similar tasks should be merged and harmonized [33].

4.9   User-Robot Relation

This dimension focuses on the role of humans during the robot's process execution. It distinguishes how humans interact with the robot and in what environment the robot operates, whether in desktop environments of workstations or on central servers. This distinction is already made for classical RPA [7] and does not directly characterize the connection with ML. However, ML influences which interactions with humans are still necessary, leading to shifts in this dimension. That is why we included it in our taxonomy. In the literature, no interaction form clearly dominates across all types of publications. Hybrid approaches are only mentioned in frameworks and concepts. Therefore, we presume that ML components enable various forms of interaction and overcome previously existing limitations.

Attended. This characteristic centers around the terms *desktop-level* automation and *human-in-the-loop*. RPA is triggered by humans for specific tasks and constantly interacts with the user on their individual desktop [7]. RPA is considered a tool for the individual user [39].

Unattended. RPA is executed on central servers and runs continuously in the background without human intervention. This is why it is also referred to as *enterprise-level* automation [7]. Humans are only required for handling exceptional cases [39]. In recent years, RPA has been shifting away from its original focus of desktop automation towards this level [21].

Hybrid. Due to the different forms of intelligence possessed by humans and machines, the connection towards a collaborative partnership between robots and users is seen as an enrichment. While humans can use social and experiential knowledge and possess creativity, intuition, and empathy, the "intelligent" robot can quickly, efficiently, and consistently process available data analytically based on underlying probability distributions [10,39].

5   Application, Limitations, and Summary

In our work, we explored the possibilities of integrating the two concepts of RPA and ML to mitigate the limitations inherent to symbolic RPA. The goal was to systematize the theoretical research landscape strewn across both communities which resulted in a nascent taxonomy comprising 2 meta-characteristics with 8 different dimensions that describe the connection of RPA and ML. Despite the high relevance of our literature corpus, we must take into account that the market develops rapidly, with recent annual growth rates ranging from 17.5% to 30.9% in the years 2021 to 2023 [13]. This leads to very dynamic developments in the RPA software products available on the market, which are increasingly equipped with more intelligent features – especially in their advertising.

Application and Evaluation. Hence, in a bid also to evaluate the usefulness and applicability of our literature-driven taxonomy, we have reviewed the current products of *UiPath*, *Automation Anywhere*, and *Microsoft Power Automate* that are leading Gartner's magic quadrant for RPA [14]. The evaluation took place from April to June 2023. We used the community editions where available and employed the taxonomy as an inspection catalog. We did not encounter any substantial concepts (viz. software functionality) that we could not classify with our taxonomy.

Our comparison of the RPA products showed that there are major similarities between all three RPA products in terms of similarities in the range of functions, the structure in similar modules, and the focus on low code. For example, all software vendors rely on a platform approach for not only ML components, but also for the integration of other APIs in marketplaces managed by them. At the same time, all RPA products offer their own ML components, especially for document processing. On all platforms, not only desktop automation, but also cloud automation via APIs is possible. The use of all capabilities of ML as mentioned in the taxonomy is either available OOTB or by third-party providers. All RPA products focus their ML components primarily on the phases of process selection (through task mining) and process execution. For ML-supported design, Automation Anywhere and Microsoft Power Automate have each recently started to offer copilot functions based on conversational agents (i.e. chatbots). However, we could not find any ML-based automatic improvement functions. Thus, we classify the RPA products as IA, but they obviously market as and aim for hyperautomation functionality. Yet, autonomous learning and autonomous adaptation are nowhere to be found as of today.

Limitations. As with all research, our taxonomy has some limitations. First, we did not perform an exhaustive structured literature review, but we opted for a hermeneutic approach to better understand the intricacies of evolving topic of RPA-ML integration. Thus, we stopped at a point of saturation to offer a nascent taxonomy for discussion rather exhaustive taxonomy that cannot yet be complete as ML continues to advances while software engineers are busy implement yesterday's advances into their products. Also, we must recognize that there are interdependencies between some of our dimensions that cannot be resolved easily: for example, the architecture significantly influences the technical depth of integration. Different levels of intelligence require varying capabilities of ML. Depending on the process context and the data being processed, various capabilities of ML from different domains are required. That is, dimensions are closely interconnected and should not be considered in isolation. Further, our capabilities dimension resides at a high level and could be detailed into concrete CV, NLP, and data analytics techniques. For the moment, we have consciously decided against detailing this dimension as it would significantly unbalance the taxonomy with respect to the other dimensions' level of detail.

Discussion and Summary. Summarizing, we found that practical applications of RPA-ML integration are primarily attributed to the intelligence level of *intelligent* RPA, which means that they work with specialized cognitive capabilities, especially in the execution phase. However, there is an effort to focus on the entire automation process and to make automation itself smarter through the application of ML, alluding to the still underspecified concept hyperautomation promising self-learning and generative robots that scale as the tasks require.

In addition, we observed that in practical reports, ML is primarily used in the execution phase with a wide range of ML capabilities. This also corresponds to the level of intelligence of *intelligent*. As narrow but superintelligent *thinking bots*, RPA robots can perform specialized tasks, such as processing unstructured data, on which they have been explicitly trained. Academic concepts predominantly focus on *hyperautomation* and self-learning robots.

Further, with the advent of generative AI we see a shift towards the process selection phase in particular. Driven by recent successes, software vendors are beginning to incorporate co-pilot functionality for the specification or automated construction of RPA robots in their products. However, contrary to some concepts from literature, human interaction is still very much necessary for this endeavor, as RPA software does not (yet) independently derive rules or analytical models from pure user observation.

The landscape of intelligent RPA is bound to evolve further especially with the recent advent of generative AI. Hence, our nascent taxonomy will require constant updates as technology advances. Nevertheless, our taxonomy can already be used today to generate archetypes of intelligent RPA robots and assist organizations in comparing the promises of next-generation RPA products in the market to assist the design of complex intelligent RPA building blocks and patterns such as those aforementioned neutral data trust models that go beyond the capabilities of symbolic RPA.

Acknowledgements. This research and development project is funded by the German Federal Ministry of Education and Research (BMBF) within the "Richtlinie zur Förderung von Projekten zur Erforschung oder Entwicklung praxisrelevanter Lösungsaspekte ("Bausteine") für Datentreuhandmodelle" (Funding No. 16DTM201B) and financed by the European Union - NextGenerationEU. The authors are responsible for the contents of this publication.

References

1. Agostinelli, S., Marrella, A., Mecella, M.: Research challenges for intelligent robotic process automation. In: Di Francescomarino, C., Dijkman, R., Zdun, U. (eds.) BPM 2019. LNBIP, vol. 362, pp. 12–18. Springer, Cham (2019). https://doi.org/10.1007/978-3-030-37453-2_2
2. Baviskar, D., Ahirrao, S., Potdar, V., Kotecha, K.: Efficient automated processing of the unstructured documents using artificial intelligence: a systematic literature review and future directions. IEEE Access **9**, 72894–72936 (2021)
3. Beerbaum, D.O.: Generative artificial intelligence (GAI) ethics taxonomy: applying chat GPT for robotic process automation (GAI-RPA) as business case. SSRN Electron. J. (4385025) (2023)
4. Benbya, H., Davenport, T., Pachidi, S.: Special issue editorial. Artificial intelligence in organizations: current state and future opportunities. MIS Q. Executive **19**(4) (2020)

5. Boell, S.K., Cecez-Kecmanovic, D.: A hermeneutic approach for conducting literature reviews and literature searches. Commun. Assoc. Inf. Syst. **34** (2014)
6. Chakraborti, T., Rizk, Y., Isahagian, V., Aksar, B., Fuggitti, F.: From natural language to workflows: towards emergent intelligence in robotic process automation. In: Marrella, A., et al. (eds.) BPM 2022. LNBIP, vol. 459, pp. 123–137. Springer, Cham (2022). https://doi.org/10.1007/978-3-031-16168-1_8
7. Chugh, R., Macht, S., Hossain, R.: Robotic process automation: a review of organizational grey literature. Int. J. Inf. Syst. Proj. Manag. **10**(1), 5–26 (2022)
8. Daase, C., Staegemann, D., Volk, M., Nahhas, A., Turowski, K.: Automation of customer initiated back office processes: a design science research approach to link robotic process automation and chatbots. In: Australasian Conference on IS (2020)
9. Dumas, M., La Rosa, M., Mendling, J., Reijers, H.A.: Fundamentals of Business Process Management. Springer, Cham (2018). https://doi.org/10.1007/978-3-662-56509-4
10. Engel, C., Ebel, P., Leimeister, J.M.: Cognitive automation. Electron. Mark. **32**, 339–350 (2022)
11. Feuerriegel, S., Hartmann, J., Janiesch, C., Zschech, P.: Generative AI. Bus. Inf. Syst. Eng. **66**(1), 111–126 (2024)
12. Gao, J., van Zelst, S.J., Lu, X., van der Aalst, W.M.P.: Automated robotic process automation: a self-learning approach. In: Panetto, H., Debruyne, C., Hepp, M., Lewis, D., Ardagna, C.A., Meersman, R. (eds.) OTM 2019. LNCS, vol. 11877, pp. 95–112. Springer, Cham (2019). https://doi.org/10.1007/978-3-030-33246-4_6
13. Gartner, Inc.: Gartner says worldwide RPA software spending to reach $2.9 billion in 2022 (2022). https://www.gartner.com/en/newsroom/press-releases/2022-08-1-rpa-forecast-2022-2q22-press-release
14. Gartner, Inc.: Magic quadrant for robotic process automation (2022). https://www.gartner.com/technology/media-products/reprints/uiPath/1-2BG6WSIY-DEU.html
15. Haenlein, M., Kaplan, A.: A brief history of artificial intelligence: on the past, present, and future of artificial intelligence. Calif. Manag. Rev. **61**(4), 5–14 (2019)
16. Herm, L.V., Janiesch, C., Helm, A., Imgrund, F., Hofmann, A., Winkelmann, A.: A framework for implementing robotic process automation projects. IseB **21**(1), 1–35 (2023)
17. Herm, L.-V., Janiesch, C., Reijers, H.A., Seubert, F.: From symbolic RPA to intelligent RPA: challenges for developing and operating intelligent software robots. In: Polyvyanyy, A., Wynn, M.T., Van Looy, A., Reichert, M. (eds.) BPM 2021. LNCS, vol. 12875, pp. 289–305. Springer, Cham (2021). https://doi.org/10.1007/978-3-030-85469-0_19
18. Hofmann, P., Samp, C., Urbach, N.: Robotic process automation. Electron. Mark. **30**, 99–106 (2020)
19. Janiesch, C., Zschech, P., Heinrich, K.: Machine learning and deep learning. Electron. Mark. **31**, 685–695 (2021)
20. Kanakov, F., Prokhorov, I.: Analysis and applicability of artificial intelligence technologies in the field of RPA software robots for automating business processes. Procedia Comput. Sci. **213**, 296–300 (2022)
21. Kedziora, D., Hyrynsalmi, S.: Turning robotic process automation onto intelligent automation with machine learning. In: Communities and Technologies, pp. 1–5. ACM (2023)
22. Kholiya, P.S., Kapoor, A., Rana, M., Bhushan, M.: Intelligent process automation: the future of digital transformation. In: System Modeling & Advancement in Research Trends, pp. 185–190. IEEE (2021)

23. Laakmann, L., Ciftci, S.A., Janiesch, C.: Dataset for Laakmann et al. "A nascent taxonomy of machine learning in intelligent robotic process automation" (2024). https://doi.org/10.23728/b2share.0cdaf090dded4be2a93679841563a5ac
24. Lacity, M., Willcocks, L.: Robotic process automation at telefonica O2. MIS Q. Executive **15**(1) (2016). https://aisel.aisnet.org/misqe/vol15/iss1/4
25. Lacity, M., Willcocks, L.: Becoming strategic with intelligent automation. MIS Q. Executive **20**(2) (2021)
26. Leopold, H., van der Aa, H., Reijers, H.A.: Identifying candidate tasks for robotic process automation in textual process descriptions. In: Gulden, J., Reinhartz-Berger, I., Schmidt, R., Guerreiro, S., Guédria, W., Bera, P. (eds.) BPMDS/EMMSAD -2018. LNBIP, vol. 318, pp. 67–81. Springer, Cham (2018). https://doi.org/10.1007/978-3-319-91704-7_5
27. Moreira, S., Mamede, H.S., Santos, A.: Process automation using RPA - a literature review. Procedia Comput. Sci. **219**, 244–254 (2023)
28. Ng, K.K., Chen, C.H., Lee, C., Jiao, J., Yang, Z.X.: A systematic literature review on intelligent automation: aligning concepts from theory, practice, and future perspectives. Adv. Eng. Inform. **47**, 101246 (2021)
29. Nickerson, R.C., Varshney, U., Muntermann, J.: A method for taxonomy development and its application in information systems. Eur. J. Inf. Syst. **22**(3), 336–359 (2013)
30. Patil, N.S., Kamanavalli, S., Hiregoudar, S., Jadhav, S., Kanakraddi, S., Hiremath, N.D.: Vehicle insurance fraud detection system using robotic process automation and machine learning. In: Conference on Intelligent Technologies, pp. 1–5. IEEE (2021)
31. Ribeiro, J., Lima, R., Eckhardt, T., Paiva, S.: Robotic process automation and artificial intelligence in industry 4.0 - a literature review. Procedia Comput. Sci. **181**, 51–58 (2021)
32. Rizk, Y., et al.: A conversational digital assistant for intelligent process automation. In: Asatiani, A., et al. (eds.) BPM 2020. LNBIP, vol. 393, pp. 85–100. Springer, Cham (2020). https://doi.org/10.1007/978-3-030-58779-6_6
33. Rizk, Y., Venkateswaran, P., Isahagian, V., Muthusamy, V., Talamadupula, K.: Can you teach robotic process automation bots new tricks? In: Marrella, A., et al. (eds.) BPM 2022. LNBIP, vol. 459, pp. 246–259. Springer, Cham (2022). https://doi.org/10.1007/978-3-031-16168-1_16
34. Rohaime, N.A., Abdul Razak, N.I., Thamrin, N.M., Shyan, C.W.: Integrated invoicing solution: a robotic process automation with AI and OCR approach. In: Student Conference on Research and Development, pp. 30–33. IEEE (2022)
35. Sudharson, D., Bhuvaneshwaran, A., Kalaiarasan, T., Satessh Kumar, D., Sushmita, V., Jyothi Lakshmi, N.: A multimodal AI framework for hyper automation in industry 5.0. In: Innovative Data Communication Technologies and Application, pp. 282–286. IEEE (2023)
36. Syed, R., et al.: Robotic process automation: contemporary themes and challenges. Comput. Ind. **115**, 103162 (2020)
37. Urabe, Y., Yagi, S., Tsuchikawa, K., Oishi, H.: Task clustering method using user interaction logs to plan RPA introduction. In: Polyvyanyy, A., Wynn, M.T., Van Looy, A., Reichert, M. (eds.) BPM 2021. LNCS, vol. 12875, pp. 273–288. Springer, Cham (2021). https://doi.org/10.1007/978-3-030-85469-0_18
38. van de Weerd, I., Nieuwenhuijs, B., Bex, F., Beerepoot, I.: Using AI to augment RPA: a conceptual framework. In: European Conference on IS (2021)
39. van der Aalst, W.M.P.: Hybrid intelligence: to automate or not to automate, that is the question. Int. J. Inf. Syst. Proj. Manag. **9**(2), 5–20 (2021)

40. van der Aalst, W.M.P., Bichler, M., Heinzl, A.: Robotic process automation. Bus. Inf. Syst. Eng. **60**(4), 269–272 (2018)
41. Viehhauser, J.: Is robotic process automation becoming intelligent? Early evidence of influences of artificial intelligence on robotic process automation. In: Asatiani, A., et al. (eds.) BPM 2020. LNBIP, vol. 393, pp. 101–115. Springer, Cham (2020). https://doi.org/10.1007/978-3-030-58779-6_7

Extending Business Process Management for Regulatory Transparency

Jannis Kiesel[(⊠)] and Elias Grünewald

Information Systems Engineering, Technische Universität
Berlin, Berlin, Germany
{kiesel,gruenewald}@tu-berlin.de

Abstract. Ever-increasingly complex business processes are enabled by loosely coupled cloud-native systems. In such fast-paced development environments, data controllers face the challenge of capturing and updating all personal data processing activities due to considerable communication overhead between development teams and data protection staff. To date, established business process management methods generate valuable insights about systems, however, they do not account for all regulatory transparency obligations. For instance, data controllers need to record all information about data categories, legal purpose specifications, third-country transfers, etc. Therefore, we propose to bridge the gap between business processes and application systems by providing three contributions that assist in modeling, discovering, and checking personal data transparency through a process-oriented perspective. We enable transparency modeling for relevant business activities by providing a plug-in extension to BPMN featuring regulatory transparency information. Furthermore, we utilize event logs to record regulatory transparency information in realistic cloud-native systems. On this basis, we leverage process mining techniques to discover and analyze personal data flows in business processes, e.g., through transparency conformance checking. We design and implement prototypes for all contributions, emphasizing the appropriate integration and modeling effort required to create business-process-oriented transparency. Altogether, we connect current business process engineering techniques with regulatory needs as imposed by the GDPR and other legal frameworks.

Keywords: business process management · transparency · privacy · cloud

1 Introduction

Ensuring transparency in business process management (BPM) is integral to supporting compliance with privacy regulations. These transparency requirements, comprising information such as data controllers, purpose specification, or personal data retention periods, necessitate systematic conformance checks against the actual execution of processes recorded in event logs [10,13,27]. To

A. Marrella et al. (Eds.): BPM 2024 Forum, LNBIP 526, pp. 337–353, 2024.
https://doi.org/10.1007/978-3-031-70418-5_20

this end, we identify a lack of legally-informed transparency enhancing tools (TETs), leading to opaque business processes.

Cloud-native architectures are subject to frequent changes through fast-paced release cycles [5]. This demands a continuous process of re-formalization of the necessary transparency information. In light of documented or discovered and executed business processes, this paper addresses the need for transparency in BPM, presenting three interconnected contributions that extend the BPMN standard, introduce a cloud-native microservices logging framework, and apply advanced conformance checking methods [3]. Our work aims to enhance transparency in BPM by offering practical solutions for modeling, configuring, enacting, and analyzing business processes while accommodating evolving compliance requirements [16].

The transparency problem poses challenges across regulatory, technical, and organizational dimensions. While existing BPM analysis methods, including process mining techniques and structured event logs, are helpful for formalizing business processes, their regulatory expressiveness is strictly limited. We differentiate the following problem areas.

First, communicating, checking, and enforcing privacy regulations is a difficult task in cloud-native system contexts. Miscommunication, out-of-date information, development costs and efforts to collect the relevant information elements, and the responsibility diffusion between the data controller (cloud consumer), data protection officer, data processor (cloud provider), and the legal and engineering staff. To overcome this challenge, we urge incorporating transparency in the business process design, that allows for automatic conformance checking [22,24].

Second, loosely coupled and evolving microservice architectures, as imposed by the prevalent cloud-native design pattern, are flexibly orchestrated and can be part of different process-related activities, associations, or gateways [5]. Meanwhile, the actual processing of personal data depends on the current execution or workflow [16]. Structured event logs help to harvest information about the actual execution but do not account for all regulatory relevant information. For a thorough legal analysis, it is, however, inevitable to describe the processing activities, including all data flows, in a structured format [1,2,25].

Furthermore, inherently complex cloud-native systems often obfuscate non-compliant personal data processing. This includes non-documented data flows, illegal storage, or data sharing with third parties without the necessary technical or organizational safeguards. We investigate the opportunities for non-compliant process discovery and conformance checking to support the data controller in executing the relevant privacy responsibilities [1,10].

As shown, the complexity of the transparency problem demands a nuanced approach. In this paper, we present three complementary contributions to advancing the BPMN standard towards legally-relevant transparency information. We undertake the task of designing and implementing these contributions to foster machine-readable transparency information integration. These contributions unfold as follows:

- First, we design and implement an *extension of the BPMN standard* using established mechanisms, mapping the language elements of a transparency information language onto BPMN. We complement this with a dedicated editor plugin on the established Camunda platform, that enables personal data processing modeling and facilitates ex ante processing transparency.
- Second, we introduce a cloud-native logging approach for *transparency - focused event logs*, inclusive of transparency information and compatible with conventional event logs. We illustrate this approach with a realistic microservice system.
- Third, we leverage the generated transparency-focused event log to enable *conformance checking, providing insightful legally relevant analysis results*. The application of process mining techniques allows for ex post processing transparency.

All in all, this work paves the way for transparency-focused BPM by providing practical solutions for transparent business process modeling, execution, and analysis.

These contributions unfold as follows: We provide background and related work in Sect. 2. Next, we present our general approach, including a set of requirements in Sect. 3. Afterward, we describe our implementation in Sect. 4. Section 5 provides a discussion, prospects of future work, and concludes.

2   Background and Related Work

In the following, we summarize related work.

2.1   Business Process Management

Business processes are characterized as activities performed and coordinated by a single organization in an organizational and technical environment that jointly realize a specific business goal and can interact with other business processes from other organizations [26]. Utilizing the BPM lifecycle with management concepts, methods, and techniques, processes can be structured, repeated, and automated easily [25,26]. The ISO/IEC 19510 standard notation for business process (meta-)models and notations is BPMN, developed by the Object Management Group, which includes core diagram types and notation elements, as well as conformance rules to allow for extensive business process modeling. It has become best practice in modeling and design approaches and supports extensions through various extension mechanisms natively [6,20]. Event data, created by application systems or technically-mediated business process execution platforms realizing business activities, are collected in event logs and then analyzed and enhanced through common data mining techniques, such as classification, clustering, or regression [2]. Process mining enables the discovery, monitoring, and improvement of real processes by extracting and enhancing information from event logs utilizing data mining techniques [1]. Utilizing process mining approaches, problems can be predicted, diagnosed, and treated [4]. These results can be harnessed to further improve the business processes under consideration.

2.2   Cloud Native Infrastructures

Modern system architectures leverage cloud native infrastructures. Components of these architectures include polyglot microservices that are loosely coupled and communicate through structured APIs, such as REST, GraphQL, or gRPC. Microservices are containerized for scalable, resilient, and dynamic container orchestration. Furthermore, these systems run on self-service, on-demand infrastructure in the cloud, typically offered by large cloud providers. While being scalable, these infrastructures are inherently complex, due to the many possible interrelations, which poses regulatory challenges.

Technical approaches to observability for detecting faults, errors, and all sorts of means for improving the quality of service, include logging, tracing, and monitoring. Coupled with business process management, respective tools are powerful resources for the detailed inspection of a system [23].

2.3   Privacy, Transparency, and BPM

Since business processes handle sensitive information, especially personal data, a range of regulatory frameworks apply. Most prominently, data protection regulations, such as the GDPR or CCPA, require data controllers to provide transparent information (Art. 12 GDPR, transparency) about all processing activities (Art. 30 GDPR) and to keep records to demonstrate compliance for accountability. However, using the BPMN standard, it is not possible to codify all regulatory-relevant information, e.g., concerning transparency obligations.

In related work, [10] proposes to re-design business processes to integrate privacy by design, yet the transparency challenges are not solved. A simple approach to integrate some form of regulatory transparency information in process diagrams is by representing data protection aspects in process models through coloring [27]. Yet, this approach underestimates the complexity of the information to be represented. Pullonen *et al.* propose PE-BPMN to enhance process diagrams by depicting dedicated privacy-enhancing technologies (PETs) activities [22]. This rightfully supports transparency but does not reveal the actual needed transparency information elements, such as purpose specifications or retention periods. Further, the Business Process eXtensions (BPeX) resemble an XML representation to incorporate full BPMN 1.0 functionality, as well as the extension with Platform for Privacy Preferences (P3P) elements [6,7,9]. Due to the limited (legal) expressiveness, and the missing link to cloud native tooling, this approach is not applicable anymore. Closely related, Jensen proposes to enhance service APIs with transparency information such as, among others, a list of countries where data is being processed, a privacy policy, and abstract service-internal business process diagram information [16]. This way, an inter-organizational, transparency-enhancing service graph can be created.

As early as the formal definition of the process mining manifest, privacy-preserving processing of event logs is a part of mining and analysis research [3]. Privacy and confidentiality are becoming an ever-increasing prerequisite for process mining [11]. So far, such related work mainly focuses on data minimization, anonymization of individuals in event logs, or re-identification and linkage threats [11,17,21].

The Transparency Information Language and Toolkit (TILT) allows for the representation and processing of transparency information in line with regulatory requirements set out by the GDPR in a well-defined and machine-readable format [13]. A corresponding TILT document contains all relevant information required for comprehensive transparency for a data controller.

Herein we address the transparency dimension and emphasize its importance for the BPM community. We argue the integration of machine-readable transparency information [13], and the development of approaches for evaluating business process in this regard seem promising.

3   General Approach

Addressing the outlined challenges, we propose a holistic approach that employs and extends existing business process management practices. Thereupon, the necessary design requirements to solve the outlined challenges are presented. Finally, we describe the mentioned challenges in a business process example.

3.1   Requirements

The general approach and all subsequent contributions are designed with the following design requirements in mind to ensure that it can be widely applied and has legal significance. These align with the requirements listed for similar endeavors of practical privacy engineering [14,15].

Regulatory expressiveness (R1): To be a practical choice, adherence to the transparency and accountability principles from the GDPR with the required expressiveness (i.e., capturing all relevant information elements) is needed. We specifically refer to Art. 12, 13, 14, 15, and 30 GDPR. Examples include information about the data controller, personal data categories, purposes, retention times, third-country transfers, legal bases for processing, etc. [13,22]

Effortless integrability (R2): Our contributions should be effortlessly integrable into existing BPM activities, BPMN modelers (incl. process and collaboration diagrams), and distributed system infrastructures [23] without overly burdensome implementation efforts for process managers, developers, or legal staff [1]. Furthermore, system design and runtime overheads must be kept reasonably low.

Compatibility (R3): To ensure compatibility, the resulting modeling and analysis artifacts must be working together with existing tools, e.g., process mining toolchains [2,4]. This is also relevant for the interplay with existing transparency enhancing technologies. Therefore, the transparency information must be machine-readable.

To the best of our knowledge, there is no related work that incorporates all of the above. Hence, we propose a novel general approach.

3.2   Transparency-Focused Business Process Management Lifecycle

To facilitate greater transparency across organizational entities, we propose to extend business process management lifecycle phases with transparency - enhancing methods and tools. Different authors propose distinct lifecycles and overarching lifecycle meta-models [18,19]. We orient our contributions around the process lifecycle of Weske [26] with its design and analysis, configuration, enactment, and evaluation phases. Accordingly, this approach is the first to enable the business-process-oriented modeling, management, and auditing of regulatory transparency. It connects privacy concerns with the design, configuration, enactment, and evaluation of business processes.

Considering the business-process lifecycle phases, our general approach comprises three cyclical integrated contributions (C1, C2, C3), as shown in Fig. 1. These contributions span eight use cases, utilize five external technology components, and create six artifacts. All contributions reinforce the communication between data controllers and development teams and can be integrated into different phases of the BPM lifecycle and agile software development practice.

Contribution 1 introduces a comprehensive BPMN 2.0 extension plug-in for transparency modeling with a prototypical implementation in the open-source Camunda Modeler modeling software.[1] The contribution maps core BPMN elements with the TILT document objects. Thus, transparency information is added to an existing business process instead of being modeled through BPMN notation elements as discussed in [22]. We opt for the Camunda platform because of its mature development state, extensibility, and open-source BPMN modeler and workflow engine (meeting R2 and R3). We choose TILT for its regulatory expressiveness (R1) fully aligned with the GDPR and beyond, and the resulting transparency toolkit compatibility (R2 and R3). The contribution entails i) an automatically generated TILT document summarizing the information in a well-defined, machine-readable format, and ii) a normative process model.

With Contribution 2, we are the first to combine transparency information fields in process event logs to enable process managers and data controllers to monitor and audit transparency. We demonstrate a flexible, universally applicable, JSON-based, transparency logging approach. We fully implement a prototype in a distributed system architecture. Utilizing C2, process managers are supported to incorporate regulatory transparency in the BPM system configuration and enactment phases. Outcomes of C2 include a syntax proposal for JSON-based, transparency-focused event logs and an example of a distributed log architecture implementation.

Contribution 3, which builds upon the outcomes of C1 and C2, leverages process discovery techniques. As such, transparency-extended business processes (based on event logs of C2) can be discovered and analyzed. In particular, a normative transparency extended business process (C1) is then used for conformance checking the discovered process. This step realizes the BPM evaluation phase. Outcomes of C3 include a transparency report, highlighted discovered

---

[1] camunda.com.

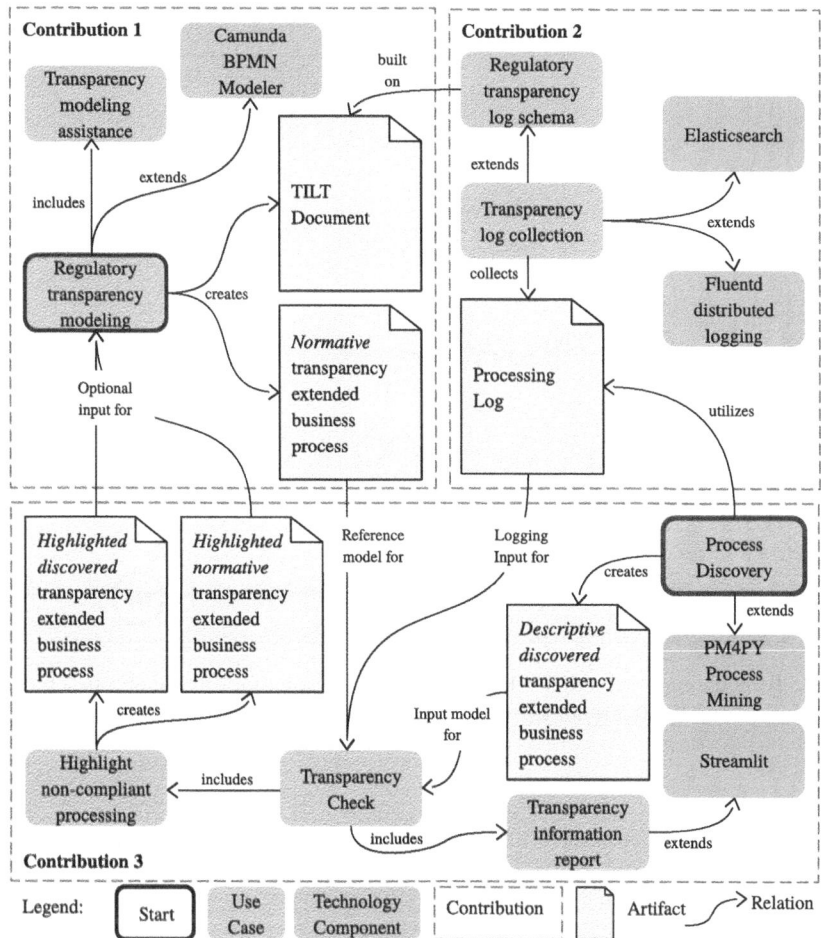

Fig. 1. Contribution overview for transparency-focused BPM.

transparency extended business processes, and highlighted normative business processes.

Figure 1 illustrates all technical contributions, their relation to each other, and their inter-dependencies. The two starts indicate the integration options of our contributions in different phases of the BPM lifecycle. First, data controllers are enabled to model normative transparency information through C1 in the design phase of creating a new business process. Secondly, C2 and C3 enable the enrichment of existing business processes with transparency information in the configuration phase, allowing for assessing the as-is conditions of transparency information in the evaluation phase. This transparency assessment, in turn, can be utilized as an input for successive lifecylce iterations.

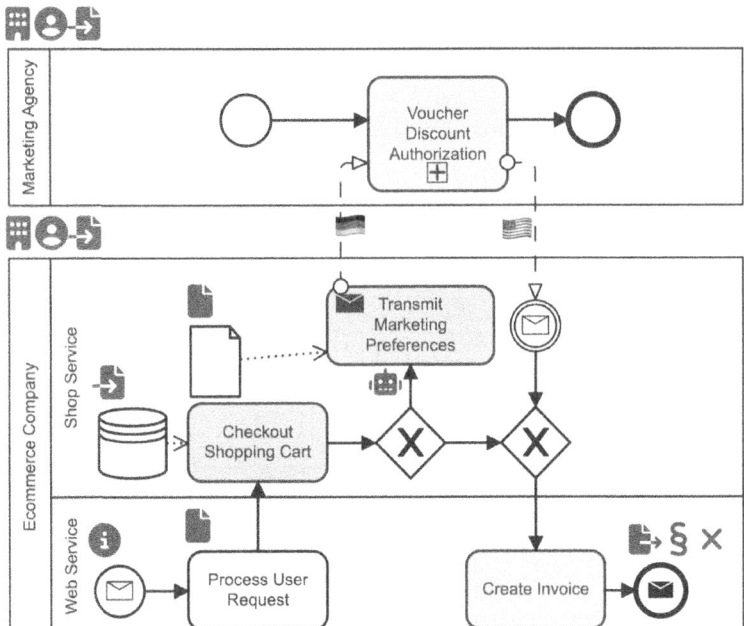

Fig. 2. Transparency-extended "shopping checkout" collaboration diagram. The icons and coloring highlights are added through our contributions C1 and C3. (Color figure online)

Due to the interconnected alignment of our contributions, we aim to integrate these well into existing BPM workflows. In the following, we illustrate a first example of applying our contributions.

3.3   Running Example Business Process

In Fig. 2, we depict a "shopping checkout" business process example, to which we already applied our transparency modeling contribution. It consists of the core notation elements of the BPMN standard and comprises two participants, five activities, eight sequence flows, two message flows, and one exclusive gateway. Furthermore, it features two participant lanes, a data store, and an object notation element. This level of complexity suffices to demonstrate several legally relevant modeling tasks.

Purple icons, as described in Table 1, and flags indicate the presence of specific transparency information, which is added through C1. Our C3 performs activity highlighting. Blue highlights indicate the absence of modeled transparency data in the discovered process. Orange highlights show the presence of discovered personal data processing activities that have not yet been modeled.

All colored elements indicate added transparency information or the results of the conformance checks. We describe them in more detail below in Sect. 4.

For instance, the flags indicate third-country transfers. We emphasize that these visualizations are much more detailed than in related work coloring privacy-related information in processes [27].

Traditionally, such a process is enacted and monitored in a workflow management system. Meanwhile, in increasingly distributed environments, new execution models can be adopted. As such, this process could be implemented as a fully autonomous enacted workflow using a polyglot microservice architecture in a cloud native setting. We develop an open-source prototypical implementation of an extended version of an example process in a containerized, Kubernetes-orchestrated microservice architecture featuring state-of-the-art cloud-native technologies. This includes the separate microservice implementations, the logging instrumentation and collection components, a load generator, etc. Following the open science principles, all subsequent artifacts are provided with extensive documentation in several repositories.[2]

4   Implementation

In the following, we explain our three contribution implementations.

4.1   Transparency Modeling in BPMN (C1)

For modeling and assessing regulatory transparency through a BPMN perspective, BPMN is required to represent relevant privacy policy information. With transparency information in BPMN, it is desirable to model as much information as possible to ensure information completeness and comprehensibility. Process and collaboration diagrams permit the modeling of multiple processes and participants in the same diagram. When modeling legally relevant information, for example, a collaboration diagram with two participants, as seen in Fig. 2, may contain multiple data controllers or data protection officers (DPO). Therefore, it is necessary to allow for the modeling of concurrent transparency information for several processes. Accordingly, we propose to include transparency information within process elements and not within metadata information.

Although BPMN comprises over 80 notation elements, only a handful are used actively [8]. While certain notation elements, such as participant and data object, may be suitable to represent aspects of a privacy policy, the limited set of notation elements subsequently prevents the regulatory expressiveness required by the GDPR. Introducing new notation elements for explicit transparency information modeling shifts the focus of process models away from the core business logic. This obscures the model for BPM stakeholders who do not have a direct interest in transparency. Therefore, this contribution proposes to extend common BPMN notation elements with TILT-based transparency information [13].

---

[2] All implementation artifacts can be referenced via doi.org/10.5281/zenodo.11396474.

Table 1. BPMN–TILT mapping overview.

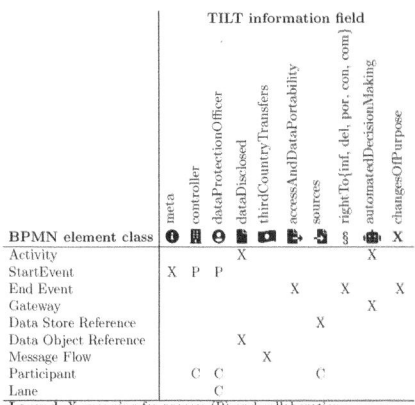

BPMN element class	meta	controller	dataProtectionOfficer	dataDisclosed	thirdCountryTransfers	accessAndDataPortability	sources	rightTo[inf, del, por, con, com]	automatedDecisionMaking	changesOfPurpose
Activity						X				X
StartEvent	X	P	P							
End Event					X	X				X
Gateway									X	
Data Store Reference						X				
Data Object Reference				X						
Message Flow					X					
Participant		C	C			C				
Lane		C								

Legend. X: mapping for process (P) and collaboration diagram (C) types.

Listing 1. BPMN 2.0 TILT extension utilizing *bpmn:extensionElements.*

```xml
<bpmn:startEvent
    id="StartEvent">
  <bpmn:extensionElements>
    <tilt:controller
      name="Chocolate Factory"
      division="Compliance"
      country="DE">
      <tilt:representative
        name="Charlie" />
    </tilt:controller>
    <tilt:dataProtectionOfficer
      name="Willy Wonka"/>
  </bpmn:extensionElements>
</bpmn:startEvent>
```

We map all TILT fields towards BPMN elements as shown in Table 1 to allow for process and collaboration diagram interoperability and regulatory transparency information completeness (R1), allowing for regulatory expressiveness as opposed to prior work [7,22]. Information fields supporting ex ante transparency are mapped towards commonly used BPMN elements and their derivatives. Ex post transparency information that occurs after the enactment of a process is mapped to its end event. This mapping facilitates the modeling and automatic generation of process-centric policies in the TILT format and subsequent presentations (e.g., a privacy policy).

The BPMN standard recommends grouping data extensions in *extensionElements*, allowing for modeling software interoperability. Listing 1 exemplifies our BPMN-compliant business process transparency extension of a process data model. Each BPMN element can carry unlimited similar or distinct TILT elements. Complex data structures such as lists are realized by element nesting.

We demonstrate the capabilities of transparency modeling through a fully-implemented Camunda modeler TILT plug-in.[3] The modeling and the graphical representation of transparency information through purple icons[4] and flags are conducted in the modeler interface through the elements properties panel.

In addition, the plug-in facilitates transparency linting, automatic transparency information adding and completion, and the process-centric TILT document creation.

We further implemented linting, a technique for analyzing, flagging, and alerting users to source code constructs that may not align with a predefined rule set.

---

[3] github.com/PrivacyEngineering/bpm-transparency-plug-in/.
[4] fontawesome.io.

To this end, the plug-in utilizes the Camunda linting *bpmnlint-utils* library to create element-specific alerts and error messages displayed in the user interface.[5]

Defined transparency linting rules are registered with a corresponding alert level. They facilitate checking complex rules, such as the order of elements, conditions under which elements can be used, and the properties of a selected element. Checking for message flows into sanctioned and otherwise disallowed third countries or for TILT information completeness are examples of rules implemented in our plug-in. These can be adapted to the compliance needs of the organization. For instance, linting rules can be used to codify data-controller-specific transparency compliance rules, facilitating straightforward organization-wide privacy-compliant process modeling, even for non-experts in the legal domain.

Assisting modeling staff with automatic transparency information filling and completion can significantly increase the overall transparency information completeness while reducing the potential necessary modeling effort (R1, R2). Our implementation features automatic third-country data transfer annotation adding and completion on message flow elements between participants in different countries as seen in Fig. 2. All message flow transparency information fields are automatically updated if a participant's country changes. Other example data fields for automatic information filling include identifier information and timestamps.

A process model is fully annotated once all applicable TILT fields are included in the process diagram. Due to the structured data model of BPMN, all transparency information can be extracted to create a process-centric TILT transparency document. The export feature of our plug-in traverses the process model and collects all transparency information in a JSON-formatted TILT document, thereby creating a process-oriented policy. Non-required and duplicate transparency elements are not exported.

These combined features enable data controllers to create normative transparency extended process models in the process design phase that allow for visually aided communication of data protection aspects between stakeholders, and the generation of regulatory expressive process-oriented policies.

4.2 Transparency Logging in Distributed Systems (C2)

Compliance auditing has been identified as a core method to ensure privacy compliance in accountable systems [12]. Auditable logs have therefore been argued to be one of the fundamental mechanisms [24]. However, existing diagnostic and functional logs are created for specific objectives and thus may be incomplete for transparency auditing [5,14]. Therefore, a dedicated transparency-focused event log is necessary to achieve the regulatory expressiveness as required by privacy regulation. We propose an expressive and easily integrable logging approach and transparency-focused processing event log for distributed systems that contains the relevant transparency information.

---

[5] Transparency linting is explained more thoroughly in the following repository: github.com/PrivacyEngineering/bpm-transparency.

Table 2. Extract of the transparency extended event log data frame.

ident: eid	time: timestamp	case: concept: name	concept: name	tilt: categories	tilt: purposes	tilt: legalBases
0	12:22:52.004	0x1234	Collect user data	[postcode, street.no, street...]	[rightToAccess]	[GDPR-15-1]

A transparency-focused event log can be realized by recording information on an event level. However, including complete TILT information within each event leads to redundant data, as most information is not exclusive to specific events. Furthermore, some ex-ante and ex-post transparency information does not exist at the design or enactment time of an activity and thus cannot be provided to individual logging instances. Updating logging components of distributed systems to include current transparency data introduces a comparatively high management overhead. Additionally, some transparency information, such as the country location of the processing service, can change at runtime. Under these considerations, we record data category, purposes, legal bases, recipients, and storage information as a grouped data-disclosed object in individual events, including case ID, timestamp, and activity identifier, in a dedicated transparency-focused event log.

Our approach is implemented in a polyglot microservice architecture.[6] We consciously avoid personal information within by utilizing trace-Ids as case identifiers that are provided by a OpenTelemetry (OTEL)[7] instrumentation. We use the Fluentd[8] logging framework that provides information about individual logging instances to filter, format, and forward logs to an Elasticsearch[9] data store. We implement a custom log format covering the relevant information items, as shown in Table 2. This can be done easily in all major programming languages supporting OTEL. Each event is logged under the standard INFO log level in JSON format.

We herein refrained from doing a structured performance assessment. Given that the performance was bound by the logging framework itself, benchmarking with a custom log format cannot yield insights beyond already known efficiency. However, had we pursued benchmarking, key variables such as log generations per second (LGPS), log size, memory, and CPU load had been suitable.

This contribution integrates into the configuration and enactment phases of the business process lifecycle.

Our implemented transparency logger component can be utilized to mitigate the logging integration effort into existing systems (R2).[10] Nevertheless, developers must still manually instrument code to log transparency information.

---

[6] github.com/PrivacyEngineering/bpm-transparency-demo.

[7] opentelemetry.io.

[8] fluentd.org.

[9] elastic.co/elasticsearch.

[10] github.com/PrivacyEngineering/bpm-transparency-logger-lib.

4.3   Transparency Information Discovery and Validation (C3)

Privacy-preserving processing of event logs is a part of process mining and analysis research as early as the formal definition of the process mining manifest [3]. Still, privacy and confidentiality are increasingly prerequisites for process mining [11]. Log-based privacy auditing and compliance checking from a systems-oriented perspective have been proposed [24]. However, to the best of our knowledge, process-oriented research focuses on the privacy-preserving analysis of traditional event logs [11,17,21] and not on regulatory transparency aspects as required by privacy legislation.

We propose to validate discovered transparency information with information derived out of normative transparency-extended process diagrams to highlight not-modeled personal data processing (orange) and not-observed personal data processing (blue) as illustrated in Fig. 2. Such a diagram visualizes personal data processing across organizational entities and systems in an aggregated overview while maintaining regulatory expressiveness (R1).

We utilize standard process mining techniques, such as process discovery, variant exploration, and rule mining, to create basic processing insights based on the transparency-focused event log, namely case ID, timestamp, and activity identifier information. Additionally, our implementation features concrete activity-specific validation of processed personal data categories, their processing purposes, and legal bases. This is realized by comparing normative data-disclosed information extracted from a normative modeled process diagram created through C1 with data-disclosed information from the transparency-focused event log created in C2.

We traverse the given data model of the normative diagram and collect all data-disclosed element information and their activity relation. These gathered objects serve as the permitted data-disclosed elements for their specified activities. As described in Sect. 4.2, recorded transparency information in an event is aggregated in grouped data-disclosed objects. Therefore, individual data fields are first separated into distinct data-disclosed elements for analysis. We utilize the inductive mining algorithm [1] to derive a BPMN diagram from the event log. Individual activities are then extended and highlighted with the discovered distinct data-disclosed elements. The resulting diagram can be used for comparison with a normative transparency extended process diagram or as a reference starting point for creating one.

This transparency analysis approach provides two main insights, which are visualized in Fig. 2: Firstly, it emphasizes activities that do not contain declared information despite being modeled to do so (blue). This includes all activities modeled to process certain personal data even though their processing is not recorded in the transparency-focused event log. Secondly, it reveals undeclared processing of personal data (orange), which is processing that is recorded in the transparency-focused event log but not modeled in the normative process diagram. Activities that process personal data conforming to the normative diagram are not highlighted.

To perform all of these analysis tasks and to visualize the results we use commonly used process mining tools, i.e., the leading open source process mining platform pm4py[11], PMtk for variant exploration[12], and Streamlit[13] for the data apps. Additional evaluation approaches, such as variant exploration and conformance coverage, illustrate our contributions' novel possibilities. Such prototypes are provided in our repository as additional artifacts.[14]

Hence, this transparency information validation approach generates expressive activity-specific personal data processing insights in an aggregated process-oriented format suitable for assessing and demonstrating personal data processing compliance. Referring to our running example, we assume strong transparency needs due to the sensitive nature of the data at hand. Other system components that do not handle personal data can do so without rich event logging or transparency-focused process mining due to the plug-in capability of our contributions (R2, R3).

5   Discussion, Future Work, and Conclusion

The interplay of our contributions and their alignment to phases of the transparency-focused BPM lifecycle (cf. Sect. 3) facilitate process-oriented regulatory transparency modeling and validation in a holistic approach. It comes with moderate integration costs into organizations with established business process management, as shown in Sect. 4. In the following, we discuss our findings.

We propose process-centric transparency modeling and auditing approaches. Our approach utilizes the transparency language TILT, which allows for fully expressive regulatory transparency required by privacy regulations, such as the GDPR, meeting our requirement R1 as opposed to comparable approaches [7,16,27]. Our contributions are easily integrable by extending the existing set of BPMN elements and not introducing explicit modeling constructs as proposed in other work [10,22] meeting requirement R2. In contrast to prior work addressing privacy considerations in the design phase of processes [7,22,27], our proposal addresses regulatory transparency across the entire BPM lifecycle, allowing for continuous regulatory transparency management and compatibility with existing process mining and BPM approaches meeting R3. To apply the presented contributions, the necessary techniques need to be implemented and kept up to date. The coordination between the technical, management, and legal stakeholders remains a challenge. Future work should streamline these tasks, for instance, to enable additional legal artifacts, such as structured data protection impact assessments.

Some existing limitations of transparency-oriented logging and complementary process mining apply. Logs (in general) may still be incomplete and rely on

---

[11] pm4py.fit.fraunhofer.de.

[12] pmtk.fit.fraunhofer.de.

[13] streamlit.io.

[14] github.com/PrivacyEngineering/bpm-transparency-demo/tree/main/src/mining-dashboard.

the willingness and skills of developers to provide them in detail. For complex compliance evaluation tasks, information flowing between different organizations must be considered. For the future, we envision extending our approaches to support cross-organizational processes and the underlying, possibly heterogeneous computing infrastructures. In any case, our contributions can reduce the risks of data breaches or data protection fines, due to increased transparency, which, in general is an indispensable precondition for ensuring all other privacy principles [15].

Additional future work shall concentrate on tight integrations with other workflow engines or service orchestrators. These may entail ex ante and ex post analyses, execution plans preventing non-modeled data processing, and automatic code instrumentation. Automatically extracted logging data, e.g., through interceptors or proxies, might be a promising research direction. Furthermore, some transparency information might be inferred or pre-filled using templates or by tapping orchestrator APIs (e.g., for data location information) or observability data.

Furthermore, in cloud-native systems, distributed tracing is increasingly used to discover data flows between services. Enhancing traces with transparency information and producing event logs might yield more comprehensive insights. However, tracing is not yet as widely established in tooling as logging. Legacy components, therefore, might not be observable. Moreover, complete automation (as in not having to instruct the creation of custom logs in code) is also not possible in all cases. Multiple business activities within a single span, including transparency information, can create undesired ambiguity.

In conclusion, this paper tackles the challenge of regulatory transparency in business processes by proposing three distinct technical contributions. On this basis, business processes could also be enhanced, particularly in accommodating forthcoming regulations on transparency in supply chains or health-data processing [21], accounting for different socio-technical aspects. This not only promises regulatory compliance but also valuable insights for people optimizing business processes in their daily routine. We illustrated the proof of concept with several example implementations.

References

1. van der Aalst, W.: Process Mining: Discovery, Conformance and Enhancement of Business Processes, vol. 136. Springer, Cham (2011). https://doi.org/10.1007/978-3-642-19345-3
2. van der Aalst, W.: Process mining. Commun. ACM **55**(8), 76–83 (2012). https://doi.org/10.1145/2240236.2240257
3. van der Aalst, W., et al.: Process mining manifesto. In: Daniel, F., Barkaoui, K., Dustdar, S. (eds.) BPM 2011. LNBIP, vol. 99, pp. 169–194. Springer, Heidelberg (2012). https://doi.org/10.1007/978-3-642-28108-2_19
4. van der Aalst, W.M.P.: Process Mining: Data Science in Action, 2nd edn. Springer, Cham (2016). https://doi.org/10.1007/978-3-662-49851-4
5. Balalaie, A., Heydarnoori, A., Jamshidi, P.: Microservices architecture enables DevOps: migration to a cloud-native architecture. IEEE Softw. **33**(3), 42–52 (2016)

6. Braun, R., Esswein, W.: Classification of domain-specific BPMN extensions. In: Frank, U., Loucopoulos, P., Pastor, Ó., Petrounias, I. (eds.) PoEM 2014. LNBIP, vol. 197, pp. 42–57. Springer, Heidelberg (2014). https://doi.org/10.1007/978-3-662-45501-2_4

7. Chinosi, M., Trombetta, A.: Integrating privacy policies into business processes. J. Res. Pract. IT **41**(2), 155–170 (2009). https://search.informit.org/doi/10.3316/ielapa.836520965194259

8. Compagnucci, I., Corradini, F., Fornari, F., Re, B.: Trends on the usage of BPMN 2.0 from publicly available repositories. In: Buchmann, R.A., Polini, A., Johansson, B., Karagiannis, D. (eds.) BIR 2021. LNBIP, vol. 430, pp. 84–99. Springer, Cham (2021). https://doi.org/10.1007/978-3-030-87205-2_6

9. Cranor, L.: Web Privacy with P3P. O'Reilly (2002)

10. Diamantopoulou, V., Karyda, M.: Integrating privacy-by-design with business process redesign. In: Katsikas, S., et al. (eds.) ESORICS 2021. LNCS, vol. 13106, pp. 127–137. Springer, Cham (2022). https://doi.org/10.1007/978-3-030-95484-0_8

11. Elkoumy, G., et al.: Privacy and confidentiality in process mining: threats and research challenges. ACM Trans. Manag. Inf. Syst. **13**(1), 1–17 (2021). https://doi.org/10.1145/3468877

12. Feigenbaum, J., Jaggard, A.D., Wright, R.N.: Towards a formal model of accountability. In: Proceedings of the 2011 New Security Paradigms Workshop, NSPW 2011, pp. 45–56. Association for Computing Machinery (2011). https://doi.org/10.1145/2073276.2073282

13. Grünewald, E., Pallas, F.: TILT: a GDPR-aligned transparency information language and toolkit for practical privacy engineering. In: Proceedings of the 2021 ACM Conference on Fairness, Accountability, and Transparency, pp. 636–646. Association for Computing Machinery (2021). https://doi.org/10.1145/3442188.3445925

14. Grünewald, E., Kiesel, J., Akbayin, S.R., Pallas, F.: Hawk: DevOps-driven transparency and accountability in cloud native systems. In: IEEE 16th International Conference on Cloud Computing (CLOUD). IEEE (2023). https://doi.org/10.1109/CLOUD60044.2023.00027

15. Grünewald, E., Wille, P., Pallas, F., Borges, M.C., Ulbricht, M.R.: TIRA: an OpenAPI extension and toolbox for GDPR transparency in RESTful architectures. In: 2021 IEEE European Symposium on Security and Privacy Workshops (EuroS&PW). IEEE (2021)

16. Jensen, M.: Towards privacy-friendly transparency services in inter-organizational business processes. In: 2013 IEEE 37th Annual Computer Software and Applications Conference Workshops, pp. 200–205 (2013)

17. Mannhardt, F., Petersen, S.A., Oliveira, M.F.: Privacy challenges for process mining in human-centered industrial environments. In: 2018 14th International Conference on Intelligent Environments (IE), pp. 64–71 (2018). https://doi.org/10.1109/IE.2018.00017

18. Macedo de Morais, R., Kazan, S., Inês Dallavalle de Pádua, S., Lucirton Costa, A.: An analysis of BPM lifecycles: from a literature review to a framework proposal. Bus. Process Manag. J. **20**(3), 412–432 (2014). https://doi.org/10.1108/BPMJ-03-2013-0035

19. Nousias, N., Tsakalidis, G., Vergidis, K.: BPM lifecycles and their core cycle steps: identification, processing and clustering. In: Matsatsinis, N.F., Kitsios, F.C., Madas, M.A., Kamariotou, M.I. (eds.) BALCOR 2020, pp. 125–132. Springer, Cham (2023). https://doi.org/10.1007/978-3-031-24294-6_13

20. Object Management Group: Business Process Model and Notation (BPMN) - Version 2.0 (2011)
21. Pika, A., Wynn, M.T., Budiono, S., ter Hofstede, A.H., van der Aalst, W., Reijers, H.A.: Privacy-preserving process mining in healthcare. Int. J. Environ. Res. Public Health **17**(5) (2020). https://doi.org/10.3390/ijerph17051612
22. Pullonen, P., Tom, J., Matulevičius, R., Toots, A.: Privacy-enhanced BPMN: enabling data privacy analysis in business processes models. Softw. Syst. Model. **18**(6), 3235–3264 (2019). https://doi.org/10.1007/s10270-019-00718-z
23. Saha, A., Agarwal, P., Ghosh, S., Gantayat, N., Sindhgatta, R.: Towards business process observability. In: Proceedings of the 7th Joint International Conference on Data Science & Management of Data, pp. 257–265. ACM (2024). https://doi.org/10.1145/3632410.3632435
24. Samavi, R., Consens, M.P.: Publishing privacy logs to facilitate transparency and accountability. J. Web Semant. **50**, 1–20 (2018). https://doi.org/10.1016/j.websem.2018.02.001
25. vom Brocke, J., Rosemann, M. (eds.): Handbook on Business Process Management 1. IHIS, Springer, Heidelberg (2015). https://doi.org/10.1007/978-3-642-45100-3
26. Weske, M.: Business Process Management: Concepts, Languages, Architectures, 3rd edn. Springer, Cham (2019)
27. Windrich, M., Speck, A., Gruschka, N.: Representing data protection aspects in process models by coloring. In: Gruschka, N., Antunes, L.F.C., Rannenberg, K., Drogkaris, P. (eds.) APF 2021. LNCS, vol. 12703, pp. 143–155. Springer, Cham (2021). https://doi.org/10.1007/978-3-030-76663-4_8

Towards Leveraging Process Mining for Sustainability – An Analysis of Challenges and Potential Solutions

Adrian Joas[1]([⊠]), Maren Gierlich-Joas[2], Charlotte Bahr[3], and Janina Bauer[4]

[1] Celonis ApS, Lautrupsgade 13, 2100 Copenhagen, Denmark
a.joas@celonis.com

[2] Department of Digitalization, Copenhagen Business School, Howitzvej 60, 2000 Frederiksberg, Denmark
mg.digi@cbs.dk

[3] Institute for Information Systems, Friedrich Alexander University, Fürther Straße 248, 90429 Nuremberg, Germany
charlotte.bahr@fau.de

[4] Celonis SE, Theresienstraße 6, 80333 Munich, Germany
j.bauer@celonis.com

Abstract. Sustainability is one of the grand challenges of our society, but organizations struggle to take a data-driven approach to identify and realize tangible value with sustainability initiatives. Due to its ability to measure and visualize actual business processes and their outcomes based on trace data, process mining (PM) can be positioned as a solution to tackle those problems. However, organizations currently lack the understanding of how to effectively leverage the technology regarding its impact and effect on sustainability initiatives' success. We apply a qualitative approach involving 28 experts to understand the crucial challenges and potential solutions in implementing PM for sustainability. Most challenges can be attributed to sustainability as an organization-wide, cross-departmental focus area, and correspondingly complex organizational setting. Our paper provides novel insights into understanding PM implementations for sustainability, especially in strategic alignment with the overall organization.

Keywords: Sustainability (transformation) · process mining · green IS · green BPM · qualitative research

1 Introduction

With the global governmental commitment to the Sustainable Development Goals (SDGs) and the Paris Agreement, sustainability is fundamentally anchored in the political agenda [12, 32]. Sustainability describes a "development that meets the needs of the present without compromising the ability of future generations to meet their own needs" [45]. The pressing need for sustainability challenges how governments develop future strategies [12], companies work and define their business model [2, 35], and individuals live their everyday lives [48]. As one of the grand challenges, sustainability demands holistic and interdisciplinary approaches [13].

A. Marrella et al. (Eds.): BPM 2024 Forum, LNBIP 526, pp. 354–371, 2024.
https://doi.org/10.1007/978-3-031-70418-5_21

The Information Systems (IS) discipline has contributed to addressing this grand challenge for more than a decade, emphasizing the role of IS in shaping the future by enabling sustainable processes and practices in organizations (Green IS), e.g.,[24], information technology (IT) being green in itself (Green IT) [7], and paving the way for research initiatives, directly and indirectly fighting this grand challenge e.g., [43]. In addition, considering the public awareness and political commitment, research on sustainability has gained momentum which is mirrored in numerous special issues and editorials (e.g., [19, 20, 33]).

In this paper, we focus on Green IS, as "configuring and applying IS to achieve environmental objectives through reducing the ecological footprint of businesses and supporting organizational decision-making toward sustainability, along with more efficient economic performance" ([19], p. 938). Investigating Green IS, industrial processes are a central lever for improvement in sustainability aspects [19]. Aiming to design sustainable organizational processes and monitor them, Green Business Process Management (BPM) is a core field of research e.g., [34]. However, a core challenge still is measuring the sustainability impact of applied process improvements [15].

In this context, process mining (PM) can be a potential solution. PM enables the discovery, analysis, and corresponding optimization of business processes based on event logs from machines, people, or software close to real-time [40]. With these unique characteristics, PM can facilitate sustainability initiatives [2, 15], e.g., on PM's use for carbon accounting highlight [4]. Leading PM vendors like Celonis also release new functionalities for their software to facilitate sustainability use cases [29]. However, as prior studies highlight, organizations still face barriers when aiming to implement PM for 'standard' use cases (an instance of using PM with a specific objective of improvement) which aim at increasing process efficiency and decreasing costs [16, 18]. The context of sustainability adds another layer of complexity to PM implementation, and it may cause organizational ambidexterity [37], which can hinder organizations from using PM to address sustainability challenges. Thus far, contributions to PM for sustainability have mainly taken a technical perspective without emphasizing the organizational requirements and challenges [15]. Hence, as part of a broader research endeavor, we pose the following research question (RQ):

What are the challenges and potential solutions for organizations when implementing PM for sustainability?

Toward answering the RQ, the remainder of the paper is structured as follows: First, we lay the foundations by synthesizing related work on sustainability and Green IT/IS, and PM in the context of sustainability. Next, we describe our method which consists of multiple qualitative data collections with experts. We derive our findings on challenges and solutions when implementing PM for sustainability. Finally, we reflect on the learnings, point out limitations, and highlight our contribution and implications.

Our paper contributes to creating an understanding of PM's role in leveraging organizational sustainability. We present challenges of PM implementations and corresponding solutions to overcome those challenges, in the context of sustainability which assists practitioners in using PM for sustainability.

2 Related Work

2.1 Sustainability and Green IT/IS

The concept of sustainability encompasses three pillars—economic, environmental, and social—also known as the "three-pillar" paradigm [28]. Economic sustainability refers to the use of IT to support long-term economic growth without negatively impacting social or environmental aspects [28]. Mostly (minimized) cost or (maximized) profitability is utilized as primary indicators for sustainable economic performance, for example, if IT decreases energy costs or improves the efficiency of business processes e.g., [19, 21]. Environmental sustainability refers to the ecological impact on the world. Its performance indicators include environmental quality function deployment (EQFD) [3], or materials, energy consumption, and toxicity (MET) matrix [5]. All indicators are aimed to be minimized for optimal environmental sustainability [14]. Lastly, social sustainability promotes social growth to fulfill the requirements of the current generation without harming the well-being of future ones [19]. The development of rural areas through the use of information and communication technologies (ICT) is an example of this level [38]. Most prior IS studies focus on one of the three levels (mostly the environmental level) [19] and quantifying those three levels is of great societal interest since the impact on the environment started to receive more attention in the late 1980s [28]. In our study, we focus primarily on economic and environmental sustainability.

The IS discipline provides a rather twofold answer regarding its substantial role in transforming towards a sustainable world and fighting the grand challenge. With Green IT and Green IS, two related yet different concepts are provided. Green IT describes the creation and usage of environmentally sustainable technology, particularly in terms of energy efficiency and carbon footprint [19]. Green IS, which is the focus of our study, is concerned with "the design and implementation of information systems that contribute to sustainable business processes" ([44], p. 2). By assisting the design and implementation of sustainable business processes [24, 43], Green IS represents a key resource to assist organizations in transforming into more sustainable entities [34]. Green IS is tackling sustainability issues from a plural focus as it aims to make an entire system, including IT, sustainable [17, 43]. Therefore, Green IS emerged to be an integral part of organizations, leading to sustainability transformations (ST) [49].

An important discipline within Green IS is Green BPM - the "sum of all IS-supported management activities that help to monitor and reduce the environmental impact of business processes in their design, improvement, implementation, and operation stages" ([27], p. 3812). The objective of Green BPM is the improvement of business processes while considering sustainability factors [26]. It enables organizations to achieve sustainability goals by guiding them in improving each process step along the BPM Lifecycle to reduce negative environmental impact [9, 36].

2.2 Foundations of Process Mining and Its Implementation in Organizations

To implement Green BPM practices within organizations, PM is a technology used to support business operations on a quantitative level, using event logs created by organizational IT systems documenting the activities carried out by people, machines, or software

[41]. In this context, an event log provides an understanding of the actual process flow and its implications on organizations, offering the opportunity to manage, support, and improve business processes [41]. PM enables the identification and evaluation of process behavior, process performance, conformance of processes to existing process models, and process improvement opportunities from event logs [6, 40]. This helps organizations to increase their operational process efficiency and achieve not only economic but also ecological sustainability.

Prior studies on PM have mostly focused on the technical side, resulting in calls for research on the organizational and ecosystem levels [42]. More recently, we see studies examining PM from a socio-technical perspective e.g., [10, 16, 42], emphasizing the organizational embedding of the technology. In this context, the implementation of PM has been a core topic of interest as it is a key prerequisite for the effective use of PM.

When implementing PM within organizations, several challenges are faced [22], especially from an end-user's or vendor's point of view. Hereby, challenges can be understood as "a difficulty or an obstacle that arises when using (or intending to use) PM in organizations, and that requires a lot of energy and determination from an individual, a team or an organization to overcome" ([23], p. 515). Our research builds upon a categorization of these challenges into 1) Strategic alignment, 2) governance, 3) people, 4) culture, and 5) methods/IT [23].

The first dimension contains all challenges that impede PM's alignment "with the overall structure of an organization" ([23], p. 516), for example, misled expectation management of customers, absence of organizational change management, or a missing long-term PM strategy or complex company politics [18]. Governance-related challenges include obstacles like insufficient department involvement during PM implementation, lacking availability of customer support, and missing engagement from the end-user organization [18]. This builds barriers toward "appropriate and transparent accountability in terms of roles and responsibilities as well as regulations regarding data collection and handling" ([18], p. 516). Furthermore, challenges can also arise on an individual level, e.g., if users' engagement in learning is insufficient [18]. Cultural challenges describe difficulties in changing the mindset of employees within the PM-implementing organization or a lack of end-user involvement [18]. This dimension of challenges incorporates all barriers that hamper the development of a facilitating environment with collective values and beliefs complementing PM initiatives [23]. Finally, challenges around methods/IT encompasses "the set of tools and techniques as well as IT-based solutions that support and enable actions [...] within PM initiatives" ([23], p. 516), including, for example, obstacles regarding data source system connections or knowledge of PM solution capabilities [18].

These challenges need to be overcome to leverage PM for 'standard' use cases of PM. Initial studies have highlighted solutions to address these obstacles, e.g., by introducing training concepts [18]. However, we are interested in examining the specificities when implementing the technology for sustainability use cases and understanding the resulting synergies and conflicts.

2.3 Process Mining Fostering Sustainability

Researchers and practitioners have already recognized PM's potential to drive sustainability [1, 2, 15]. For example, PM can enhance value creation and operational activities regarding economic and ecological efficiency [2]. More specifically, PM's potential as an auxiliary technique to support carbon accounting has been examined to support the assessment of the impact on its environment [15]. Moreover, contributing from a technical perspective, there is a research focus on investigating concrete technical implementations of PM fostering sustainability. Integrating sustainable performance measures in well-known benchmark PM methods aims to raise awareness among developers of PM methods about the ecological impact of their solutions [8, 15]. A rather indirect application of PM was evaluated by Acerbi [1] who aimed to foster a circular economy and thus reduce material use, and redesign materials, products, and services to be less resource-intensive by implementing PM.

However, as PM is a relatively new field for researchers and practitioners, there is a lack of substantial research on how PM is fostering sustainability in organizations, as recent calls for papers indicate [42]. This is where our research contributes.

3   Methodology

3.1   Methodological Approach

Towards answering the RQ, we applied a qualitative approach which allowed us to collect rich insights from real-life experiences [31]. As we investigate the phenomenon of using PM for sustainability in the complex socio-technical context, we aim to integrate diverse perspectives. Our study is embedded in the ecosystem of Celonis, a leading PM vendor so we were able to collect insights from its customers, partners, and internal experts. When approaching experts, we ensured that they had experience in PM and were at different stages of using it for sustainability use cases.

Our qualitative data contains insights from semi-structured expert interviews and focus groups. The interviews aimed to unpack individuals' experiences with PM sustainability solutions and perceived challenges. They covered questions like "How can process mining help you in achieving the 'how' of sustainability solutions?" and "What challenges do you face when implementing process mining for sustainability?". We followed the guidelines by Myers and Newmann [25] for qualitative interviews to ensure a valid study design and avoid interview biases.

With the focus groups, we intended to broaden the perspective and "elicit people's understandings, opinions and views, or to explore how these are advanced, elaborated and negotiated" ([47], p. 187). The focus groups were exploratory and had a designated moderator [39]. They included the three building blocks: 1) Introduction to PM for sustainability, 2) discussion of the participants along several predefined questions, and 3) closing.

3.2 Data Collection and Data Analysis

The data was collected between December 2022 and September 2023. In sum, we engaged with 28 purposefully sampled experts on PM and sustainability (either in person or via video-conferencing tools). 20 of them were customers from various industries, six were partners and two were employees of Celonis (see Table 1).

Table 1. Overview of interview partners and focus group participants

ID	Industry	Role	ID	Industry	Role
1	Manufacturing	Senior transformation manager	15	Manufacturing	Digitalization and process optimization
2	Manufacturing	Manager sales, planning & logistics	16	Automotive	Sustainability manager
3	Energy	Program manager logistics	17	Healthcare	Specialist trade pricing, value chain & key account
4	Logistics	Area operations director	18	Private Equity	Manager business analytics
5	Industry retail	Sustainability manager	19	Partner	Strategic Advisor
6	Manufacturing	PM data analyst	20	Partner	Analyst
7	Construction	Global category analyst	21	Airline	Process analyst
8	PM vendor	Sustainability team	22	Airline	Data analyst
9	PM vendor	Sustainability team	23	Manufacturing	Consultant business analyst
10	Construction materials	Product owner PM & robotics	24	Manufacturing	Internal control analyst
11	Construction materials	Process analyst data mining & robotics	25	Partner	Consultant
12	Construction materials	Manager ESG reporting	26	Partner	CEO
13	Manufacturing	Project & process manager	27	Manufacturing	Senior manager B2C operations
14	Partner	Machine learning engineer	28	Partner	SR Consultant

In December 2022, we started with seven initial semi-structured interviews to investigate how PM customers explored the technology for their first sustainability use cases. In May, we built upon these initial insights in the context of a sustainability hackathon that was organized by Celonis for its customers and partners. We conducted five additional semi-structured interviews and an exploratory focus group with seven participants to deepen our understanding of challenges related to the implementation of sustainability use cases. Finally, in September, the challenges were refined in a focus group with ten participants, and potential solutions to overcome them were discussed.

Table 2. Overview of data collection, analysis, and outcome

Time	Type of data	Data analysis	Outcome
Dec 2022	7 semi-structured interviews (IDs 1–7)	Videos and transcripts; iterative coding cycles	Initial understanding of current PM sustainability use cases and related challenges
May 2023	5 semi-structured interviews (IDs 14–18)	Detailed thought protocols; qualitative content analysis	Initial understanding of challenges related to PM implementations for sustainability use cases
May 2023	Focus group with 7 participants (IDs 7–13)	Detailed thought protocols; qualitative content analysis	
Sep 2023	Focus group with 10 participants (IDs 19–28)	Videos and transcripts, additional notes on virtual whiteboard; deductive coding cycles	Refined understanding of challenges related to PM implementations for sustainability use cases and potential solutions

Whenever possible, the interviews and focus group discussions were recorded, transcribed verbatim, and anonymized. Otherwise, we relied on detailed thought protocols and written output from the focus group participants (virtual whiteboard) (see Table 2).

For the data analysis, we followed iterative coding cycles with multiple researchers involved in the coding [30]. We used deductive coding, taking the dimensions of PM challenges as initially introduced by Martin et al. [23] and extended by Joas and Matzner [18] as a basis for the development of our second-order themes. This approach resulted in 20 second-order themes and five aggregated dimensions which we use to describe our results in the following.

4   Findings

4.1   Current Challenges Around the Use of Process Mining for Sustainability

We were able to cluster challenges from the expert interviews and the focus groups into "strategic alignment", "governance", "people", "culture", and "methods/IT" [23] (Summary see Table 3). We will elaborate on the challenges for sustainability PM use cases in the following by pointing out how they differ from "traditional" PM implementations that aim at cost reduction and business optimization.

Strategic Alignment. The dimension of strategic alignment was discussed most controversially, and many challenges can be grouped into this dimension. With "strategic alignment", we refer to the organizational process of initiating PM projects in the field of sustainability, convincing stakeholders, measuring the progress, and managing contradicting business goals.

Users stated that "it is very difficult to find a good use case you can start on because for several topics you have great ideas." (ID 21). Moreover, the business value of PM sustainability solutions is difficult to determine: "Looking at the initiation, one problem that we saw was the KPI that we want to track. Because we don't have an easy KPI that we can track apart from the CO_2 pollution. It's hard to quantify how we can reach our goals. [...] In the sustainability case, this is harder for me than for the other process mining cases." (ID 28).

With the *unclear use cases* and the *elective business value*, we observed a *lack of management support*. The experts discussed that "sponsors and business don't see the value" (participants' notes in the focus group) and they were concerned that "without specific targets, nothing will happen" (ID 15). Furthermore, they stressed that "an external push is mandatory for sustainability initiatives to be successful" (ID 14). Hence, for some organizations, external factors like legal regulations were considered a crucial driver for implementing PM sustainability solutions (ID 9, ID 13). For others, the drivers were investors' demand or customers' demand (ID 13). Only in some cases, the drivers were internal with intrinsically motivated individuals driving sustainability projects (ID 17).

But even if managers understand the value of PM for sustainability, they face the burden of *organizational ambidexterity*. As participants in the focus group highlighted, "sustainability is not the core business and does not have the prioritization." (participants' notes in the focus group). If sustainability and cost reduction goals have the same effective direction, such as "in most cases, cost and CO_2 emissions are closely combined" (ID 3), organizations adopt PM for sustainability as it is a win-win situation (ID 8, ID 13). However, if this leads to higher costs and causes ambidexterity between competing business goals, organizations carefully navigate between these tensions: "Our sole reason to look into this topic are [legal regulations], as we're otherwise only focusing on monetary KPIs. Our key element is being competitive in monetary KPIs and not all sustainability activities lead to higher willingness to pay—therefore we're very selective." (ID 18). Contrarily, some started to prioritize PM recommendations that tackle sustainability over those that foster business value: "So we've actually made the decision to accept an increase in cost, as we're not using those [environmentally unfriendly] suppliers anymore." (ID 13). Finally, even if PM is implemented to support sustainability goals, it is difficult to determine which prerequisites are needed and the *success factors are unclear* (participants' notes in focus group).

Governance. Some challenges can be categorized as governance challenges. These cover the unavailability of data, constraining data access barriers, lack of interdisciplinary and cross-functional teams, and unclear organizational anchoring.

As many experts stressed, they consider it challenging that some *data for sustainability use cases are lacking*. In both focus groups, experts discussed their "inability to measure certain sustainability indicators like carbon footprint" (participants' notes in focus group) and found it difficult to establish a measurable baseline (ID 11, ID 13).

But even if data exist and are of decent quality, with dispersed local entities, data silos emerge, which *constrain data access* and are harmful when aiming to pursue global sustainability targets (ID 15). The reason is, as one expert observed, a *lack of interdisciplinary and cross-functional teams* as "[t]he challenge is to include the different departments to be able to actually bring sustainability initiatives into life." (ID 8). Closely related, sustainability initiatives are mostly *not anchored and well-organized in organizations* yet. This leads to unclear responsibilities and long approval flows (ID 17), hindering the effective usage of PM for sustainability. Toward defining suitable governance structures, some interviewees reported on hiring sustainability experts on a global level who can own the topic (ID 18). However, these positions are not necessarily staffed with tech-affine people.

People. A central success factor for PM's usage for sustainability is a skilled workforce. However, experts problematized insufficient domain expertise, analytical skills, and technical skills.

As sustainability addresses all organizational functions, close collaboration is required. Individuals using PM for sustainability use cases do not only need to have *expertise in sustainability* and PM but also *in the different domains* (e.g., manufacturing, supply chain) that are to be optimized. Hence, PM subject matter experts and sustainability experts "need support, as [they] don't know what [they] don't know" (ID 15). Furthermore, most sustainability teams are not data-driven yet, which limits their efficiency and ability to make the best decisions (ID 8). They *lack technical skills* "and it is a challenge to explain technical capabilities to sustainability teams—which is even more difficult than explaining sustainability to tech teams." (ID 14). Therefore, the experts stressed the importance of *technical resources* to enable sustainability teams (ID 10).

Culture. Next to the individuals' capabilities, the organization's culture plays a significant role in implementing sustainability solutions based on PM. The mentioned barriers range from lacking orientation toward sustainability and lack of trust in PM to resistance to change and unwillingness to share the domain knowledge.

We observed a *lack of orientation towards sustainability* for some organizations (participants' notes in focus group) which is mirrored in the lack of managerial support and the overall strategic alignment (see section above). But even if the impetus for sustainability initiatives is given, sustainability teams tend to *question data-driven decisions based on PM* (ID 15) as PM is not yet a commonly used and accepted best-practice method. If pursuing sustainability targets implies a "change or adaptation of your processes or tools to get more sustainable, this is tricky" (participants' notes in focus group). For example, "[employees] support the initiatives that do not have a direct impact on them, but especially if we're talking about direct restrictions, there is push-back. For example, with reducing company travel." (ID 8). Consequently, interviewees underlined the complexity of external and internal transformational change being required to align with the organization's values and employees' goals (ID 8, ID 12).

Finally, also the culture regarding *sharing domain knowledge* needs to be adjusted. Despite the increased need to share knowledge in the case of sustainability, organizational departments are reluctant to collaborate on sustainability use cases: "They consciously decide not to involve any other departments, because they fear their ideas might get stolen. Therefore, it is very difficult to implement sustainability measures as they need to run on a global level." (ID 17).

Methods/IT. While the four previously described themes stress the social, and organizational side, the articulated challenges on the technical side are not too different to those observed for standard use-cases. The experts mentioned barriers related to data preparation, bridging different systems, and missing advanced sustainability features.

First, *complex data preparation and mapping* are needed as sustainability KPIs are difficult to calculate. For example, it requires lots of effort to break the used material down into unique bills of material or to determine the energy used by buildings or employees working from home (ID 14).

Second, this complexity is increased using *fragmented solutions*. Not all needed data can be retrieved from organizational IT, such as SAP, but dispersed solutions have to be integrated (participants' notes in focus group, ID 15). One interviewee described how their use of PM to make their supply chain management more sustainable is hindered as the relevant sensory data from their rented trucks are not centrally available and they have to coordinate their entire ecosystem to optimize their supply chain (ID 7). This leads to inefficient practices such as: "Our current 'solution' is sending Excel files back and forth and having the individual organizations fill it out." (ID 18) which prevents the usage of PM. Another interviewee agrees that they are "trying to bring in both types of data: So, the key right now is to create a data set that has [their] IDs and the secondary keys for procurement and the order-to-cash process, […] but also maintaining the sustainability information" (ID 3).

Third, current PM solutions are only starting to incorporate *advanced features* for sustainability use cases. One interviewee explained how they are using simulations to optimize material selection based on the emission factor (ID 16). As the technology is only developing in this direction, further features might follow.

Table 3. Overview of challenges discovered

Category	Challenges when using process mining for sustainability
Strategic Alignment	Ambiguity in the definition of use cases
	Elective business value
	Lack of engagement
	Lack of management support
	Insufficient clarification of success factors
Governance	Lack of usable data
	Limited data access (data silos)
	Lack of interdisciplinarity and cross-functionality in teams
	Lack of anchor & structurization within organizations
People & Culture	Lack of support
	Lack of technical skills
	Lack of orientation toward sustainability
	Criticality toward data-driven decisions
	Low-level maturity in a culture of sharing and exchanging knowledge
Methods/IT	Complexity in data preparation and mapping
	Fragmented solutions
	Incorporation of advancements in PM technology (Features)

4.2   Towards Solutions Facilitating the Use of Process Mining for Sustainability

Despite the mentioned challenges, the interviewees emphasized the potential they see in PM for sustainability, stating e.g., that "the hypothesis for today is, that [they] use process mining to access the data and use this, to actually compute CO2 emissions, taking the emission factor on top of the data [they] already have and get a pretty comprehensive and transparent picture [...]" (ID 1). Hence, they are eager to overcome the stated challenges. Their approaches for solutions can also be grouped according to the aggregated dimensions [23] (Summary see Table 4).

Strategic Alignment. The area of sustainability makes it possible to get *executive management support*, which, supported by a strong *business case calculation*, can act as a door opener case for [use of PM in] other areas of the company to see the value and in turn increase the willingness to invest (participants' notes in focus group). One interviewee even mentioned that in a group of multiple organizations, a central sustainability responsible was hired to drive the agenda on the C-level (ID 18). The willingness to invest in balancing the KPIs can be further facilitated by "hold[ing] joint workshops in-between departments to jointly define KPIs and optimize them respectively" (ID 15). A strategic step many organizations try to aim at is "the decoupling of economic growth from emissions growth to build growing economies" (ID 8) but currently, they lack the option to see both dimensions in one joint view. PM might be a potential technology that enables organizations to see those KPIs or *success factors* and their interdependence in one view and understand processes and performance holistically. Given *clear objectives on what the goal* is, they acknowledge that they need help to define the strategic objective at the current stage, to bring their ideas into a consistent setting and deliver tangible value: "We need some help. We have a lot of good ideas, but we need some kind of global focus to figure out where do we start [...]. We need data or we need models to help us find our way." (ID 4).

Governance. Solution proposals mostly revolve around *implementation guidance*, *organizational anchoring*, and *interdisciplinary and cross-functional teams*. Similar to the support needed to define the strategic objective for PM in sustainability, several interviewees also mentioned that they are looking for guidance around how others are using PM in a sustainability context and from an implementation perspective, either in general around structuring a PM sustainability project or specifically on solutions like the emission reduction. This guidance might come both from PM vendors as well as directly from other organizations (ID 2, ID 17, ID 18). The idea of establishing sustainability departments as internal champions to ensure organizational alignment and internal distribution proposes a central, yet cross-functional team to lead efforts around sustainability and PM (ID 10).

People & Culture. The experts are looking for guidance not only for the implementation but also for individuals with *domain*, *analytical*, and *technical* expertise (ID 15). However, the experts found it difficult to derive solutions for people-related and cultural challenges. The only solution proposal mentioned is closely tied to a change in the organizations' communication culture to support both the creation of transparency and the sharing of insights for continuous improvement. This specifically includes *technical* and *domain knowledge*: "I don't only speak with the sustainability people from my team, but also with the more tech people and we can make something together out of it" (ID 5).

Methods/IT. Lastly, focusing on the technical side, the interviewees mentioned some initial ideas for solutions: "We're using simulations to optimize material selection based on the emission factor. This is our priority as it is the biggest impact we see in manufacturing. As we have limited runs, we need to calculate the emissions value per car, as those are very individual in the manufacturing process. We look into the emission level on product level through the bill of material and aggregate to corporate" (ID 16), which combines *complex data preparation* and *advanced features* to create sustainability-focused solutions with PM.

These solutions were framed by the interviewees with an overall optimistic outlook on the challenges: "So it is actually not that difficult to measure [emissions], because we have the data, we just have to automate the process, and combine different data sets. I see a lot of potential there" (ID 5). This creates a positive environment in which PM can develop as a driver for sustainability initiatives.

Table 4. Overview of potential solutions facilitating the use of PM for sustainability

Category	Solution towards using process mining for sustainability
Strategic Alignment	Provide strong business calculation Establish measures (KPIs, success factors) Clarify and align objectives/ goals Anchor executive management support Establishment of measures (KPIs, Success Factors)
Governance	Provide implementation guidance Establish organizational anchoring Establish interdisciplinary and cross-functional teams
People & Culture	Ensure availability and establishment of domain knowledge Establish analytical and technical expertise
Methods/IT	Reduce complexity in the data preparation process Enable access to new and advanced PM features Invest in feature development

5 Discussion

5.1 Connecting the Findings to Prior Literature: Process Mining as a Facilitator for Sustainability

ST is considered one of the biggest organizational challenges today, calling for researchers to join forces [13]. Sustainability research with its two perspectives—Green IT and Green IS—has a long tradition in the IS field e.g., [19, 24, 33, 43]. Building upon this body of knowledge, our empirical findings indicate that PM is a promising technology to facilitate sustainability initiatives at an organizational level. Therefore, we discuss our findings in the context of Green IS.

Prior studies on Green BPM have focused on modeling and deploying greener and more sustainable business processes e.g., [11, 27, 34, 44]. As an extension of this research, initial contributions on PM for sustainability have illustrated PMs' ability to leverage sustainability from a conceptual perspective [17]. We build upon these studies with our empirical investigation with PM experts and sustainability experts. Our findings are in line with Graves et al. [17] as we could confirm PM's overall potential to foster sustainability. Looking at the three pillars of sustainability [19], we found mainly improvements on the ecological and economic level, whereas social sustainability is currently difficult to quantify using PM.

To leverage the potential of PM for sustainability, it needs to be implemented successfully. As prior studies have highlighted the difficulty of implementing PM due to its socio-technical nature and embedding in ecosystems [18, 23], we examine the implementation for sustainability use cases. Comparing our findings on the implementation of PM to the challenges identified by Martin et al. [23] and extended by Joas and Matzner [18], we find that the number of mentions of the dimensions differs. In Martin et al. [23], most of the challenges belonged to the category of methods/IT. With the focus on PM ecosystems, Joas and Matzner [18] already stressed the socio-side. In our observations, we can see an even greater increase in perceived challenges on the socio-side, compared to the technical focus in more traditional usages of PM.

The empirical data suggest that most mentions from the interviewees can be grouped into the dimension of strategic alignment, both from a challenge and solution-proposal perspective. This emphasizes the relevance of the socio-side of ST in organizations. We observe a high level of attention from top-level management on sustainability cases for the organization-wide strategy, especially compared to more common PM use cases [29], where most perceived solutions are at the department level. Given the German background of some of our interviewees, this place on the top-level agenda could be due to the "German Supply Chain Act"—a legal requirement for companies to follow, which includes several sustainability-related aspects organizations need to fulfill [46]. This highlights the intervention from a political viewpoint that can be observed in the sustainability space. Both investors as well as customers may refer to this legislation in gathering information about the sustainability of an organization's sustainability KPIs and its processes. Hence, organizations see a clear relevance of PM to support their sustainability initiatives but are just starting to see initial and easy use cases. The same legislative push cannot be seen in more traditional use cases, which are motivated by the organization's desire to improve revenue or decrease cost. Despite the external drivers to implement PM for sustainability, organizational as well as individual uncertainty are burdens. Externally available guidance and success stories around PM for sustainability use cases are the most mentioned solution proposals to remedy these challenges.

We observe three organization-wide trends in the solution proposals and challenges of the interviewees. First, organizational KPI ambidexterity is considered highly influential, with the overall aim of decoupling economic growth from environmental impact, while the challenge of measuring the impact of sustainability initiatives is a current struggle for most interviewees. Second, given the organization-wide implications sustainability initiatives have, organizations aim to reduce silos and create a culture of data-driven decision-making, sustainable growth, and interdisciplinary teams. Third, using PM for sustainability cases might act as a door opener for continued use of PM, leading to an increase in digital transformation in tandem with sustainability transformation in organizations. Summarized by an interviewee: "We follow the hypotheses internally, that digitalization is very closely related to sustainability. We want to use existing structures and tools to ensure we also achieve our sustainability targets. Therefore, we want to optimize our process based on this foundation." (ID 15).

5.2 Limitations and Outlook

Our study has certain limitations that can guide future work. First of all, this paper is based on the understanding that PM is a strong lever for organizational sustainability, following the narrative of prior studies in the field. Therefore, the selected technology for our investigation is case-centric PM. However, in further research, our findings could be extended by looking at object-centric PM, a novel approach to PM allowing organizations to retrieve and analyze comprehensive, and easy-to-understand process models with multiple case notations. Furthermore, considering different technologies like robotic process automation, simulation, or AI for predictive analyses and how their potential to foster sustainability is utilized, could serve as a general guideline on how to develop PM's full potential for sustainability initiatives. While we consider sustainability as an overarching goal to achieve using PM, our insights on how PM impacts the distinct pillars of sustainability—economical, ecological, and social—are limited, as we mainly examine the ecological and economical dimension. Further investigations into how PM can facilitate social sustainability may be required.

Furthermore, narrowing the topic down to PM implementation for sustainability use cases (Green IS), we take a socio-technical perspective and highlight the socio-side. However, it would also be a valuable contribution to explore the technical side more in-depth by researching the sustainability of PM itself (Green IT).

Looking at our selected method, we intentionally collect qualitative data at one point in time within the ecosystem of Celonis. We argue that we selected an information-rich, representative case and followed the guidelines for rigorous qualitative research. However, some of the derived challenges and solutions might also apply to ST in general and could be related to certain regulatory framing conditions. Even though the challenges were derived based on an established framework, they are not mutually exclusive. Also, most of the experts were only in the initial stage of using PM for sustainability, and we were not able to investigate cases from a longitudinal perspective. Hence, we advise researchers to conduct further studies in different countries and industries to explore PM implementation journeys from a holistic, longitudinal perspective.

6 Conclusion

This paper explores the potentials and challenges of PM for sustainability, guided by the RQ—*What are the challenges and potential solutions when implementing PM for sustainability?* Collecting data from 28 experts in the PM ecosystem of Celonis, we derive challenges in the fields of "strategic alignment", "governance", "culture", "people", and "methods/IT", as well as potential solutions.

We contribute to understanding the specificities of PM implementations for sustainability use cases. We highlight the importance of the socio-side and thereby extend the body of knowledge that has mainly been informed by technical-focused studies. Furthermore, we position PM as an impactful lever for organizational sustainability. PM can assist in automatizing ST and making it measurable, and thereby, it can overcome many shortfalls of BPM solutions. We contribute to the realm of Green IS by highlighting one specific IS that can make ST quantifiable.

Our study holds implications for practice as well. With the growing importance of sustainability, organizations are searching for IS to support them in reaching their goals. Due to its complexity, sustainability also presents risks—especially when there is little focus on profitable use cases. Since the countervalue of sustainability initiatives is already difficult to identify, there is a risk of diluting it even further. PM for sustainability offers using an existing technical solution for another strategic topic. However, it is crucial to understand the reasons and barriers to adoption.

Beyond that, PM for sustainability holds enormous potential for PM software providers. Additional use cases increase the value potential and the scope of the application. In this way, new business potential can be leveraged. However, this requires investment in research, expertise in the sustainability area, and product development. A better understanding of customer challenges, as well as adaptation drivers, will help vendors to develop and deploy profitable solutions with focus.

References

1. Acerbi, F., Polenghi, A., Quadrini, W., Macchi, M., Taisch, M.: Fostering circular manufacturing through the integration of genetic algorithm and process mining. In: Kim, D.Y., von Cieminski, G., Romero, D. (eds.) APMS 2022. IFIPAICT, vol. 664, pp. 407–414. Springer, Cham (2022). https://doi.org/10.1007/978-3-031-16411-8_47
2. Böttcher, T.P., Empelmann, S., Weking, J., Hein, A., Krcmar, H.: Digital sustainable business models: Using digital technology to integrate ecological sustainability into the core of business models. Inf. Syst. J. 1–26 (2023). https://doi.org/10.1111/isj.12436
3. Bovea, M.D., Pérez-Belis, V.: A taxonomy of ecodesign tools for integrating environmental requirements into the product design process. J. Clean. Prod. 20(1), 61–71 (2012). https://doi.org/10.1016/j.jclepro.2011.07.012
4. Brehm, L., Slamka, J., Nickmann, A.: Process mining for carbon accounting: an analysis of requirements and potentials. In: Dibbern, J., Förderer, J., Kude, T., Rothlauf, F., Spohrer, K. (eds.) Digitalization Across Organizational Levels. PROIS, pp. 209–244. Springer, Cham (2022). https://doi.org/10.1007/978-3-031-06543-9_9
5. Byggeth, S., Hochschorner, E.: Handling trade-offs in Ecodesign tools for sustainable product development and procurement. J. Clean. Prod. 14(15–16), 1420–1430 (2006). https://doi.org/10.1016/j.jclepro.2005.03.024
6. de Leoni, M., van der Aalst, W.M.P., Dees, M.: A general process mining framework for correlating, predicting and clustering dynamic behavior based on event logs. Inf. Syst. 56, 235–257 (2016). https://doi.org/10.1016/j.is.2015.07.003
7. Dedrick, J.: Green IS: concepts and issues for information systems research. Commun. Assoc. Inf. Syst. 27 (2010). https://doi.org/10.17705/1CAIS.02711

8. Delgado, A., García, F., Moraga, M.Á., Calegari, D., Gordillo, A., Peña, L.: Adding the sustainability dimension in process mining discovery algorithms evaluation. In: Di Francescomarino, C., Burattin, A., Janiesch, C., Sadiq, S. (eds.) BPM 2023. LNBIP, vol. 490, pp. 163–177. Springer, Cham (2023). https://doi.org/10.1007/978-3-031-41623-1_10
9. Dumas, M., La Rosa, M., Mendling, J., Reijers, H.A.: Fundamentals of Business Process Management. Springer, Heidelberg (2018). https://doi.org/10.1007/978-3-662-56509-4
10. Eggers, J., Hein, A., Böhm, M., Krcmar, H.: No longer out of sight, no longer out of mind? How organizations engage with process mining-induced transparency to achieve increased process awareness. Bus. Inf. Syst. Eng. **63**(5), 491–510 (2021). https://doi.org/10.1007/s12599-021-00715-x
11. Fritsch, A., von Hammerstein, J., Schreiber, C., Betz, S., Oberweis, A.: Pathways to greener pastures: research opportunities to integrate life cycle assessment and sustainable business process management based on a systematic tertiary literature review. Sustainability **14**(18), 11164 (2022). https://doi.org/10.3390/su141811164
12. Fuldauer, L.I., Thacker, S., Haggis, R.A., Fuso-Nerini, F., Nicholls, R.J., Hall, J.W.: Targeting climate adaptation to safeguard and advance the Sustainable Development Goals. Nat. Commun. **13**(1), 3579–3594 (2022). https://doi.org/10.1038/s41467-022-33518-z
13. Gholami, R., Watson, R., Hasan, H., Molla, A., Bjorn-Andersen, N.: Information systems solutions for environmental sustainability: how can we do more? J. Assoc. Inf. Syst. **17**(8), 521–536 (2016). https://doi.org/10.17705/1jais.00435
14. Gohar, S.R., Indulska, M.: Business process management: saving the planet? In: ACIS 2015 Proceedings (2015)
15. Graves, N., Koren, I., van der Aalst, W.M.P.: ReThink your processes! A review of process mining for sustainability. In: International Conference on ICT for Sustainability, pp. 164–175 (2023). https://doi.org/10.1109/ICT4S58814.2023.00025
16. Grisold, T., Mendling, J., Otto, M., vom Brocke, J.: Adoption, use and management of process mining in practice. Bus. Process. Manag. J. **27**(2), 369–387 (2020). https://doi.org/10.1108/BPMJ-03-2020-0112
17. Jenkin, T.A., Webster, J., McShane, L.: An agenda for 'Green' information technology and systems research. Inf. Organ. **21**(1), 17–40 (2011). https://doi.org/10.1016/j.infoandorg.2010.09.003
18. Joas, A., Matzner, M.: Unleashing digital process innovation with process mining: designing a training concept with action design research. In: European Conference on Information Systems, vol. 1829 (2022)
19. Kotlarsky, J., Oshri, I., Sekulic, N.: Digital sustainability in information systems research: conceptual foundations and future directions. J. Assoc. Inf. Syst. **24**(4), 936–952 (2023). https://doi.org/10.17705/1jais.00825
20. Lehnhoff, S., Staudt, P., Watson, R.T.: Changing the climate in information systems research. Bus. Inf. Syst. Eng. **63**(3), 219–222 (2021). https://doi.org/10.1007/s12599-021-00695-y
21. Ma, J., Harstvedt, J.D., Jaradat, R., Smith, B.: Sustainability driven multi-criteria project portfolio selection under uncertain decision-making environment. Comput. Ind. Eng. **140**, 106236 (2020). https://doi.org/10.1016/j.cie.2019.106236
22. Mamudu, A., Bandara, W., Wynn, M.T., Leemans, S.J.J.: A process mining success factors model. In: Di Ciccio, C., Dijkman, R., del Río Ortega, A., Rinderle-Ma, S. (eds.) BPM 2022. LNCS, vol. 13420, pp. 143–160. Springer, Cham (2022). https://doi.org/10.1007/978-3-031-16103-2_12
23. Martin, N., et al.: Opportunities and challenges for process mining in organizations: results of a Delphi study. Bus. Inf. Syst. Eng. **63**(5), 511–527 (2021). https://doi.org/10.1007/s12599-021-00720-0
24. Melville. Information systems innovation for environmental sustainability. MIS Q. **34**(1), 1 (2010). https://doi.org/10.2307/20721412

25. Myers, M.D., Newman, M.: The qualitative interview in IS research: examining the craft. Inf. Organ. **17**(1), 2–26 (2007). https://doi.org/10.1016/j.infoandorg.2006.11.001

26. Nowak, A., Leymann, F., Schleicher, D., Schumm, D., Wagner, S.: Green business process patterns. In: Proceedings of the 18th Conference on Pattern Languages of Programs, pp. 1–10 (2011). https://doi.org/10.1145/2578903.2579144

27. Opitz, N., Krup, H., Kolbe, L.M.: Green business process management – a definition and research framework. In: 2014 47th Hawaii International Conference on System Sciences, pp. 3808–3817 (2014). https://doi.org/10.1109/HICSS.2014.473

28. Purvis, B., Mao, Y., Robinson, D.: Three pillars of sustainability: in search of conceptual origins. Sustain. Sci. **14**(3), 681–695 (2019). https://doi.org/10.1007/s11625-018-0627-5

29. Reinkemeyer, L.: How Process Mining can turn Sustainability Targets into Action (2022). https://www.linkedin.com/pulse/how-process-mining-can-turn-sustainability-targets-lars-rei nkemeyer/

30. Saldaña, J.: The Coding Manual for Qualitative Researchers. SAGE Publications Ltd. (2009)

31. Sarker, S., Xiao, X., Beaulieu, T.: Editorial: qualitative studies in information systems: a critical review and some guiding principles. MIS Q. **37**(4), iii–xviii (2013)

32. SDGs: Sustainable Development Goals (2015). https://sdgs.un.org/goals

33. Seidel, S., et al.: The sustainability imperative in information systems research. Commun. Assoc. Inf. Syst. **40**, 40–52 (2017). https://doi.org/10.17705/1CAIS.04003

34. Seidel, S., Recker, J., vom Brocke, J.: Green Business Process Management. Springer, Heidelberg (2012). https://doi.org/10.1007/978-3-642-27488-6

35. Seidler, A.-R., Henkel, C., Fiedler, M., Kranz, J.: Greening the organisation: an institutional logics approach to corporate pro-environmentalism. In: British Academy of Management Conference (2017)

36. Sohns, T.M., Aysolmaz, B., Figge, L., Joshi, A.: Green business process management for business sustainability: a case study of manufacturing small and medium-sized enterprises (SMEs) from Germany. J. Clean. Prod. **401**, 136667 (2023). https://doi.org/10.1016/j.jclepro.2023.136667

37. Thambusamy, R., Salam, A.F.: Corporate ecological responsiveness, environmental ambidexterity and IT-enabled environmental sustainability strategy. In: International Conference on Information Systems (2010)

38. Tim, Y., Cui, L., Sheng, Z.: Digital resilience: how rural communities leapfrogged into sustainable development. Inf. Syst. J. **31**(2), 323–345 (2021). https://doi.org/10.1111/isj.12312

39. Tremblay, M.C., Hevner, A.R., Berndt, D.J.: Focus groups for artifact refinement and evaluation in design research. Commun. Assoc. Inf. Syst. **26** (2010). https://doi.org/10.17705/1CAIS.02627

40. van der Aalst, W.: Data science in action. In: van der Aalst, W. (eds.) Process Mining, pp. 3–23. Springer, Heidelberg (2016). https://doi.org/10.1007/978-3-662-49851-4_1

41. van der Aalst, W., et al.: Process mining Manifesto. In: Daniel, F., Barkaoui, K., Dustdar, S. (eds.) BPM 2011. LNBIP, vol. 99, pp. 169–194. Springer, Heidelberg (2012). https://doi.org/10.1007/978-3-642-28108-2_19

42. vom Brocke, J., Jans, M., Mendling, J., Reijers, H.A.: A five-level framework for research on process mining. Bus. Inf. Syst. Eng. **63**(5), 483–490 (2021). https://doi.org/10.1007/s12599-021-00718-8

43. Watson, R.T., Boudreau, M.-C., Chen, A.J.: Information systems and environmentally sustainable development: energy informatics and new directions for the IS community. MIS Q. **34**(1), 23 (2010). https://doi.org/10.2307/20721413

44. Watson, R.T., Boudreau, M.C., Chen, A., Chen, A.: Green IS: building sustainable business practices **17**, 1–17 (2008)

45. WCED: Our Common Future (1987)
46. Weihrauch, D., Carodenuto, S., Leipold, S.: From voluntary to mandatory corporate account-
    ability: the politics of the German Supply Chain Due Diligence Act. Regul. Gov. **17**(4),
    909–926 (2023). https://doi.org/10.1111/rego.12501
47. Wilkinson, S.: Focus group methodology: a review. Int. J. Soc. Res. Methodol. **1**(3), 181–203
    (1998). https://doi.org/10.1080/13645579.1998.10846874
48. Wunderlich, P., Veit, D.J., Sarker, S.: Adoption of sustainable technologies: a mixed-methods
    study of German households. MIS Q. **43**(2), 673–691 (2019). https://doi.org/10.25300/MISQ/
    2019/12112
49. Zimmer, M.P., Järveläinen, J.: Digital–sustainable co-transformation: introducing the triple
    bottom line of sustainability to digital transformation research. In: Kreps, D., Davison, R.,
    Komukai, T., Ishii, K. (eds.) HCC 2022. IFIPAICT, vol. 656, pp. 100–111. Springer, Cham
    (2022). https://doi.org/10.1007/978-3-031-15688-5_10

Author Index

© The Editor(s) (if applicable) and The Author(s), under exclusive license
to Springer Nature Switzerland AG 2024
A. Marrella et al. (Eds.): BPM 2024 Forum, LNBIP 526, pp. 373–374, 2024.
https://doi.org/10.1007/978-3-031-70418-5

GPSR Compliance

The European Union's (EU) General Product Safety Regulation (GPSR) is a set of rules that requires consumer products to be safe and our obligations to ensure this.

If you have any concerns about our products, you can contact us on ProductSafety@springernature.com

In case Publisher is established outside the EU, the EU authorized representative is:

Springer Nature Customer Service Center GmbH
Europaplatz 3
69115 Heidelberg, Germany

The manufacturer's authorised representative in the EU is Springer
Nature Customer Service Centre GmbH, Europaplatz 3, 69115 Heidelberg,
Germany. If you have any concerns regarding our products, please
contact ProductSafety@springernature.com

Printed and bound by CPI Group (UK) Ltd, Croydon, CR0 4YY
24/04/2026
02096351-0010